Tourism, Religion and Spiritual Journeys

Religion and spirituality are still among the most common motivations for travel. Many major tourism destinations have developed largely as a result of their connections to sacred people, places and events.

Tourism, Religion and Spiritual Journeys provides a comprehensive assessment of the primary issues and concepts related to the intersection of tourism and religion. The book provides a balanced discussion of both theoretical and applied subjects that destination planners, religious organizations, scholars and tourism service providers must deal with on a daily basis. The volume draws together a distinguished list of contributors to examine the global phenomenon of religious tourism. Included are examples from most parts of the world, with substantial empirical cases being taken from Hinduism, Islam, Judaism, Roman Catholicism, Mormonism (Church of Jesus Christ of Latter-day Saints), the New Age movement, Sikhism, Buddhism and the spiritual philosophies of East Asia. Also included are many allusions to other important religious traditions throughout the world and their treatment of pilgrimage travel.

In this vibrant collection of essays, *Tourism, Religion and Spiritual Journeys* discusses many important ideas, paradigms and problems that are currently being examined and debated at the forefront of scholarly research on religious tourism the world over.

Dallen J. Timothy is Associate Professor of Community Resources and Development at Arizona State University and Visiting Professor of Heritage Tourism at the University of Sunderland. Dr Timothy is editor of the Journal of Heritage Tourism and has published numerous books and articles on heritage, planning, borders and religious tourism.

Daniel H. Olsen is a PhD candidate in the Department of Geography at the University of Waterloo. His research interests include religious heritage tourism, cultural geography and the geography of religion. He has published several articles and book chapters on various aspects of heritage tourism, religious tourism and cultural geography.

Contemporary Geographies of Leisure, Tourism and Mobility
Series Editor: C. Michael Hall
Professor at the Department of Tourism, University of Otago, New Zealand

The aim of this series is to explore and communicate the intersections and relationships between leisure, tourism and human mobility within the social sciences.

It will incorporate both traditional and new perspectives on leisure and tourism from contemporary geography, e.g. notions of identity, representation and culture, while also providing for perspectives from cognate areas such as anthropology, cultural studies, gastronomy and food studies, marketing, policy studies and political economy, regional and urban planning, and sociology, within the development of an integrated field of leisure and tourism studies.

Also, increasingly, tourism and leisure are regarded as steps in a continuum of human mobility. Inclusion of mobility in the series offers the prospect to examine the relationship between tourism and migration, the sojourner, educational travel, and second home and retirement travel phenomena.

The series comprises two strands:

Tourism, Religion and Spiritual Journeys

Edited by
**Dallen J. Timothy and
Daniel H. Olsen**

Routledge
Taylor & Francis Group

LONDON AND NEW YORK

First published 2006
by Routledge
2 Park Square, Milton Park, Abingdon, Oxon OX14 4RN

Simultaneously published in the USA and Canada
by Routledge
270 Madison Ave, New York, NY 10016

Routledge is an imprint of the Taylor & Francis Group

Typeset in Times by
Florence Production Ltd, Stoodleigh, Devon
Printed and bound in Great Britain by
Antony Rowe Ltd, Chippenham, Wiltshire

Transferred to Digital Printing 2006

British Library Cataloguing in Publication Data
A catalogue record for this book is available from the British Library

Library of Congress Cataloging in Publication Data
   Tourism, religion and spiritual journeys/edited by Dallen J. Timothy and
   Daniel H. Olsen.
       p.cm – (Routledge studies in contemporary geographies of leisure,
       tourism, and mobility)
   Includes bibliographical references.
   1. Tourism – Religious aspects.   2. Sacred space.   I. Timothy,
   Dallen J.   II. Olsen, Dallen H., 1973–   III. Series
   G156.5.C5T68 2005
   201′.691 – dc22   2005017079

ISBN10: 0–415–35445–5
ISBN13: 9–78–0–415–35445–5

Contents

Illustrations

Figures

Tables

Contributors

Thomas S. Bremer is an Assistant Professor of Religious Studies at Rhodes College, Memphis, Tennessee, USA. As a specialist in American religions, his research has concentrated on the intersections of religion and tourism at such locations as Temple Square in Salt Lake City, the Lorraine Motel, site of the National Civil Rights Museum, in Memphis, and pilgrimage destinations in Mexico. His book, *Blessed with Tourists: The Borderlands of Religion and Tourism in San Antonio* (2004, University of North Carolina Press), includes both historical and contemporary views of tourists at religious places in San Antonio, Texas. He recently began work on a history of religious life in America's national parks.

Justine Digance is Senior Lecturer in Tourism Management, in the Department of Tourism, Leisure, Hotel and Sport Management at Griffith University's Gold Coast Campus, Brisbane, Australia. Her research interests are tourism management, business event tourism, and pilgrimage. Her doctorate (University of Sydney) was based upon research on modern secular pilgrimage, and she has published extensively on the subject of religion and tourism, with a recent article in *Annals of Tourism Research* exploring pilgrimage at contested sites.

Mara W. Cohen Ioannides is a Lecturer in the English Department at Missouri State University, Springfield, Missouri, USA. Her area of expertise is Jewish American History, focusing primarily, but not exclusively, on the Ozarks region of the United States. In 1999, she guest edited a special double issue of *OzarksWatch* on documenting Jews of the Ozarks. She has published articles in Jewish studies and tourism journals and co-authored book chapters with Dimitri Ioannides on the topic of Jewish tourism. She has also been active in recent years in presenting her work at conferences, such as the meetings of the Midwest Jewish Studies Association, Western Jewish Studies Association, Midwest Popular Culture Association, the Missouri Folklore Association, and the Association of American Geographers. In addition, Cohen Ioannides co-directed and wrote the documentary, *Home, Community, Tradition: The Women of Temple Israel*.

Erik H. Cohen, a sociologist, is a senior lecturer in the Institute for the Study and Advancement of Religious Education, School of Education, Bar Ilan University, Ramat Gan, Israel. He is the director of Research and Evaluation, an independent group of researchers that has conducted numerous international studies. Cohen is also a member of the Facet Theory Association and served two years as its secretary (2001-2003). He serves on a number of scientific committees and is consulting editor for two scholarly journals. His research interests include cross-cultural studies, tourism, youth culture, informal education, research methods, and data analysis. He has published many articles on these subjects and authored three books: *L'Etude et l'éducation juive en France* (1991); *Israel Experience: Educational and Policy Analysis* (with Eynath Cohen, in 2000 in Hebrew) and *Entre France et Israël: la jeunesse juive* (with Maurice Ifergan 2005).

Paul J. Conover is a Park Ranger for the city of Phoenix at the South Mountain Environmental Education Center, Phoenix, Arizona, USA. He recently completed a master's degree at Arizona State University with a focus on parklands management. In addition, he has personal and research interests in spiritual tourism in the southwestern United States.

Chao Guo is an Assistant Professor in the School of Community Resources and Development at Arizona State University, Tempe, Arizona, USA. He obtained his PhD in public administration from University of Southern California. His research interests include non-profit govern-ance, representation and responsiveness in non-profit organizations, collaboration within and across sectors, stakeholder theory, and the crossover between eastern spiritual philosophies and tourism. He has published several articles in public administration and non-profit studies journals. Guo has also presented his work at national and international conferences, including meetings of the Association for Research on Non-profit Organizations and Voluntary Action, American Society of Public Administration, and the Academy of Management.

C. Michael Hall is a Professor and the Head of the Department of Tourism at the University of Otago, Dunedin, New Zealand and a Docent in the Department of Geography at the University of Oulu, Finland. He is a co-editor of *Current Issues in Tourism* and has published widely on the topics of tourism and human mobility, peripheral regional development, and environmental history. He serves on numerous editorial boards and is a commissioning editor/co-editor for two tourism book series at Routledge and Channel View Publications. His research at the time of writing relates primarily to the role of human mobility in regional development, competitiveness and innovation; lifecourses and human mobility; foodways; and Nordic tourism.

Dimitri Ioannides is Professor of Tourism Planning and Development in the Department of Geography, Geology and Planning at Missouri State University, Springfield, Missouri, USA. Since 2003 he has also been a Senior Research Fellow at the Center for Regional and Tourism Research on the island of Bornholm, Denmark. He has co-edited two books, including the acclaimed *Economic Geography of the Tourist Industry* (Routledge, 1998). In addition, he has written a number of journal articles and book chapters on tourism planning, tourism and sustainable development, tourism and entrepreneurial activity, and tourism in borderlands. In 2003 Ioannides received the Roy Wolfe Award for Outstanding Contributions to the Recreation, Tourism, and Sport Specialty Group of the American Association of Geographers.

Thomas Iverson is a Professor of Economics and the Executive Director of the Sustainable Development Institute at the Micronesian Area Research Center of the University of Guam, Mangilao, Guam, USA. Dr. Iverson received his PhD in Economics from the University of Texas. Prior to moving to Guam in 1988, he was an Associate Professor of Management Information Systems at the Graduate Center of Kentucky State University. Iverson has published articles, notes, and book reviews in most of the major tourism journals. His work centers on sustainable development and the Japanese, Korean, and Chinese outbound markets. In 2001 Tom converted to Islam and became Abdul Salaam but continues to publish under his Christian/professional name. Still a neophyte, he has not yet undertaken the Hajj and continues to study and practice Islam.

Rajinder S. Jutla is Associate Professor of Community and Regional Planning in the Department of Geography, Geology and Planning at Missouri State University, Springfield, Missouri, USA. His graduate work focused on urban planning and landscape architecture. His undergraduate education was in architecture. Prior to working at Missouri State University, he taught environmental design and regional and community planning at Kansas State University. His research interests include urban tourism, historic preservation, urban design, and environmental perception and preferences. Jutla has published articles and presented papers at national and international conferences on these subjects, as well as Sikh pilgrimage. Presently he is researching tourism in the context of urban space from the perspective of social, cultural and visual dimensions.

Lutz Kaelber is Associate Professor of Sociology at the University of Vermont, Burlington, Vermont, USA. He researches and teaches social theory, sociology of religion, and comparative historical sociology with a special focus on the work of Max Weber and medieval heresy. He authored *Schools of Asceticism: Ideology and Organization in Medieval*

Religious Communities (Penn State University Press, 1998), which received the 1999 best book award from the American Sociological Association's Sociology of Religion section. Additionally, he translated and edited Max Weber's *History of Commercial Partnerships in the Middle Ages* (Rowman and Littlefield, 2003) and co-edited a collection of essays on Weber's *Protestant Ethic*, titled *The Protestant Ethic Turns 100: Essays on the Centenary of the Weber Thesis* (Paradigm Publishers, 2005). He has published widely on these subjects and is currently working on new projects related to Weber's 'adventure capitalism' and a sociological account of travel.

Daniel H. Olsen is a Lecturer and PhD student in the Department of Geography, University of Waterloo, Waterloo, Ontario, Canada. The focus of his dissertation is religious heritage tourism with an emphasis on both religious tourism theory and the practical side of managing sacred sites. He has published a number of articles and book chapters, as well as presented numerous conference papers, on the topics of religious tourism, the geography of tourism, and tourism in peripheral regions. He presently serves on the executive board of the Tourism, Recreation and Sport Specialty Group of the Association of American Geographers.

Myra Shackley is Professor of Culture Resource Management in the Nottingham Business School, Nottingham Trent University, UK. She has a particular interest in the management of visitors to historic sites and protected areas and has published thirteen books, the latest being *Managing Sacred Sites: Service Provision and Visitor Experience* (Continuum, 2001). She has also authored many journal articles, book chapters and reports. Her most recent work deals with issues affecting World Heritage sites, the management of sacred sites, and developments in cultural tourism.

Rana P.B. Singh is Professor of Geography at Banaras Hindu University, Varanasi, UP, India. He is the Founding President of the Society of Pilgrimage Studies, and also of the Society of Heritage Planning and Environmental Health. His work involves heritage planning and spiritual tourism in the Varanasi region. He has lectured at various centres in the United States, Europe, East Asia, and Australia and published more than 100 papers and six books on the geography of pilgrimage. His books include titles such as *Trends in the Geography of Pilgrimages* (1987, with R.L. Singh); *Banaras (Varanasi), Cosmic Order, Sacred City, Hindu Traditions* (1993); *The Spirit and Power of Place* (1994); *Banaras Region: A Spiritual and Cultural Guide* (2002, with P.S. Rana); *Panchakroshi Yatra of Banaras* (2002); *Where the Buddha Walked: A Companion to the Buddhist Places of India* (2003); *Literary Images of Banaras* (2004).

Dallen J. Timothy is Associate Professor in the School of Community Resources and Development at Arizona State University, Tempe, Arizona, USA. He is the editor-in-chief of the *Journal of Heritage Tourism* and Visiting Professor of Heritage Tourism at the University of Sunderland, England. Timothy serves on the editorial boards of nine international journals and is reviews editor of *Tourism Geographies*. He is also the secretary of the Tourism Commission of the International Geographical Union and co-editor of the Aspects of Tourism books series published by Channel View Publications in the UK. His research in tourism focuses primarily on heritage, religion, ethnicity, consumption, political boundaries, peripheral and developing regions, and grassroots empowerment. He has published several books and numerous articles and chapters on these subjects.

Boris Vukonić is Professor of Economics at the University of Zagreb, Zagreb, Croatia. He has conducted research throughout the world on many aspects of tourism economics and religious tourism, including working closely with the World Tourism Organization (WTO) and the United Nations Development Program (UNDP). He is a member of the International Academy for Tourism, Vice President of the International Management Development Association, and serves on the Education Council of the WTO. He is the editor of the longstanding journal, *Acta Turistica*, and serves on the editorial board of *Annals of Tourism Research* and several other scientific journals. Vukonić has authored numerous articles and books about various aspects of tourism, including the foundational text, *Tourism and Religion* (Elsevier, 1996).

Preface

The foundations of this book stem from our professional interest in tourism and our own spiritual backgrounds and religious belief systems, which guide our everyday lives. After having travelled for religious purposes ourselves on many occasions, we have realized the important position that spiritually and religiously motivated travel holds within the realm of global tourism. Millions of people travel each year for various reasons that essentially boil down to spiritual foundations, whether or not they adhere to an 'official' religion. In addition to volume, this form of travel is geographically widespread with most regions of the world being home to sacred spaces that appeal to religious adherents or other people who visit sacredscapes out of sheer curiosity.

Despite the pervasiveness and volume of religious tourism throughout the world, relatively little has been said about it in the religion or tourism literatures. A nascent collection of journal articles has appeared during the past 15 years, most in two special issues on the subject in *Annals of Tourism Research* (1992) and *Tourism Recreation Research* (2001), and a few books have been published since 2000 that outline the history of pilgrimage and the transformation of pious journeys into modern-day tourism. Given this dearth of information compared to other forms of tourism and the fact that scholars, religious leaders and pilgrimage destination officials continue to debate the nature of tourism, religion, pilgrimage and religious tourism, the idea for this book developed.

The aim of this volume is twofold. The first purpose is to examine the theoretical bases that underlie the fusion between religion and tourism by highlighting conceptual issues that guide scholarly understanding of spiritually minded or religiously motivated people to venerate certain spaces and visit them. Scholars of religion and scholars of tourism are brought together in this volume to examine the social, economic, historical and political elements of the religion and tourism crossover. Second, in light of the fact that most recent works on the subject focus either on a specific religion or particular patterns of pilgrimage, the second part of the book aims to create empirically an overview of various world religions and spiritual movements and their dealings, requirements, practices and traditions in relation to various forms of travel. Unfortunately, space and time constraints necessarily limited the range and scope of faiths that could have been analysed. The decision to include a specific belief system was based on availability of information and expertise. In this approach, efforts were made to include contributors who are both academic scholars and where possible adherents of the belief systems under examination. In all but a few cases, this was a notable success.

Our hope is that this volume will provide first-hand insight into the spiritual and religious elements of tourism and provide a conceptual foundation for increased research into this pervasive but understudied global phenomenon.

Dallen J. Timothy
Gilbert, Arizona, USA

Daniel H. Olsen
Kitchener, Ontario, Canada

Acknowledgements

The authors would like to express their gratitude to the staff at Routledge, who have made this book possible. Andrew Mould, among the most skilful and insightful commissioning editors in the field, was instrumental in making this book a reality. Zoe Kruze's patience and perseverance at the time of delivery are much appreciated – she has been a pleasure to work with. Michael Hall's interest in the book from the beginning has made this endeavour a success and has facilitated its timely production. We also wish to thank the contributors to this volume who graciously withstood our relay races back and forth for clarifications and additional information.

Finally and more personally, we both would like to thank our wives and children who have put up with our absences during the final months of the book's preparation. Your patience and understanding are truly appreciated – thanks for shouldering so much responsibility alone during our late nights on the computer.

1 Tourism and religious journeys

Daniel H. Olsen and Dallen J. Timothy

Religious travel is not a new phenomenon. Religion has long been an integral motive for undertaking journeys and is usually considered the oldest form of non-economic travel (Jackowski and Smith 1992). Every year millions of people travel to major pilgrimage destinations around the world, both ancient and modern in origin. Jackowski (2000) estimates that approximately 240 million people a year go on pilgrimages, the majority being Christians, Muslims, and Hindus. Religiously or spiritually motivated travel has become widespread and popularized in recent decades, occupying an important segment of international tourism, having grown substantially in recent years both in proportional and absolute terms. A continued increase in this market segment seems to be a foreseeable trend in the future as well (Bywater 1994; Holmberg 1993; Olsen and Timothy 1999; Post *et al.* 1998; Russell 1999; San Filippo 2001; Singh 1998).

Increases in spiritually motivated travel have coincided with the growth of tourism in the modern era (Lloyd 1998), and even though the industry and its "associated practices interact with religious life and the institutions of religion in virtually every corner of the world" (Bremer 2005: 9260), religious tourism is one of the most understudied areas in tourism research (Vukonić 1998). This is particularly so when compared to other aspects of the tourism system and their associated markets. This is surprising because religion has played a key role in the development of leisure over the centuries and has influenced how people utilize their leisure time (Kelly 1982). As such, modern travel patterns and activities cannot be fully understood unless religion is also considered (Mattila *et al.* 2001). Only recently have scholars, governments, and tourism agencies taken notice of the increasing numbers of religiously motivated travelers, or at least the increase in visitation to sacred sites in conjunction with the general growth of cultural and heritage tourism. This public interest has arisen mainly owing to the economic potential of religious tourists. As a result, venerated places are now being seen as tourism resources that can be commodified for travelers interested in cultural and historic sites. Mosques, churches, cathedrals, pilgrimage paths, sacred architecture, and the lure of the metaphysical are used prominently in tourism promotional literature, as evidenced in the recent marketing efforts surrounding the year 2000 and its millennial religious connotations (Olsen and Timothy 1999). As a result of marketing and a growing general interest in cultural tourism, religious sites are being frequented more by curious tourists than by spiritual pilgrims, and are thus commodified and packaged for a tourism audience (Vukonić 1996; Shoval 2000; Shackley 2001a; Olsen 2003).

Researchers from a variety of disciplines have considered different aspects of the relationships between religion and tourism (e.g. Turner and Turner 1978; Morinis 1992; Vukonić 1996; Stoddard and Morinis 1997; Olsen and Timothy 1999, 2002; Schelhe 1999; Swatos and Tomasi 2002; Timothy and Boyd 2003). These and other observers have tended to focus on a number of theoretical and practical concerns, including critiquing the paradigms, theories, definitions, and characteristics of religious travel, the planning of pilgrimages, the management and interpretation of sacred sites, and the training of tour guides (Kaszowski 2000). Bremer (2005) notes three broader approaches in which researchers have considered the inter-sections of religion and tourism: the spatial approach (pilgrims and other tourists occupying the same space with different spatial behaviors), the historical approach (the relationship between religious forms of travel and tourism), and the cultural approach (pilgrimage and tourism as modern practices in a (post)modern world).

The only attempt to date to examine religion and tourism in a holistic manner has been by Vukonić (1996) in his book *Tourism and Religion*. However, he mainly documented his observations and reflections on the topic rather than critically evaluating the existing literature within any of the theoretical or paradigmatic frameworks that have guided most research on tourism and religion. Messenger (1999) investigated early Methodist theologies of leisure at Ocean Grove, a religious seaside resort in nineteenth-century America. More recently, Swatos and Tomasi (2002) compiled a collection of essays focusing on the shift from medieval peni-tential pilgrimage to modern tourism, a special issue of *Tourism Recreation Research* (2002) was dedicated to spiritual journeys, the Religious Tourism and Pilgrimage Research Group of the European Association for Tour-ism and Leisure Education (ATLAS) has published a book based on papers presented at its first expert meeting (Fernandes *et al.* 2003), and Badone and Roseman (2004) edited a volume discussing the intersections of religion and tourism from an anthropological perspective. As well, there are a number of recent journal articles and book chapters that relate to various aspects of religion and tourism (e.g. Cohen 1999; Tweed 2000; Cai *et al.* 2001; Epstein and Kheimets 2001; Jacobs 2001; Joseph and Kavoori 2001; Mattila *et al.* 2001; Shackley 2001b; Tilson 2001; Collins-Kreiner and Kliot 2000; Collins-Kreiner 2002; Koskansky 2002; Bar and Cohen-Hattab 2003; Digance 2003; McNeill 2003; Poria *et al.* 2003; Coleman and Eade 2004; Schramm 2004). However, the literature is still fragmented and lacks synthesis and holistic conceptualization.

The aim of this book is to contribute to the growing literature on religiously motivated travel by reviewing and challenging existing paradigms, concepts, and practices related to pilgrimage and other forms of religious travel. It focuses on a number of theoretical and practical perspectives related to spiritual journeys, including the nature, creation, and management of hallowed place, the pilgrim–tourist dichotomy, the economics of religious tourism, and the educational implications of religious tourism. In addition, much of the book focuses on the intersections of religion and tourism from a perspective that has been little studied, even the perspective of religion and spirituality, whether it is religious leaders commenting on tourism as a social phenomenon, or the perspectives of religious adherents who travel to various destinations in search of truth and enlightenment.

Religion, tourism, and spirituality

Traditionally and historically, pilgrimage has been defined as a physical journey in search of truth, in search of what is sacred or holy (Vukonić 1996: 80). It is where people are drawn to sacred places "where divine power has suddenly burst forth" (Sallnow 1987: 3) as a result of their spiritual magnetism (Preston 1992). This search for truth, enlightenment, or an authentic experience with the divine or holy leads people to travel to sacrosanct sites that have been ritually separated from the profane space of everyday life. Religiously motivated travel, including pilgrimage, has grown tremendously during the past fifty years, surprising many who conjecture that religious pilgrimage is losing social and institutional significance. Modern religious pilgrimage "appears to be at odds with our widely held belief in the progressive development of the West into a complex modern civilization based on science, technology, and reason, rather than on magic, religion, and irrationality" (Campo 1998: 41). As well, postmodern culture, the privatization of religion, and the ability to take cyber-pilgrimages (Heelas 1998; Inoue 2000; Kong 2001; MacWilliams 2002), it could be argued, should lead people to participate in unmediated, reflexive forms of spiritual travel rather than pilgrimages where authenticity of experience is dependent in part on ecclesiastical institutions and structures (Sharf 1998; Rountree 2002; York 2002). Kosti (1998: 5) observes, however, that pilgrimage is increasing at a rapid rate rather than diminishing. This can be seen in Europe, where visitation to religious sites has been increasing while regular church attendance has been on the decline (Nolan and Nolan 1992).

This global resurgence of religious pilgrimage has occurred for many reasons, including the rise of fundamentalism (Friedland 1999; Riesebrodt 2000; Stump 2000), the retreat of some religious faiths into traditional forms of medieval spirituality and religious ritual (Post *et al.* 1998), the increasing investment in mass transportation infrastructure (Griffin 1994), the globalization of the local through the mass media (termed *mediascapes* by Vásquez and Marquandt 2000; see Koskansky 2002), and the recent turn of the millennium (Olsen and Timothy 1999). This indicates in part the increasing numbers of people who are searching for the answers to basic questions of human existence, including "What is the meaning of life?" or, more specifically, "What is the meaning of my life?" (Olsen and Guelke 2004; Clark 1991). This parallels some scholars' view that a growing number of people experience feelings of dislocation and root-lessness, particularly those immersed in western, postmodern social life (MacCannell 1976; Lowenthal 1997), who "search to be themselves, to be givers of sense" (Voyé 2002: 123). According to Nuryanti (1996: 25), the twentieth century has been characterized by the heritage movement, where people actively search out their ancestral roots. In essence, this quest for an understanding of the past involves the asking of the question "Who am I?" in terms of "Who was I?" (Dann 1998: 218) through searching for "new points of orientation . . . strengthen[ing] old boundaries and . . . creat[ing] new ones" (Paasi 2003: 475).

The popularity of religious travel can be seen not only in the increase of religiously motivated travel to sacred sites but also in the combining of New Age spirituality with pilgrimage travel (Rountree 2002). The concept of religion has shifted with the advent of modern secularizing trends, such as post-industrialism, cultural pluralism, and scientific rationality (Baum 2000), which have, according to some social commentators, led to decreasing significance of religious institutions and their associated prac-tices (Houtman and Mascini 2002). As such, the term "religion" is used in everyday public discourse to refer to things outside the realm of tradi-tional religious institutions. This may come in part, as Williams (2002: 603) suggests, in that the use of such terms in everyday vernacular "reflect[s] the deep penetration of New Age ideas of 'spirituality' into . . . culture" (see also Marler and Hadaway 2002). As a result of both of these secularizing trends and the changing use of the word religion, religion is being seen more and more as a privatized and pluralized experience where the "spiritual" and the "religious" are separate. As Heelas (1998: 5) notes, "people have what they take to be 'spiritual' experiences without having to hold religious *beliefs*." In other words, spirituality is an individual experi-ence that is outside "preconstituted discourse[s] of meaning" (Hervieu-Léger 1999 quoted in Voyé 2002: 124), where experimenting with the mixing of various religious traditions – both traditional and alternative – is seen as both accepted and encouraged; thus making "real" religion identifiable with personal faith outside of religious institutions (Tilley 1994: 185). Thus, many people who consider themselves spiritual would not see themselves as religious and vice versa. In fact, atheists and agnostics may also have deep spiritual experiences in relation to nature and their own self-consciousness without believing in god or any organized religious affiliation.

This separation of the spiritual from the religious has led to a re-interpretation of what constitutes the "sacred," where people are no longer constrained by religion in interpreting what spaces they view as sacred (Hammond 1991: 118). While historically the search for the metaphysical or the supernatural has led people to sites where, in their minds, there was the potential to commune with the "holy" (Hauser-Schäublin 1998), Tomasi (2002: 1; italics in original) argues that today "the search for the *super-natural* [has been replaced] by [the] search for *cultural-exotic* and the *sacred.*" Because of this, modern society has expanded what it defines as sacred, bringing about the creation of new sites of sacrality, with travel to these sites being termed pilgrimage in its own right because rather than pilgrimage being travel to sites where heaven and earth converge, it is considered to be journeys "undertaken by a person in quest of a place or a state that he or she believes to embody a *valued ideal*" (Morinis 1992: 4; italics added). In this light, then, pilgrimage has also been extended beyond the "religious" realm to include travel to places symbolizing nationalistic values and ideals (Zelinsky 1990; Guth 1995), disaster sites, such as Ground Zero in New York and the site of the Oklahoma City bombing (Foote 1997; Blasi 2002; Conran 2002), war memorials and ceme-teries (Johnstone 1994; Lloyd 1998; Gough 2000; Seaton 2002), New Age and pagan sacred sites (Attix 2002; York 2002), places related to the lives of literary writers and the settings of their novels (Herbert 2001), places associated with music stars, such as Elvis Presley's mansion (Graceland) in Memphis, Tennessee, or John Lennon's monument, 'Strawberry Fields' in New York City (Alderman 2002; Davidson *et al.* 1990; King 1993; Kruse 2003), nostalgic tourist attractions (e.g. Walt Disney World) (Knight 1999), sporting events (Gammon 2004), genealogical trips (Kurzwell 1995), shopping malls (Pahl 2003), and even cyberspace (Kong 2001; MacWilliams 2002).

Likewise, many people travel to a widening variety of sacred sites not only for religious or spiritual purposes or to have an experience with the sacred in the traditional sense, but also because they are marked and marketed as heritage or cultural attractions to be consumed (Timothy and Boyd 2003). They may visit because they have an educational interest in learning more about the history of a site or understanding a particular religious faith and its culture and beliefs, rather than being motivated purely by pleasure-seeking or spiritual growth. While not necessarily motivated by their own religious beliefs, tourists who belong to a particular religious tradition may visit a site associated with their faith out of a sense of obligation to do so while in the area, for nostalgic reasons, or to educate their family members about their religious beliefs. Tourists also visit sacred sites seeking authentic experiences, whether through watching religious leaders and pilgrims perform rituals or by experiencing a site's "sense of place" or sacred atmosphere (Shackley 2001a, 2002).

Present inquiries

This "blurring of the lines" (Kaelber 2002) between pilgrimage and other forms of travel traditionally viewed as part of tourism has led to an increase in religiously and spiritually motivated travel to a wide range of sacred sites around the world. This has opened up many avenues of research into the area of religion and tourism. However, most research and writing on the topic has centered on four distinct themes of inquiry: distinguishing the pilgrim from the tourist; the characteristics and travel patterns of religious tourists; the economics of religious tourism; and the negative impacts of tourism on religious sites and ceremonies. These four themes will now be discussed.

Pilgrim–tourist dichotomy

The primary focus of research and debate among scholars examining religious travel has been on the tourist and the pilgrim, the main players in the relationship between religion and tourism (Cohen 1998). In much of the literature, tourists and pilgrims are viewed from two distinct perspectives. The first and more popular view is that tourists and pilgrims are similar, if not one and the same, for "even when the role[s] of tourist and pilgrim are combined, they are necessarily different but form a continuum of inseparable elements" (Graburn 1983: 16).

From the perspective of tourism, pilgrims and tourists are structurally and spatially the same or forms of one another, and the very fact that people address the issue as "pilgrim and/or tourist" with dichotomous undertones reflects a lack of understanding about the meaning of a tourist and tourism. Motivations and activities undertaken by travelers have little to do with whether a person is a tourist or not. Tourists are not by their very nature pleasure-seeking hedonists; such a stereotype is a very narrow and erroneous view of tourists as noted in the words of Digance and Cusack (2002: 265): "the superficial view that tourists travel solely for pleasure has been questioned for several decades, and it is now acknowledged that there are many complex reasons why people . . . elect to travel." Simply speaking, a tourist is someone who travels temporarily (at least 24 hours but less than a year) away from home to another region. This is the most widely accepted definition of a tourist among the World Tourism Organization, tourism scholars, and government agencies charged with data collection and promotion and reflects the modern paradigm that there are multitudinous underlying motives for people traveling away from home to become tourists. Motivation for travel is not integral to the definition of tourist or the meaning of tourism. People on short-term humanitarian missions from the developed world offering their services to improve the lives of people of the developing world in areas of health, hygiene, and education are tourists, typically referred to as volunteer tourists (Wearing 2001; Wearing and Neil 2000). Similarly, business people traveling to attend intense marketing meetings or business conventions away from their home areas are tourists (business tourists) the same way a person vacationing on the beach in Jamaica is a tourist (beach tourist). Thus, we speak of types of tourists rather than whether or not one motivation is more important than another in defining a tourist. From this perspective, then, a "pilgrim" is a tourist (religious tourist) who is motivated by spiritual or religious factors.

This notion is important in examining pilgrimage as a form of tourism. Most researchers today do not distinguish between pilgrims and tourists or between pilgrimage and tourism. Rather, pilgrimage is typically accepted as a form of tourism (Fleischer 2000), for it exhibits most of the same general characteristics in terms of travel patterns and the use of trans-portation, services, and infrastructure. For example, Gupta (1999: 91) notes that "apart from the devotional aspect, looked at from the broader point of view, pilgrimage involves sightseeing, travelling, visiting different places and, in some cases, voyaging by air or sea etc. and buying the local memorabilia." This is also noted by Eade (1992) who, in discussing pilgrimage and tourism at Lourdes, France, observed that pilgrims partici-pate in typical tourist activities, dress like tourists, make similar purchases, and cannot be differentiated from their tourist counterparts in the manner they relax at night. Smith (1992) argues that tourists and pilgrims share the fundamentals of travel – leisure time, income, and social sanctions for travel – and in most instances the same infrastructure. Some researchers even view tourism as a quasi-pilgrimage in which travelers associate with other cultures seeking authentic experiences they feel they cannot gain at home (Graburn 1989; MacCannell 1973). In this vein Turner and Turner (1978: 20) write, "a tourist is half a pilgrim, if a pilgrim is half a tourist."

The other, opposing perspective is that pilgrims are not tourists. In this vein, travelers motivated by deep spiritual or religious convictions (i.e. pilgrims) are seen as somehow being different from those motivated by pleasure, education, curiosity, altruism, and relaxation. Typically, it is various religious organizations that hold to this viewpoint. Thus, from a religious institutional perspective, tourists are seen as vacationers while pilgrims are seen as religious devotees (de Sousa 1993). Cohen (1992) differentiates between pilgrims, who journey towards the center of their world, and tourists, who travel away from their center to a pleasurable periphery. They are seen as having different virtues – the pilgrim being pious and humble and sensitive to the host culture, while the tourist is hedonistic, demanding and far more taxing on the destination in terms of services, needs, and desires (de Sousa 1993; Gupta 1999). In sum, those who hold to this perspective essentially define pilgrims and tourists based on their motivation rather than activities and travel patterns.

Smith (1992), rather than viewing pilgrims and tourist as two distinct groups, has placed them both on a continuum with pilgrims at one end and tourists at the other. Situated between the two are infinite possibilities of sacred–secular combinations (de Sousa 1993). Based on Smith's continuum, Fleischer (2000) argues that religious tourism and religious heritage tourism would lie in the middle of this spectrum. Graham and Murray (1997: 401), while accepting the continuum as valid and useful for classifying various sub-markets of contemporary pilgrimage, argue that the scale does not reflect the multi-layered meanings of pilgrimage in the modern era, as "'holy' and 'pious' no longer define the 'spiritual.' Thus the search for personal consciousness and meaning far transcends the realm of the religious, and pilgrimage becomes the product of a multiplicity of motivations, attitudes and behavioural mindsets."

Also based on this continuum is the notion that travelers can swing from one extreme to the other. Schlehe (1999: 8) rightly notes that "a considerable number of pilgrims will perform the proper rituals first, and then enjoy their nocturnal jaunt to the beach." Jackson and Hudman (1995), in discussing tourist visits at English cathedrals, note that while cultural heritage and architecture are the main motivations for visitation, visitors generally are touched by religious feelings when they are on site. Along these lines Eade (1992) comments that while tourists may appear different from pilgrims, they can be moved by religious emotion just as well as pilgrims. Likewise, de Sousa (1993) argues that people can switch from being a pilgrim to a tourist without the individual being aware of the change from one to the other. Hunt (1984: 408) states that in ancient Rome journeys undertaken out of curiosity were "often inseparable from . . . the element of piety and devotion which might be aroused in the traveler when he arrived at places highly charged with the legacy of the past. The tourist, in other words, might easily become the pilgrim." The difficulty of distinguishing between pilgrims and other tourists can be seen in the official statistics of many countries, where existing figures tend to combine pilgrimage and religious tourism with cultural or heritage tourism (Russell 1999).

Characteristics and travel patterns of religious tourists

There is a small but important literature that focuses on the characteristics and travel patterns of religiously motivated tourists. Rinschede (1992) differentiates between different forms of religious tourism based on time involved and distance traveled, namely short- and long-term religious tourism. The short-term type involves travel to nearby pilgrimage centers or religious conferences, while long-term religious tourism involves travel to religious sites and conferences around the world. Rinschede also distinguishes between the two forms of religious tourism using other variables such as number of participants, seasonality of demand, social structure, and choice of transportation. Bhardwaj and Rinschede (1988) distinguish between the periodic pilgrimage and the uninterrupted pilgrimage, which refers to a lifelong journey towards perfection.

Much has been made in the anthropological literature over the experiences of religious travelers and why they travel. The religious motives literature was initiated primarily by Eliade (1961) who noted that religions have sacred centers that people desire to visit. Turner (1973) argued that sacred sites were typically on the periphery of society away from the profane social world. Cohen (1992) sees pilgrims as traveling to the center of their religious world, and Eade (1992: 129) suggests that religious travelers go on pilgrimages to gain "emotional release ... from the world of everyday structure." Turner (1984) terms this emotional release *liminality*, where during a journey travelers experience a temporary release from social ties and may "expect things of themselves and others which they may not expect while they are at home" (Holmberg 1993: 23). Travelers expect a spiritual experience and a world where they meet with other travelers in a classless society called *communitas* in Turnerian lexicon (Turner 1973; Turner and Turner 1978). In a sense then they expect hyperreal experiences where they are "betwixt and between periods of normal, everyday life" (Holmberg 1993: 23). With no formal ties to the outside world, inner reflection and meditation is easier for many pilgrims (Turner 1987). While this notion of *communitas* has been contested recently (Coleman and Elsner 1991; Eade and Sallnow 1991), the principal of *liminality* as a hyperreal experience is useful for explaining the experiences of religiously motivated travelers.

In terms of behavior patterns, the literature shows that religious tourists tend to be focused in their itineraries and areas of interest. Sizer (1999), in arguing for increased contact with indigenous Arab Christian groups during group pilgrimages to the Holy Land, discusses the different categorizations of Protestant pilgrimage to the Holy Land, dividing it into three groups: Evangelicals, Fundamentalists, and Living Stones. Evangelicals focus on educational tours of sites of biblical significance, while fundamentalists travel for similar reasons but include eschatological motivations for travel. The third group focuses more on experiences with indigenous Christian groups in their travels. Russell (1999) suggests that destinations that attract the largest number of religious tourists are those either associated with sites from the Bible, Quran, or other sacred texts, or with spiritual happenings, such as miracles and visions.

Fleischer (2000) and Collins-Kreiner and Kliot (2000) note that there are differences between the visitation patterns of Catholic and Protestant pilgrims (those who listed themselves as pilgrims on an official government exit survey) in Israel. Fleischer (2000) discovered that Catholic pilgrims tended to travel solely to biblical and historic sites, whereas Protestants were more interested in the land of the Bible as a whole. Catholic tours were strictly organized in terms of number of sites to be visited and the requirement to participate in daily prayers and mass, while Protestant tours were more flexible with their requirements for prayer. Collins-Kreiner and Kliot (2000) suggest that this has everything to do with the religious beliefs and background of the tour participants. Fleischer (2000) noted that the Catholic and Protestant visitors stayed in Israel a relatively short period of time, visited many sites with emphasis on holy and historical sites, traveled within organized tours, and preferred to visit during the main Christian holidays of Christmas and Easter and some during the off-season.

Another issue in discussing behavior patterns and characteristics of religious tourists is that of tour guides. According to Wilkinson (1988), modern religious travelers prefer to have as their tour guide someone who has the same views of Christianity they can respect and relate to. This is because most religious travelers, while acknowledging the presence of other churches and belief systems, are more concerned with their own orthodoxy and beliefs in terms of how religious sites are interpreted to them.

Economics of religious tourism

Economics and religion have been influential forces in shaping world history. However, according to Vukonić (2002), the economic aspects of religious travel have been the least studied topic in relation to the religion–tourism crossover, only being of interest to researchers when a single sacred site is under consideration. Religious pilgrimage has a history of being an economic generator in the areas pilgrims visited, as services developed to cater to their needs. This is much the same today, where in many places religious sites are the main tourist attractions and sometimes anchor entire economies, such as in Santiago de Compostela, Medjugorje, Lourdes, and Mecca. In many countries and localities, tourism is seen as a way either to diversify or rescue a struggling economy, especially with current tourism forecasts, as mentioned earlier, showing that religious tourism will increase in the near future (Russell 1999). Jackowski and Smith (1992) give the example of pilgrimage sites in Poland where because of Second World War damage and communist repression, a tourism infrastructure was virtually non-existent in the early 1990s. They point out that this lack of infrastructure limited tourists' length of stay, subsequently limiting opportunities for local residents to gain from the economic benefits of pilgrimage tourism. Jackowski and Smith argue for the potential of religious tourism to become an important source of income and employment in Poland. Tourism development at the El Rocio shrine in Spain has also played a central economic role in increasing employment and local revenues (Crain 1996).

In some areas the demand for services from tourists and pilgrims has changed the landscape and urban land use patterns at pilgrimage centers. Two examples are Lourdes, France, and Fatima, Portugal, where the pilgrimage zone can be divided into the profane or commercialized area (restaurants, shops, hotels, etc.) and the sacred (pilgrimage shrines, churches, etc.) (Gesler 1996; Rinschede 1990). The urban morphology of Lourdes has changed in that there is an intensive agglomeration of souvenir shops and restaurants along the roads leading from the commercialized part of town to the pilgrimage areas (Rinschede 1986). Similarly, Gupta (1999) points out that almost every pilgrimage place in India has shops catering to tourists and pilgrims seeking special handicrafts.

In many cases the economic benefits of tourism to sites of religious importance are seen to outweigh the negative impacts. This has been illustrated in the literature focusing on less-developed areas of the world. Baedcharoen (2000) illustrates how in Thailand residents are ambivalent about the negative impacts of tourism on religious sites because of the economic benefits gained. Shackley (2002: 349) notes that

> although excessive numbers of visitors may adversely affect the fabric
> of a cathedral (by theft, vandalism, graffiti, erosion, etc.), as well as
> diminishing the quality of the experience of visitors by excessive noise,
> and crowding, visitors undeniably make a positive economic input into
> cathedral finances.

Shackley (1999) elsewhere has discussed how because of the potential revenue, Buddhist monks allow their religious festivals to be interrupted by tourists. These examples are parallel to a comment by Fleischer (2000), who argues that the economic impacts associated with religious tourism are greater than those associated with other market segments, because pilgrims and other religious travelers are avid buyers of religious souvenirs. Because the economics of religious tourism can be so lucrative, many religious groups are willing to put up with ignorant tourists who visit and disturb their religious centers (Baedcharoen 2000).

From a religious perspective, while "there are probably no orthodox theologians or other theorists who would deny the economic impacts of religious tourism" (Vukonić 1998: 86), there is some perceived incompatibility between religion and economics and between the material and the spiritual (immaterial) (Vukonić 2002; Zaidman 2003). Some theologians argue that the consumption or indulgence associated with tourism corrupts the spiritual nature of traveling, changing tourism from a process of accumulating knowledge and wisdom and appreciation for the creations of God to a forum for hedonistic indulgence (Vukonić 1998). As a result, some religious groups distance themselves from the idea of overtly benefiting from tourism, though many site managers encourage visitation as a way of funding maintenance and preservation (Griffin 1994; Stevens 1988; Willis 1994). While religious items have been sold at holy sites for millennia (Evans 1998; Houlihan 2000), religious authorities and some academics have criticized vendors and entrepreneurs who reproduce versions of devotional items and other religious products for sale, preferring to see them as tourist trash or religious kitsch. This raises issues of commodification, religious authenticity, and dilution of spiritual value (Vukonić 2002; Zaidman 2003).

Negative impacts of tourism on religious sites/ceremonies

While religious tourism has some positive impacts in an economic sense, much of the literature focuses on the negative side of religious tourism in relation to sites and ceremonies. While tourism is seen in many circles as a way of contributing to the preservation of heritage and religious sites and to bolster sagging economies, most observers feel it is a destructive force in terms of cultural unity and degradation of the natural and built environment. Gupta (1999) suggests that the only enduring difference between pilgrimage and tourism is that pilgrimage has not produced the negative cultural, environmental and social impacts associated with mass tourism. However, this is not entirely correct. In fact there are many recorded instances where pilgrims, those who should respect holy places the most, are just as culpable for breaking off pieces of shrines, churches, mosques, and natural sites as non-religious tourists are (Powell 2003; Timothy 1994, 1999). Mass tourism to sites of religious significance has caused some structures to be closed owing to mismanagement and overuse. Overcrowding leads to the local populations having little room to enjoy their own spiritual environment (Fish and Fish 1993). Mass tourism, the media, and various social groups have taken traditional pilgrimages and transformed them from a cult ritual to a festival status with an international and secular flavor (Crain 1992).

Some commentators have noted that tourism "erodes the penitential dimension of pilgrimage" (Preston 1992: 36) and disengages the religiously motivated traveler from the "moving experiences" he or she should feel (Pfaffenberger 1983: 60). Mass tourism has in many instances violated the sanctity of sacred places, for example at Jebel Musa (Mt Sinai) in Egypt, (Hobbs 1992), and at Buddhist Himalayan monasteries in Tibet and Nepal (Kaur 1985; Singh 2004; Shackley 1999). In many European countries mass tourism has almost completely taken over Christian religious sites, causing these places of worship to cease their normal functions to some degree (Cohen 1998). Attitudes of modern tourists and their lack of education about proper behavior and the meanings behind sacred rituals also downplay the religious significance of sacred sites. For many tourists travel to pilgrimage places is combined with other educational and holiday-type activities found in the immediate area (Rinschede 1992). Many people travel to religious sites to fulfill social goals and are described in a heritage context by Moscardo (1996) as "mindless visitors." This can cause problems for those who attend religious sites to worship and meditate or fulfill obligations (Timothy 1994), for pilgrims and other religiously motivated tourists visit sacred sites to worship because they believe they can become closer to deity, more so than worshipping in regular churches, temples, or synagogues (Marshall 1994).

Cohen (1998: 7) argues that mass tourism has a negative effect on the religiosity or level of spirituality of people who live in tourist destinations. He suggests that the impact is "generally a secularizing one – a weakening of the local adherence to religion and of the beliefs in the sacredness and efficacy of holy places, rituals, and customs." This pattern may be seen in the secularization of religion in general, which involves "a gradual narrowing down, if not the elimination, of the role of religious beliefs, practices and institutions in everyday life" (Madan 1986: 257). Beckerleg (1995) and Sindiga (1996) provide examples from East Africa where tourism influences young Muslim boys negatively by leading to irregular and infrequent mosque attendance, as well as drug use, drinking, and sexual activity. Another example of this is found at Mount Sinai, where monks at St Katherine's Monastery complain that the sacred mountain has lost its inviolability because 300–500 people camp on the mountain every night. The monks believe this activity detracts from the spirit of the place, especially the resultant litter, pollution, and noise (Shackley 1998). Nolan and Nolan (1992) note the impact of tourism on religious festivals in Europe. Similarly, Shackley (1999: 96) argues that the character of religious dance rituals and festivals in the Himalayas has changed as a result of the large number of non-Buddhist tourists who neither understand the rituals nor reverence the religious meanings behind them. These tourists' behavior excludes local residents, decreases grassroots participation, and alters the function of the festivals as a basis for social cohesion.

The present volume

This book, as noted earlier, aims to examine the crossover between religion and tourism. The discussion so far in this chapter only begins to scratch the surface of what has been said or what could be said about the matter; these issues and many others are discussed in the chapters that follow. The book is divided into two sections. Part I, "Concepts, concerns and management issues," deals primarily with substantial conceptual problems, theoretical concerns, and some practical implications. In Chapter 2, Thomas S. Bremer examines the meaning of place and the role of sacred and tourist spaces in creating personal identity and meaning among visitors to religious sites. The third chapter (Justine Digance) examines the traditional meanings and views of religious pilgrimage based on Christian, Hindu, Islamic, and Buddhist traditions, and the contemporary practice of secular pilgrimage that occurs outside of the major religious traditions. In Chapter 4, Lutz Kaelber confirms that pilgrimages are alive and well throughout the world today. He traces the transformation of ancient pilgrimage traditions to modern-day versions, including secularized religious journeys, and introduces the idea of "virtual pilgrimages," which allow people to participate in pilgrim voyages by proxy, never having to leave home. In his piece on religious humanism (Chapter 5) C. Michael Hall speaks to the idea that life itself may be a pilgrimage in which people's faith may be deepened, spirits may grow, yet simultaneously, doubts about the existence of deity may deepen. Spiritual experiences, he suggests, can come from within rather than from without and may not necessarily involve a belief in deity. In the sixth chapter, Erik H. Cohen examines the junction of religious tourism, educational tourism, and religious education as manifested in people's quests for knowledge and information, identity, and community formation. Myra Shackley (Chapter 7) looks at the commodification of religious material culture as represented in the development of religious souvenirs for sale at sacrosanct attractions. Finally, in Chapter 8, Daniel H. Olsen reviews the main concerns and issues related to religious site management at the internal level (e.g. crowding, conservation, finances) and at the external level, dealing with issues such as stakeholder involvement, regional planning, and destination marketing.

The chapters comprising Part II, "Religious traditions and tourism," address individual religions and their treatment of tourism and pilgrimage based on doctrines, scriptures/holy writings, traditions, and common practices associated with travel. With clear space constraints in mind, and being cognizant of the vast array of world religions, the following religious or spiritual belief systems were selected for inclusion in this book: the spiritual philosophies of East Asia (Chapter 9, Chao Guo), paganism and the New Age (Chapter 10, Dallen Timothy and Paul Conover), Judaism (Chapter 11, Mara Cohen Ioannides and Dimitri Ioannides), Buddhism (Chapter 12, Michael Hall), Islam (Chapter 13, Dallen Timothy and Tom Iverson), Sikhism (Chapter 14, Rajinder Jutla), Hinduism (Chapter 15, Rana P.B. Singh), Catholicism (Chapter 16, Boris Vukonić) and the Church of Jesus Christ of Latter-day Saints (Chapter 17, Daniel H. Olsen). The concluding chapter consolidates many of the main issues and concepts touched upon throughout the book and suggests areas of weakness in the study of religion and tourism that could fruitfully be brought to bear on future research.

References

Alderman, D.H. (2002) "Writing on the Graceland Wall: on the importance of authorship in pilgrimage landscapes," *Tourism Recreation Research* 27(2): 27–33.

Attix, S.A. (2002) "New Age-oriented special interest travel: an exploratory study," *Tourism Recreation Research* 27(2): 51–58.

Badone, E. and Roseman, S.R. (2004) *Intersecting Journeys: The Anthropology of Pilgrimage and Tourism*, Chicago: University of Illinois Press.

Baedcharoen, I. (2000) "Impacts of Religious Tourism in Thailand," unpublished Masters thesis, Department of Tourism, University of Otago, Dunedin, New Zealand.

Bar, D. and Cohen-Hattab, K. (2003) "A new kind of pilgrimage: the modern tourism pilgrim of nineteenth-century and early twentieth-century Palestine," *Middle Eastern Studies* 39(2): 131–148.

Baum, G. (2000) "Solidarity with the poor," in S.M.P. Harper (ed.) *The Lab, the Temple, and the Market*, Ottawa: International Development Research Centre.

Beckerleg, S. (1995) "'Brown sugar' or Friday prayers: youth choices and community building in coastal Kenya," *African Affairs* 94(374): 23–38.

Bhardwaj, S.M. and Rinschede, G. (1988) "Pilgrimage: a world-wide phenomenon," in S.M. Bhardwaj and G. Rinschede (eds) *Pilgrimage in World Religions*, Berlin: Dietrich Reimer Verlag.

Blasi, A.J. (2002) "Visitation to disaster sites," in W.H. Swatos and L. Tomasi (eds) *From Medieval Pilgrimage to Religious Tourism*, Westport, CT: Praeger.

Bremer, T.S. (2005) "Tourism and religion," in L. Jones (ed.) *Encyclopedia of Religion*, Detroit: Macmillan.

Bywater, M. (1994) "Religious travel in Europe," *Travel & Tourism Analyst* 2: 39–52.

Cai, L.A., Schwartz, Z., and Cohen, E. (2001) "Senior tourists in the Holy Land," *Journal of Teaching in Travel and Tourism* 1(4): 19–33.

Campo, J.E. (1998) "American pilgrimage landscapes," *Annals of the American Academy of Political and Social Science* 558: 40–56.

Clark, M. (1991) "Developments in human geography: niches for a Christian contribution," *Area* 23(4): 339–345.

Cohen, E. (1992) "Pilgrimage centers: concentric and excentric," *Annals of Tourism Research* 19: 33–50.

Cohen, E. (1998) "Tourism and religion: a comparative perspective," *Pacific Tourism Review* 2: 1–10.

Cohen, E.H. (1999) "Informal marketing of Israel Experience educational tours," *Journal of Travel and Tourism Marketing* 37: 238–243.

Coleman, S. and Eade, J. (2004) "Introduction: reframing pilgrimage," in S. Coleman and J. Eade (eds) *Reframing Pilgrimage: Cultures in Motion*, London: Routledge.

Coleman, S. and Elsner, J. (1991) "Contesting pilgrimage: current views and future directions," *Cambridge Anthropology* 15(3): 63–73.

Collins-Kreiner, N. (2002) "Is there a connection between pilgrimage and tourism? The Jewish religion," *International Journal of Tourism Sciences* 2(2): 1–18.

Collins-Kreiner, N. and Kliot, N. (2000) "Pilgrimage tourism in the Holy Land: the behavioural characteristics of Christian pilgrims," *GeoJournal* 50: 55–67.

Conran, T. (2002) "Solemn witness: a pilgrimage to Ground Zero at the World Trade Center," *Journal of Systemic Therapies* 21(3): 39–47.

Crain, M.M. (1992) "Pilgrims, 'Yuppies', and media men: the transformation of an Andalusian pilgrimage," in J. Boissevain (ed.) *Revitalizing European Rituals*, New York: Routledge.

Crain, M.M. (1996) "Contested territories: the politics of touristic development at the Shrine of El Rocío in Southwestern Andalusia," in J. Boissevain (ed.) *Coping with Tourists: European Reactions to Mass Tourism*, Providence, RI: Berghahn Books.

Dann, G.M.S. (1998) "'There's no business like old business': tourism, the nostalgia industry of the future," in W.F. Theobald (ed.) *Global Tourism*, Boston: Butterworth Heinemann.

Davidson, J.W., Hecht, A., and Whitney, H.A. (1990) "The pilgrimage to Graceland," in G. Rinschede and S.M. Bhardwaj (eds) *Pilgrimage in the United States*, Berlin: Dietrich Reimer Verlag.

de Sousa, D. (1993) "Tourism and pilgrimage: tourists as pilgrims?," *Contours* 6(2): 4–8.

Digance, J. (2003) "Pilgrimage at contested sites," *Annals of Tourism Research* 30: 143–159.

Digance, J. and Cusack, C. (2002) "Glastonbury: a tourist town for all seasons," in G.M.S. Dann (ed.) *The Tourist as a Metaphor of the Social World*, Wallingford: CABI.

Eade, J. (1992) "Pilgrimage and tourism at Lourdes, France," *Annals of Tourism Research* 19: 18–32.

Eade, J. and Sallnow, M.J. (1991) *Contesting the Sacred*, London: Routledge.

Eliade, M. (1961) *The Sacred and the Profane: The Nature of Religion*, New York: Harper & Row.

Epstein, A.D. and Kheimets, N.G. (2001) "Looking for Pontius Pilate's footprints near the Western Wall: Russian Jewish tourists in Jerusalem," *Tourism, Culture & Communication* 3: 37–56.

Evans, G. (1998) "Mementos to take home: the ancient trade in souvenirs," in J.M. Fladmark (ed.) *In Search of Heritage as Pilgrim or Tourist?*, Shaftesbury: Donhead.

Fernandes, C., McGettigan, F., and Edwards, J. (2003) *Religious Tourism and Pilgrimage*, Fatima, Portugal: Tourism Board of Leiria/Fatima.

Fish, J.M. and Fish, M. (1993) "International tourism and pilgrimage: a discussion," *The Journal of East and West Studies* 22(2): 83–90.

Fleischer, A. (2000) "The tourist behind the pilgrim in the Holy Land," *Hospitality Management* 19: 311–326.

Foote, K.E. (1997) *Shadowed Ground: America's Landscapes of Violence and Tragedy*, Austin: University of Texas Press.

Friedland, R. (1999) "When God walks in history," *Tikkun* 14(3): 17–22.

Gammon, S. (2004) "Secular pilgrimage and sport tourism," in B.W. Ritchie and D. Adair (eds) *Sport Tourism: Interrelationships, Impacts and Issues*, Clevedon: Channel View.

Gesler, W. (1996) "Lourdes: healing in a place of pilgrimage," *Health & Place* 2(2): 95–105.

Gough, P. (2000) "From heroes' groves to parks of peace: landscapes of remembrance, protest and peace," *Landscape Research* 25(2): 213–228.

Graburn, N.H.H. (1983) "The anthropology of tourism," *Annals of Tourism Research* 10: 9–33.

Graburn, N.H.H. (1989) "Tourism: the sacred journey," in V.L. Smith (ed.) *Hosts and Guests: The Anthropology of Tourism*, Philadelphia: University of Pennsylvania Press.

Graham, B. and Murray, M. (1997) "The spiritual and the profane: the pilgrimage to Santiago de Compostela," *Ecumene* 4(4): 389–409.

Griffin, J. (1994) "Order of service," *Leisure Management* 14(12): 30–32.

Gupta, V. (1999) "Sustainable tourism: learning from Indian religious traditions," *International Journal of Contemporary Hospitality Management* 11(2/3): 91–95.

Guth, K. (1995) "Pilgrimages in contemporary Europe: signs of national and universal culture," *History of European Ideas* 20(4–6): 821–835.

Hammond, P.E. (1991) "How to think about the sacred in a secular age," in D.G. Bromley (ed.) *Religion and the Social Order*, Greenwich, CT: JAI Press.

Hauser-Schäublin, B. (1998) "Temples and tourism: between adaptation, resistance and surrender?," *Rima* 32(1): 144–178.

Heelas, P. (1998) "Introduction: on differentiation and dedifferentiation," in P. Heelas (ed.) *Religion, Modernity and Postmodernity*, Oxford: Blackwell.

Herbert, D. (2001) "Literary places, tourism and the heritage experience," *Annals of Tourism Research* 28: 312–333.

Hervieu-Léger, D. (1999) *Le Pèlerin et Le Converti: La Religion en Mouvement*, Paris: Flammarion.

Hobbs, J.J. (1992) "Sacred space and touristic development at Jebel Musa (Mt. Sinai), Egypt," *Journal of Cultural Geography* 12(2): 99–112.

Holmberg, C.B. (1993) "Spiritual pilgrimages: traditional and hyperreal motivations for travel and tourism," *Visions in Leisure and Business* 12(2): 18–27.

Houlihan, M. (2000) "Souvenirs with soul: 800 years of pilgrimage to Santiago de Compostela," in M. Hitchcock and K. Teague (eds) *Souvenirs: The Material Culture of Tourism*, Aldershot: Ashgate.

Houtman, D. and Mascini, P. (2002) "Why do churches become empty, while New Age grows? Secularization and religious change in the Netherlands," *Journal for the Scientific Study of Religion* 41(3): 455–473.

Hunt, E.D. (1984) "Travel, tourism and piety in the Roman Empire: a context for the beginnings of Christian pilgrimage," *Echos Du Monde Classique* 28: 391–417.

Inoue, N. (2000) "From religious conformity to innovation: new ideas of religious journey and holy places," *Social Compass* 47(1): 21–32.

Jackowski, A. (2000) "Religious tourism: problems with terminology," in A. Jackowski (ed.) *Peregrinus Cracoviensis*, Krakow: Institute of Geography, Jagiellonian University.

Jackowski, A. and Smith, V.L. (1992) "Polish pilgrim-tourists," *Annals of Tourism Research* 19: 92–106.

Jackson, R.H. and Hudman, L. (1995) "Pilgrimage tourism and English cathedrals: the role of religion in travel," *The Tourist Review* 4: 40–48.

Jacobs, C.F. (2001) "Folk for whom? Tourist guidebooks, local color, and the spiritual churches of New Orleans," *Journal of American Folklore* 114: 309–330.

Johnstone, I.M. (1994) "A visit to the Western Front: a pilgrimage tour of World War I cemeteries and memorials in northern France and Belgium, August–September 1993," *Australian Folklore* 9: 60–76.

Joseph, C.A. and Kavoori, A.P. (2001) "Mediated resistance: tourism and the host community," *Annals of Tourism Research* 28: 998–1009.

Kaelber, L. (2002) "The sociology of Medieval pilgrimage: contested views and shifting boundaries," in W.H. Swatos and L. Tomasi (eds) *From Medieval Pilgrimage to Religious Tourism*, Westport, CT: Praeger.

Kaszowski, L. (2000) "Methodology of pilgrimaging," in A. Jackowski (ed.) *Peregrinus Cracoviensis*, Krakow: Institute of Geography, Jagiellonian University.

Kaur, J. (1985) *Himalayan Pilgrimages and the New Tourism*, New Delhi: Himalayan Books.

Kelly, J.R. (1982) *Leisure*, Englewood Cliffs, NJ: Prentice Hall.

King, C. (1993) "His truth goes marching on: Elvis Presley and the pilgrimage to Graceland," in I. Reader and T. Walter (eds) *Pilgrimage in Popular Culture*, London: Macmillan.

Knight, C.K. (1999) "Mickey, Minnie, and Mecca: destination Disney World – pilgrimage in the twentieth century," in D. Perlmutter and D. Koppman (eds) *Reclaiming the Spiritual in Art*, Albany: State University of New York Press.

Kong, L. (2001) "Religion and technology: refiguring place, space, identity and community," *Area* 33(4): 404–413.

Koskansky, O. (2002) "Tourism, charity, and profit: the movement of money in Moroccan Jewish pilgrimage," *Cultural Anthropology* 17(3): 359–400.

Kosti, F. (1998) "The 'enchantment' of religious tourism: the case of Mount Athos, Greece," *Vrijeti Studies* 15(4): 5–13.

Kruse, R.J. (2003) "Imagining Strawberry Fields as a place of pilgrimage," *Area* 35(2): 154–162.

Kurzwell, A. (1995) "Genealogy as a spiritual pilgrimage," *Avotaynu* 11(3): 15–20.

Lloyd, D.W. (1998) *Battlefield Tourism*, New York: Berg.

Lowenthal, D. (1997) *The Heritage Crusade and the Spoils of History*, London: Viking.

MacCannell, D. (1973) "Staged authenticity: arrangements of social space in tourist settings," *American Journal of Sociology* 79(3): 589–603.

MacCannell, D. (1976) *The Tourist: A New Theory of the Leisure Class*, New York: Schocken Books.

McNeill, D. (2003) "Rome, global city? Church, state and the Jubilee 2000," *Political Geography* 22: 535–556.

MacWilliams, M.W. (2002) "Virtual pilgrimages on the Internet," *Religion* 32: 315–335.

Madan, T.N. (1986) "Secularization and the Sikh religious tradition," *Social Compass* 33(2/3): 257–273.

Marler, P.L. and Hadaway, C.K. (2002) "'Being religious' or 'being spiritual' in America: a zero-sum proposition," *Journal for the Scientific Study of Religion* 41(2): 289–300.

Marshall, J. (1994) "The Mosque on Erb Street: a study in sacred and profane space," *Environments* 22(2): 55–66.

Mattila, A.S., Apostolopoulous, Y., Sönmez, S., Yu, L., and Sasidharan, V. (2001) "The impact of gender and religion on college students' spring break behavior," *Journal of Travel Research* 40: 193–200.

Messenger, T. (1999) *Holy Leisure: Recreation and Religion in God's Square Mile*, Minneapolis: University of Minnesota Press.

Morinis, E.A. (1992) "Introduction: the territory of the anthropology of pilgrimage," in A. Morinis (ed.) *Sacred Journeys: The Anthropology of Pilgrimage*, Westport, CT: Greenwood.

Moscardo, G. (1996) "Mindful visitors: heritage and tourism," *Annals of Tourism Research* 23: 376–397.

Nolan, M.L. and Nolan, S. (1992) "Religious sites as tourism attractions in Europe," *Annals of Tourism Research* 19: 68–78.

Nuryanti, W. (1996) "Heritage and postmodern tourism," *Annals of Tourism Research* 23(2): 249–260.

Olsen, D.H. (2003) "Heritage, tourism, and the commodification of religion," *Tourism Recreation Research* 28(3): 99–104.

Olsen, D.H. and Guelke, J.K. (2004) "Spatial transgression and the BYU Jerusalem Center controversy," *Professional Geographer* 56(4): 503–515.

Olsen, D.H. and Timothy, D.J. (1999) "Tourism 2000: selling the Millennium," *Tourism Management* 20: 389–392.

Olsen, D.H. and Timothy, D.J. (2002) "Contested religious heritage: differing views of Mormon heritage," *Tourism Recreation Research* 27(2): 7–15.

Paasi, A. (2003) "Region and place: regional identity in question," *Progress in Human Geography* 27(4): 475–485.

Pahl, J. (2003) "Pilgrimage to the Mall of America," *Word and World* 23(3): 263–271.

Pfaffenberger, B. (1983) "Serious pilgrims and frivolous tourists: the chimera of tourism in the pilgrimage of Sri Lanka," *Annals of Tourism Research* 10: 57–74.

Poria, Y., Butler, R.W., and Airey, D. (2003) "Tourism, religion and religiosity: a holy mess," *Current Issues in Tourism* 6(4): 340–363.

Post, P., Pieper, J., and van Uden, M. (1998) *The Modern Pilgrim: Multidisciplinary Explorations of Christian Pilgrimage*, Leuven: Uitgeverij Peeters.

Powell, E.A. (2003) "Solstice at the stones," *Archaeology* 56(5): 36–41.

Preston, J.J. (1992) "Spiritual magnetism: an organizing principle for the study of pilgrimage," in A. Morinis (ed.) *Sacred Journeys: The Anthropology of Pilgrimage*, Westport, CT: Greenwood.

Putter, J.A. (1998) "Two pilgrim towns: a quest for St Andrew and St Margaret," in J.M. Fladmark (ed.) *In Search of Heritage: as Pilgrim or Tourism?*, Shaftesbury: Donhead.

Riesebrodt, M. (2000) "Fundamentalism and the resurgence of religion," *Numen* 47: 266–287.

Rinschede, G. (1986) "The pilgrimage town of Lourdes," *Journal of Cultural Geography* 7(1): 21–34.

Rinschede, G. (1990) "Religionstourismus," *Geographische Rundschau* 42(1): 14–20.

Rinschede, G. (1992) "Forms of religious tourism," *Annals of Tourism Research* 19: 51–67.

Rountree, K. (2002) "Goddess pilgrims as tourists: inscribing their body through sacred travel," *Sociology of Religion* 63(4): 475–496.

Russell, P. (1999) "Religious travel in the new millennium," *Travel & Tourism Analyst* 5: 39–68.

Sallnow, M.J. (1987) *Pilgrims of the Andes: Regional Cults in Cusco*, Leiden: Koninklijke Brill.

San Filippo, M. (2001) "The religious niche," *Travel Weekly* 60(18): 12.

Schlehe, J. (1999) "Tourism to holy sites," *International Institute for Asian Studies Newsletter* 19: 8.

Schramm, K. (2004) "Coming home to the motherland: pilgrimage tourism in Ghana," in S. Coleman and J. Eade (eds) *Reframing Pilgrimage: Cultures in Motion*, London: Routledge.

Seaton, A.V. (2002) "Thanatourism's final frontiers? Visits to cemeteries, church-yards and funerary sites as sacred and secular pilgrimage," *Tourism Recreation Research* 27(2): 73–82.

Shackley, M. (1998) "A golden calf in sacred space? The future of St. Katherine's Monastery, Mount Sinai (Egypt)," *International Journal of Heritage Studies* 4(3/4): 123–134.

Shackley, M. (1999) "Managing the cultural impacts of religious tourism in the Himalayas, Tibet and Nepal," in M. Robinson and P. Boniface (eds) *Tourism and Cultural Conflicts*, New York: CABI.

Shackley, M. (2001a) *Managing Sacred Sites: Service Provision and Visitor Experience*, London: Continuum.

Shackley, M. (2001b) "Sacred world heritage sites: balancing meaning with management," *Tourism Recreation Research* 26(1): 5–10.

Shackley, M. (2002) "Space, sanctity and service: the English cathedral as heterotopia," *International Journal of Tourism Research* 4: 345–352.

Sharf, R.H. (1998) "Experience," in M.C. Taylor (ed.) *Critical Terms for Religious Studies*, Chicago: University of Chicago Press.

Shoval, N. (2000) "Commodification and theming of the sacred: changing patterns of tourist consumption in the 'Holy Land'," in M. Gottiener (ed.) *New Forms of Consumption: Consumers, Culture, and Commodification*, Oxford: Rowman & Littlefield.

Sindiga, I. (1996) "International tourism in Kenya and the marginalization of the Waswahili," *Tourism Management* 17: 425–432.

Singh, Sagar (2004) "Religion, heritage and travel: case references from the Indian Himalayas," *Current Issues in Tourism* 7(1): 44–65.

Singh, Shalini (1998) "Probing the product life cycle further," *Tourism Recreation Research* 23(2): 61–63.

Sizer, S.R. (1999) "The ethical challenges of managing pilgrimages to the Holy Land," *International Journal of Contemporary Hospitality Management* 11(2/3): 85–90.

Smith, V.L. (1992) "Introduction: the quest in guest," *Annals of Tourism Research* 19: 1–17.

Stevens, T. (1988) "The ministry of welcome: tourism and religious sites," *Leisure Management* 8(3): 41–44.

Stoddard, R.H. and Morinis, E.A. (eds) (1997) *Sacred Places, Sacred Spaces: The Geography of Pilgrimages*, Baton Rouge: Louisiana State University Press.

Stump, R.W. (2000) *Boundaries of Faith: Geographical Perspectives on Religious Fundamentalism*, Lanham, MD: Rowman & Littlefield.

Swatos, W.H. and Tomasi, L. (eds) (2002) *From Medieval Pilgrimage to Religious Tourism*, Westport, CT: Praeger.

Tilley, T.W. (1994) "The institutional element in religious experience," *Modern Theology* 10(2): 185–212.

Tilson, D.J. (2001) "Religious tourism, public relations and church-state partnerships," *Public Relations Quarterly* 46(3): 35–39.

Timothy, D.J. (1994) "Environmental impacts of heritage tourism: physical and socio-cultural perspectives," *Manusia dan Lingkungan* 2(4): 37–49.

Timothy, D.J. (1999) "Built heritage, tourism and conservation in developing countries: challenges and opportunities," *Journal of Tourism* 4: 5–17.

Timothy, D.J. and Boyd, S.W. (2003) *Heritage Tourism*, Harlow: Prentice Hall.

Tomasi, L. (2002) "Homo viator: from pilgrimage to religious tourism via the journey," in W.H. Swatos and L. Tomasi (eds) *From Medieval Pilgrimage to Religious Tourism*, Westport, CT: Praeger.

Turner, E. (1987) "Pilgrimage," in M. Eliade (ed.) *The Encyclopedia of Religion*, New York: Macmillan.

Turner, V. (1973) "The center out there: pilgrim's goal," *History of Religions* 12(3): 191–230.

Turner, V. (1984) "Liminality and the performance genres," in J.J. MacAloon (ed.) *Rite, Drama, Festival, Spectacle: Rehearsals Toward a Theory of Cultural Performance*, Philadelphia, PA: Institute for the Study of Human Issues, pp. 19–41.

Turner, V. and Turner, E. (1978) *Image and Pilgrimage in Christian Culture*, New York: Columbia University Press.

Tweed, T.A. (2000) "John Wesley slept here: American shrines and American Methodists," *Numen* 47(1): 41–68.

Vásquez, M.A. and Marquandt, M.F. (2000) "Globalizing the rainbow Madonna: old time religion in the present age," *Theory, Culture & Society* 19(4): 119–142.

Voyé, L. (2002) "Popular religion and pilgrimages in Western Europe," in W.H. Swatos and L. Tomasi (eds) *From Medieval Pilgrimage to Religious Tourism*, Westport, CT: Praeger.

Vukonić, B. (1996) *Tourism and Religion*, Oxford: Elsevier.

Vukonić, B. (1998) "Religious tourism: economic value or an empty box?," *Zagreb International Review of Economics & Business* 1(1): 83–94.

Vukonić, B. (2002) "Religion, tourism and economics: a convenient symbiosis," *Tourism Recreation Research* 27(2): 59–64.

Wearing, S. (2001) *Volunteer Tourism: Experiences that Make a Difference*, Wallingford: CABI.

Wearing, S. and Neil, J. (2000) "Refiguring self and identity through volunteer tourism," *Society and Leisure* 23(2): 389–419.

Wilkinson, J. (1998) "In search of holy places: then and now," in J.M. Fladmark (ed.) *In Search of Heritage as Pilgrim or Tourist?*, Shaftesbury: Donhead.

Williams, P.W. (2002) "Sacred space in North America," *Journal of the American Academy of Religion* 70(3): 593–609.

Willis, K.G. (1994) "Paying for heritage: what price for Durham Cathedral?," *Journal of Environmental Planning and Management* 37(3): 267–278.

York, M. (2002) "Contemporary pagan pilgrimages," in W.H. Swatos and L. Tomasi (eds) *From Medieval Pilgrimage to Religious Tourism*, Westport, CT: Praeger.

Zaidman, N. (2003) "Commercialization of religious objects: a comparison between traditional and New Age religions," *Social Compass* 50(3): 345–360.

Zelinsky, W. (1990) "Nationalistic pilgrimages in the United States," in G. Rinschede and S.M. Bhardwaj (eds) *Pilgrimage in the United States*, Berlin: Dietrich Reimer Verlag.

Part I

Concepts, concerns and management issues

2 Sacred spaces and tourist places

Thomas S. Bremer

Tourists encounter places of religion in nearly every corner of the world. Wherever religious people have made places of interaction with sacred powers, touristic practices are able to establish a site worthy of visitation. At religious sites that host significant numbers of tourists, a simultaneity of places emerges in parallel geographies of both the sacred and the touristic. Moreover, the meaningful content of these overlapping places relies upon narratives of identity that consolidate the bonds that religious adherents and tourist visitors feel toward the place. In short, religious tourist sites tell stories that articulate coherent identities.

The stories that make these sites meaningful relate first of all to a heuristic distinction between space and place. Space refers to an undifferentiated expanse lacking in meaningful content. On the other hand, place distinguishes particular locales by punctuating the meaningless expanse of space with meaningfulness. As the cultural geographer Yi-Fu Tuan notes, "'Space' is more abstract than 'place.' What begins as undifferentiated space becomes place as we get to know it better and endow it with value" (Tuan 1977: 6). Indeed, the value that distinctive communities or particular individuals attribute to a site differentiates it from other places and from the monotonous expanse of space in general. This in turn emphasizes the particularity of place in contrast to the homogeneity of space. This can be visualized in terms of maps. A map is not so much a representation of space as it is a guide to the meaningful content that differentiates space into particular places. Michel de Certeau (1984: 116–132) discusses a "bipolar distinction between 'map' and 'itinerary'"; he demonstrates how historically "the map has slowly disengaged itself from the itineraries that were the condition of its possibility." Thus, we do not see space on a map; a map of space would be nothing more than a blank sheet of paper. Instead, we see the configuration of particular places in relation to each other in a given segment of space. Map is not territory, as Jonathan Z. Smith (1993: 309) correctly points out, but as artifacts of past itineraries, maps do provide a visual image of the places that make a territory meaningful and therefore recognizable to particular communities and individuals.

Meaningful places, however, do not remain static. In acknowledging the contingencies and dynamics apparent in all places, the author has discussed elsewhere what he calls the itinerant dimensions of place in contrast to the locative (Bremer 2004: 12). A locative view of place emphasizes the power, stability, and permanence of location. Material markers of a particular place, whether they be naturally occurring topographic features such as mountains, rivers, lakes, and caves, or built environments of human interactions such as buildings, shrines, enclosures, and fortifications, announce an air of enduring stability that makes the place distinctly recognizable. Try imagining, for instance, the island of Manhattan without visualizing its stunning skyline highlighted with the towering presence of the Empire State building, the Chrysler building, and other distinctive landmarks that make its breathtaking materiality so memorable and lasting.

But the locative stability of the Manhattan skyline disguises its itinerant reality. In contrast to the apparent permanency of the locative, the itinerant dimensions of place are the ephemeral and unstable realities that belie the illusion of permanence. The itinerant draws attention to the contingencies of place that inevitably erode whatever enduring qualities a particular place may possess. Not too long ago it would have been difficult to imagine the Manhattan skyline without the imposing presence of the Twin Towers at the World Trade Center. Today their absence reminds us that the permanency of place is always a contingent stability; the itinerant nature of place inevitably brings change to otherwise enduring landscapes.

Features of place

The tension between the enduring stability of the locative and the forces of change that characterize the itinerant draw attention to two features that every place displays. First, all places are social. In other words, places involve relationships between various individuals and groups of people. It is impossible to think of a particular place without inferring a social dimension. The meaningfulness of a given place derives to a large extent from its status in the relations that people have with each other. Even sites of solitude or social estrangement contain a social dimension, perhaps more so than those that thrive on contact and interaction with others. Indeed, the absence of society can be a powerful social force.

The importance of place in social relationships goes beyond its role as the site where people interact. In fact, place serves as an integral element in all social relations, both as a determinant of those relations and as a product of them as well. This is especially evident in places of religion. The special character of a holy site endows its occupants with a degree of social prestige. Priests, ritual specialists, and others who inhabit and control sacred sites command the respect of religious adherents who recognize their importance in part because of their association with the auspicious nature of the site. At the same time, the prestige of the place itself gains recognition due to the presence and activities of religious leaders and specialists who practice there. Hence, the social and spatial dimensions reinforce each other and sustain their special character in an inextricable relationship of reciprocal meaningfulness.

Because places are social, they are never unchanging. The meaningfulness of a particular place derives from practices that establish them and maintain them, and from the discursive force of those practices in the communities that regard them as special or peculiar. These practices and the rhetorical powers they deploy change over time in the ongoing ebb and flow of the social relations where they occur. As a result, the place itself changes as its meanings shift for the individuals and communities who find it distinctive.

Because places continually reflect the dynamic processes of shifting meanings, they are vulnerable to contestation by various interested parties, a point aptly demonstrated in regard to sacred sites in particular (Chidester and Linenthal 1995: 1–42). Various groups have strongly vested interests in particular places. These interests may be economic, political, religious, moral, aesthetic, nostalgic, or some combination of these. When the interests of different groups are at odds, places can become the focus of intense conflicts. These struggles can involve questions of actual ownership or control of a particular site, or they may involve nothing more than a rhetorical battle over the specific meanings of the place. The social aspect of place thus plays itself out in the discursive outcome of these never-ending attempts to define and control the site.

Besides their inherently social nature, the second feature that all places display is a temporal dimension. Places occur at the intersection of the spatial and the temporal domains. Or, to put it another way, the meaningfulness of a particular place relies on cultural assumptions about time. Spatial location has little significance without temporal location; places include pasts, presents, and futures. The history of the place and its commemorative value help to maintain personal and collective memories that in turn contribute to contemporary perceptions of self and other. Current practices that sustain the place's meanings rely on the calendrics of "social time," a term that Emile Durkheim (1965: 23) coined in reference to a category of time that expresses "a time common to the group." This contrasts with the personal experience of time that he describes as "sensations and images which serve to locate us in time." Finally, the perceptions of place also coincide with assumptions about how time will operate in (or on) the future.

Of course, not all groups regard time as a continuity of past, present, and future. But other cultural perceptions of time likewise impact understandings of place. For instance, a cyclical notion of time coincides with a view of place as cyclically significant. In ancient Mesoamerica, for instance, the unification of the double count calendar system relied on a cosmological union of time and space; consequently, the ritual "binding of the years" practiced in Mesoamerican religions involved an elaborate linking of significant places into a cosmological whole. These places would reemerge periodically as auspicious sites in the recurring rounds of the calendar cycles (Aveni 1980: 154–158; Carrasco 1999: 88–114; Greenhouse 1996: 144–174).

But whether a particular culture perceives time as a linear continuity, a cyclical repetition, or some other configuration, its temporal assumptions figure into its understandings of place. Moreover, as Carol Greenhouse (1996: 4) demonstrates, the primary significance of time relates to cultural formulations of agency that give legitimacy to the dominant political order within a particular society. These formulations of agency likewise create the meaningful significance that defines particular places.

The formulation of subjective agency through cultural assumptions about time suggests that time is in fact part of the social character of place. Thus, the two fundamental features of place – the social and the temporal – are closely related. And both in turn relate to the locative and itinerant dimensions of place. Locative claims tend to regard place as timeless; the perception of permanence implies that a particular site escapes the forces of historical change. These sorts of claims also serve as rhetorical strategies in the politics of place; they operate discursively in the relations of power between and among communities that find the site meaningful in some way. Indeed, a declaration of permanence imposes a particular group's interpretation of the site and helps to consolidate its control by foreclosing any competing claims on the place. In contrast, the itinerant dimensions of place are embedded in the dynamic forces of time; the contingencies of place rely on cultural assumptions about how time operates. Moreover, the itinerant quality of places derives in part from social contention over the control and significance of the site. The unstable and changing nature of a place reflects social differences and consequent struggles over actual ownership and ultimate interpretation of the place.

Ownership and interpretation of a particular site involves control over narratives of place. Places are made in human spatial experience, and the stories that people tell of those experiences give places their meanings and sustain them as distinct locales in the landscape. In other words, place gains its meaningfulness and retains its importance in narratives, a fact that Michel de Certeau (1984: 115) draws attention to when he notes regarding stories that "every day, they traverse and organize places; they select and link them together; they make sentences and itineraries out of them. They are spatial trajectories." These spatial trajectories are evident, for example, in the images of Manhattan's skyline. Certainly, the narration of the human experiences of the place reveals Manhattan's most profound significance. The stories of human drama, triumph, tragedy, or simply of the mundane unfolding of daily lives invest the places of the city with a myriad of meanings that pervade its character. Among the most powerful of those stories relate to the absence of the Twin Towers at the

World Trade Center. The downtown site evokes the horror that struck the island and its occupants on September 11, 2001. The narration of those events and their aftermath are profound elements in the meaning of Manhattan as a place.

Travel practices in discourses of place

The stories that articulate the meaningful value of a particular locale are a dimension of the travel practices that make places. An important element in the itinerant nature of all places is the travel practices that make places possible to begin with. The undifferentiated expanse of space undergoes its transformation into distinct places in the human itineraries that experience the landscapes they traverse and then narrate those experiences to themselves and to others. Places thus emerge in human practices of travel.

Travel practices encompass all sorts of human translocal movements, as the author has discussed elsewhere (Bremer 2004: 12–13; 2005: 9262). These include migration, pilgrimage, business and trade, tourism, military deployments, research excursions, family visitations, and many other manners of travel. The practices that constitute these movements go beyond actual travel itself. Certainly travel practices involve various modes of transportation, most commonly airlines, trains, buses, and automobiles. But they also include communication networks that facilitate travel, including telecommunications and the Internet, as well as media that make travel desirable and feasible such as radio and television broadcasts, newspapers and magazines, and other forms of mass communication. Modern travel practices also include banking facilities that allow convenient and trustworthy currency exchange; accommodations for lodging and food services; and any other services or products that meet the needs and desires of modern travelers.

Yet travel practices go beyond modern infrastructures and services that make global travel possible, convenient, and comfortable. They include any practice, discourse, or circumstance that either necessitates translocal movements or that generates desires for and encourages people to travel. In the context of market capitalism, the construction and promotion of travel destinations constitutes important travel practices. Making a locale desirable for travelers to visit or settle there, and participating in discourses of place that travelers pay heed to, all are part of travel practices in the modern world. Likewise, the exclusion of populations deemed undesirable, both resident and transient, must be counted among modern travel practices. Thus, border controls and the policing of sites qualify as well.

Like other places, religious sites that host tourist visitors are sustained with travel practices. But although religious adherents and non-adherent tourists both occupy the same site at the same times, their respective practices and resulting interpretations of its significance make distinct places of it. Thus, religious sites of tourism maintain what Bremer (2004: 4) has called a simultaneity of places. Neither their religious quality nor their touristic character can make a total claim on these places – they remain both religious and touristic, occupied by both religious adherents and other tourists whose respective experiences of the site are quite different from each other. True, the experiences of particular individuals may cross between the religious and the touristic, as when tourists participate in religious activities or when religious followers indulge in the touristic attractions of their sacred precinct (pilgrims, for instance, often indulge in touristic practices in the course of their religious journeys). Yet the hybridity of these sites does not allow a seamless blending of the two; careful observations reveal clear distinctions between the places of religion and the places of tourism, despite the very real overlaps that inevitably constitute religious sites of tourism.

The narratives of sacred places occur within the discourses of particular religious traditions. This is not to say that all sacred places must be associated with a recognized religious body. It simply means that the meaningful experience articulated in the narrative of the place must have some larger discursive framework that orients an understanding of sacrality. To deem a place sacred, there must be an understanding of what the sacred entails. This understanding, I contend, relies on discourses of particular religious traditions. Such reliance usually takes a positive form. Christians who regard a site as sacred because it is a place of God participate positively in a Christian theological discourse; they reinforce what they regard as the accepted Christian tradition. On the other hand, the narrative of a sacred place may pose a negative counter-discourse to the religious tradition. An auspicious gathering place for a band of Satanists, for instance, poses a threatening alternative to normative Christian discourse. As Elaine Pagels (1995) demonstrates, the discursive origins of Satan in Jewish apocalyptic sources became part of normative Christian traditions that vilify anyone who pays respect to the demonic figure. Yet, the Satanists' claims of sacrality nevertheless rely on normative Christian traditions to suggest their own criteria and parameters of sacred experience.

The experience of the sacred in particular places tends to reinforce locative claims, especially in monotheistic traditions. Thus, places regarded as sacred often have an air of permanency that transcends the contingencies of human mortality; the sacred places of a religious tradition will survive long after individual religious adherents have passed away. In fact, sacred sites serve as an enduring symbol of, and orientation for, the continuance of the religious community itself. Their importance also makes them geographical centers that orient the spatial worlds of religious communities. In many cases they become literal centers of the world. As an example, Mircea Eliade (1959: 23), a great advocate of locative claims of sacred places, argues: "Revelation of a sacred space makes it possible to obtain a fixed point and hence to acquire orientation in the chaos of homogeneity, to 'found the world' and to live in a real sense." He consistently emphasizes the role of sacred places in orienting religious communities, as when he describes "techniques of *orientation*" as "techniques for the *construction* of sacred space" (Eliade 1959: 29). To Eliade, a sacred space represents the center of the world; it is, in his famous term, an *axis mundi*. Even moveable objects can operate in this locative, orienting manner. Eliade's example of the *kauwa-auwa*, or sacred pole, of the Achilpa people of Australia (an example that Jonathan Z. Smith (1992: 1–23) critiques), demonstrates how an itinerant object can serve locative ends. According to Eliade:

> This pole represents a cosmic axis, for it is around the sacred pole that territory becomes habitable, hence is transformed into a world. . . . During their wanderings the Achilpa always carry it with them and choose the direction they are to take by the direction toward which it bends. This allows them, while being continually on the move, to be always "in their world" and, at the same time, in communication with the sky into which Numbakuyla [their creator god] vanished.
>
> (Eliade 1959: 33)

Sacred space, in Eliade's estimation, organizes around particular sites, or even particular objects, that become orientational centers in the religious community's world. These *axes mundi* decree a locative stability that allows the community to endure in a changing world.

The rhetorical force of these locative claims seeks to squelch the contested, itinerant nature of these places. The assertion of an enduring, unchanging place that serves as the center of a community's world attempts to foreclose counterclaims that would undermine the authority of those who control the *axis mundi* and the normative tale of its significance. Individuals and groups vested in maintaining the status quo will resist the itinerant aims of a counter-discourse that seeks to transform the sacred geography according to alternative mythic and social realities. To give one example, this resistance was apparent in the persecution of the Church of Jesus Christ of Latter-day Saints in nineteenth-century America. Part of the Mormon revision of Christianity involved a reconfiguration of the sacred geography; fundamental to the Church's efforts was the establishment of a New Jerusalem on the American continent. The tenth article of faith of the Latter-day Saints states in part: "We believe . . . that Zion will be built upon this [the American] continent" (Talmage 1949: 2, 345–355). This New Jerusalem, according to the words of Jesus in *The Book of Mormon*, will be a place where "the powers of heaven shall be in the midst of this people; yea, even I will be in the midst of you" (3 Nephi 20:22). Their struggle to build the place of Zion in the American west posed a discursive threat to more established Christian sects who responded with a vicious campaign of persecution against the Mormons. The conflict resulted in the execution of their founder and the removal of the Saints to a new homeland in Utah. Indeed, the itinerant force of an alternative voice usually elicits a strong response from those with a vested interest in the locative claims of established sacred geographies.

Places of tourism

In contrast to the struggles that establish and sustain religious sites, places of touristic appeal demonstrate other discursive concerns. The conventional practices of tourism participate most acutely in discourses on modernity (Adler 1998). In fact, tourists can be regarded as exemplary modern subjects. They seek out the experience of unfamiliar places by leaving the familiarity of home, but the conventions of tourism tend to domesticate these novel places by commodifying the experiences that visitors have of them. Thus, touristic experience conforms to the forces of modern capitalism, especially in its globalizing capacity to encompass all places in the marketplace of consumer demand. Tourists, then, as they participate in this marketplace, serve as practitioners of modernity (Bremer 2005: 9262).

The specific practices that typically constitute modern tourism include specialized forms of travel practices, including activities such as sightseeing, picture taking, and souvenir shopping. But tourists partake in far more than what the stereotypes of touristic behavior suggest. Tourism, as Franklin and Crang (2001: 15) point out, involves "material practices that serve to organize and support specific ways of experiencing the world." Besides the commodification of places and experiences brought about by modern capitalism, the touristic way of experiencing the world also relies on a modern aesthetic sense. Consequently, tourists serve as consumers in a marketplace of aesthetically pleasing experiences.

Fundamental to tourists' experiences is an aesthetic obsession with authenticity (Redfoot 1984; Olsen 2002). Tourist discourses attribute significant value to authenticity; the most authentic experiences are the most aesthetically pleasing. This explains tourists' desire for destinations "off the beaten path." The glee expressed in finding that out-of-the-way place where there are no tourists employs a touristic discourse on authenticity that bolsters one's own esteem as a traveler by ironically disparaging others as tourists. By denying one's own status as a tourist, one gains esteem at the expense of others less adept in the practice of authentic travel. In a fundamental paradox of touristic practice, contrasting oneself with lowly "tourists" makes one a better tourist (Culler 1981: 130–131; MacCannell 1999: 91–107).

In places of religion, the touristic discourse on authenticity coincides with religious discourses of place. Indeed, both pleasure tourists and religious adherents have a keen interest in the authentic experience of place. To some extent, touristic concerns and religious interests respond to and reinforce each other to produce a meaningful sacred site. On the one hand, the authentically sacred character of a religious place makes it an appealing attraction for traditional tourists. At the same time, tourist attention helps to sustain the locative claims of the place as a meaningful site in the religious tradition. Together, the touristic and religious perspectives engage in a symbiotic reciprocity that makes certain sites both sacred religious places and appealing travel destinations. They are both sacred spaces and tourist places.

Identity and place

The narratives produced in the discourses of these places also articulate particular identities. There persists a close relationship between place and identity. As I have argued elsewhere:

> The making of place, sacred or otherwise, always involves the making of identities; conversely, the construction of identity always involves the construction of places. On the one hand, the meaningful content of a particular place relies on the production of both subjects who inhabit the place and subjects who observe, comment on, and interpret the place, including both insiders and outsiders. On the other hand, all subjects are situated; to attain subjective agency means at the most fundamental level to occupy a particular place, both in the physical and social senses. Thus, place and identity emerge together in a relationship of simultaneity.
>
> (Bremer 2003: 73–74)

This simultaneous relationship between place and identity relies on a reciprocal meaningfulness that binds people and places together. If one's identity involves a "meaningful self-image of affiliation with an imagined community" (Bremer 2004: 144), or, as Moya (2000: 8) puts it, "our conceptions of who we are as social beings," then identity, like place, is foremost a matter of social relationships. The meaningful content of an individual's being depends on her or his association with various social groups as imagined in her or his own cultural and historical contexts. At the same time, an affinity of meanings provides a basis for imagining a distinct community to identify with.

At religious places of tourism, visitors attribute meanings according to their own social affiliations. Religious people understand such places as sacred in accord with their own affiliation with a particular religious community. The meaningfulness of the place for such visitors derives from their community's understanding of the religious tradition. In contrast, tourists who do not regard themselves as members of the religious community ascribe different meanings to the place. The touristic propensity for aesthetic pleasures creates an authentic place of beauty and power in conformity with modern aesthetic tastes. This in turn offers an aesthetic dimension to tourists' modern identities.

Conclusion

The religious understandings of a site create one set of places, while touristic interpretations produce a different set of places. Certainly, this simultaneity of places offers an abundant opportunity for overlap and convergence. But the social realities of identity require a certain distinctiveness that allows individuals to imagine the appeal of the site in terms of their own sense of self. The stories that visitors hear at religious sites of touristic appeal, and the tales they recite to themselves and others, make distinctive places in a landscape of coherent identities. Sacred for the religious, aestheticized and commodified for the tourists, these places contribute decisively to the social affiliations and personal identities of those who enter their precincts.

References

Adler, J. (1998) "Origins of sightseeing," in C.T. Williams (ed.) *Travel Culture: Essays on What Makes Us Go,* Westport, CT: Praeger.

Aveni, A.F. (1980) *Skywatchers of Ancient Mexico*, Austin: University of Texas Press.

The Book of Mormon: An Account Written by the Hand of Mormon, upon Plates Taken from the Plates of Nephi (1987) (J. Smith, Trans.), Salt Lake City: Church of Jesus Christ of Latter-day Saints.

Bremer, T.S. (2003) "Il Genius Loci Ignotus di Eranos e la Creazione di un Luogo Sacro," in E. Barone, M. Riedl, and A. Tischel (eds) *Eranos, Monte Veritá, Ascona,* Pisa, Italy: Edizioni ETS.

Bremer, T.S. (2004) *Blessed with Tourists: The Borderlands of Religion and Tourism in San Antonio*, Chapel Hill: University of North Carolina Press.

Bremer, T.S. (2005) "Tourism and religion," in L. Jones (ed.) *The Encyclopedia of Religion*, New York: Macmillan.

Carrasco, D. (1999) "The New Fire Ceremony and the binding of the years," in D. Carraso (ed.) *City of Sacrifice: The Aztec Empire and the Role of Violence in Civilization*, Boston: Beacon Press.

Certeau, M. de (1984) *The Practice of Everyday Life* (S. Rendall, Trans.), Berkeley: University of California Press.

Chidester, D. and Linenthal, E.T. (eds.) (1995) *American Sacred Space*, Bloomington: Indiana University Press.

Culler, J. (1981) "Semiotics of tourism," *American Journal of Semiotics* 1(1–2): 127–140.

Durkheim, E. (1965) *The Elementary Forms of the Religious Life* (J.W. Swain, Trans.), New York: The Free Press.

Eliade, M. (1959) *The Sacred and the Profane: The Nature of Religion* (W.R. Trask, Trans.), New York: Harvest/HBJ.

Franklin, A. and Crang, M. (2001) "The trouble with tourism and travel theory?," *Tourist Studies* 1: 5–22.

Greenhouse, C.J. (1996) *A Moment's Notice: Time Politics Across Cultures*, Ithaca: Cornell University Press.

MacCannell, D. (1999) *The Tourist: A New Theory of the Leisure Class* (3rd edn), Berkeley: University of California Press.

Moya, P.M.L. (2000) "Introduction: reclaiming identity," in P.M.L. Moya and M.R. Hames-García (eds) *Reclaiming Identity: Realist Theory and the Predicament of Postmodernism*, Berkeley: University of California Press.

Olsen, K. (2002) "Authenticity as a concept in tourism research: the social organization of the experience of authenticity," *Tourist Studies* 2(2): 159–182.

Pagels, E.H. (1995) *The Origin of Satan*, New York: Random House.

Redfoot, D.L. (1984) "Touristic authenticity, touristic angst, and modern reality," *Qualitative Sociology* 7(4): 291–309.

Smith, J.Z. (1992) *To Take Place: Toward Theory in Ritual*, Chicago: University of Chicago Press.

Smith, J.Z. (1993) *Map Is Not Territory: Studies in the History of Religions*, Chicago: University of Chicago Press.

Talmage, J.E. (1949) *A Study of the Articles of Faith; Being a Consideration of the Principal Doctrines of the Church of Jesus Christ of Latter-day Saints* (28th edn), Salt Lake City: The Church of Jesus Christ of Latter-day Saints.

Tuan, Y.-F. (1977) *Space and Place: The Perspective of Experience*, Minneapolis: University of Minnesota Press.

3 Religious and secular pilgrimage
Journeys redolent with meaning

Justine Digance

The word 'pilgrimage' usually conjures up images of travellers undertaking long arduous journeys to religious shrines around the world. Most of this imagery is predicated upon notions and concepts drawn from medieval pilgrimage, a social movement that occurred between 500 and 1500 CE. Pilgrimage offered a temporary escape from the generally harsh existence in an agrarian-based society, so much so that medieval pilgrimage is generally given as the first example of mass tourism as we know it today. However, by the end of the fifteenth century the motivation for pilgrimage had changed from the spiritual to be one of curiosity, the desire to see new places and experience new things reflecting broader changes that were occurring in European society around that time (Sumption 1975).

The medieval roots of pilgrimage are also reflected in the word itself, which is derived from the Latin *peregrin-um*, meaning one that comes from foreign parts, linked with its usual meaning as denoting a journey (usually involving a long distance) to a sacred place to undertake demonstrations of religious devotion. While this definition may have been apt for pilgrimage up until the beginning of the twentieth century, the term has now been adopted into much more popular parlance with corresponding secular overtones. Today, the journey may be short in both distance travelled and duration of stay, or there even may be no physical travel involved at all if the pilgrim is surfing cyberspace. Therefore this chapter proposes that the twenty-first century calls for a newer and more flexible definition of pilgrimage, namely 'undertaking a journey that is redolent with meaning'.

This chapter focuses on traditional religious pilgrimage, which evolved from the world's main religious traditions such as Christianity, Hinduism, Buddhism and Islam, and the more relatively contemporary process of secular pilgrimage or contemporary pilgrimage occurring outside the context of the main religious traditions. A third pilgrimage tradition – prehistoric and/or tribal pilgrimage (e.g. Egypt, Babylonia, Meso-American, Australian Aboriginal and pre-Christian Europe) – is not covered in detail; however, its sites and/or customs play an important role when discussing some aspects of secular pilgrimage today. When the author first commenced her doctoral research some years ago (Digance 2000), a search of the term 'secular pilgrimage' revealed no results, but when researching for this chapter, the writer was rewarded with four results via ProQuest: Latin American women's literature (Fahey 2003), American planters on a nineteenth-century European tour (Kilbride 2003), American Jews visiting the Soviet Union in the 1920s and 1930s (Soyer 2000), and Donald Horne's commentary on tourism (Hollinshead 1999).

It is interesting to note at the turn of the millennium that this distinction is starting to be made by academics despite the long-popular usage of describing virtually any journey (touristic or otherwise) as a 'pilgrimage'. For example, the headline of *The Wall Street Journal* (Carlton 2000) 'EBay CEO touts the Company in a rare Wall Street pilgrimage', which utilized the word 'pilgrimage' to draw readers' attention. Nonetheless, despite outward appearances, this author suggests that there are many shared similarities or common threads found in both traditions – faithful and secular forms of travel. These similarities are discussed by following a typical pilgrimage sequence as set out later in the chapter. While there are many similarities, the conclusion looks at distinguishing between the two by proposing that being motivated to undertake a pilgrimage as 'an act of faith' is fundamental to traditional religious pilgrimage, and is lacking in modern secular pilgrimage. However, that distinction also creates some difficulties, leaving one to ponder if there is any meaningful real difference at all.

The quest

Central to pilgrimage is the view that the pilgrim is on a spiritual quest in search of meaning (often simply described as 'the quest'), but in archaic pilgrimages this was not always so. For example, oracles and festivals in Hellenic times are sometimes described in the literature as constituting pilgrimage, but many of these events were also times of celebration, such as during the games at ancient Olympia. While modern society may evince an increasingly secularized world, many commentators emphasize that we live in a re-sacralized world with a blurring of spirituality between the religious and secular domains. Traditional religious pilgrimage is far from diminishing in popularity with age-old centres such as Rome, Jerusalem and Lourdes still attracting the faithful, with newer sites such as Medjugorje and Sri Sathya Sai Baba's palatial Ashram at Puttaparthi in India, proving to be popular pilgrimage sites today. A veritable cornucopia of secular pilgrimages abounds, far too numerous to cover in detail in this chapter.

This chapter suggests that there are two groups of modern secular pilgrims: those who still claim traditional religion as meeting their spiritual needs but for whom the journey fulfils a deep personal meaning, and those who could loosely be grouped under the broad 'New Age' banner.

Much has been written on the New Age, and suffice it to say that much of the 1970s counterculture has been absorbed into many mainstream activities so that the term itself is somewhat misleading. New Age followers emphasize transformation of the Self, which Heelas (1996: 18) calls 'self-spirituality', with an emphasis on the experiential and looking within for an inner spirituality. In today's consumer society, religion is just another marketable commodity or meaning system (Olsen 2003), with individuals being able to choose packaged meaning systems, with 'Buy this product and change your life' being a common marketing theme (Aldred 2000). McColl (1989) called this 'spiritual smorgasbording', while Solomon (1999) views it as 'spiritual promiscuity'. Caplan (2001: 51) sees seeking spirituality as a fad and 'a commodity that is bought and sold for millions of dollars, an identity, a club to belong to, an imagined escape'.

All pilgrims share the common trait in that they are searching for, and expect also to be rewarded with, a mystical or magico-religious experience – a moment when they experience something out of the ordinary that marks a transition from the mundane secular humdrum world of our everyday existence to a special and sacred state. The moment may be fleeting so that many may not even instantly recognize what has occurred, but perhaps only afterwards when back in their quiet, crowd-free hotel room, and reflecting on their experiences will they realize that they have experienced an encounter with the Other. These experiences can be described in any number of ways, such as transformation, transcendence, life- and/or consciousness-changing event, hierophany, enlightenment and so on, but words seem somewhat inadequate to describe experiences that often are not amenable to reason. Brainard (1996) categorizes these mystical experiences as a spiritual awakening, notable because they are both profound and extraordinary. Again, many of the descriptive terms used flow from the literature on traditional religious pilgrimage, and thus the suggested definition of pilgrimage used in the title of this chapter, journeys redolent with meaning (cf. Morinis 1992), is preferred as it allows secular pilgrimage to be accorded similar status to religious pilgrimage. 'Magical experiences' are commonly used to describe many of today's experiences that stand out as unforgettable, memorable moments and thus the proposed definition broadens the scope of the word 'pilgrimage' and all that it denotes. Also worthy of discussion at this point is the term 'spiritual', which is often associated with the pilgrimage experience. Definitions of spiritual or spirituality very often see this as the sole preserve of religion, pertaining to things that are not worldly or material. However, secular pilgrimage journeys largely fall outside this ambit, for example shopping pilgrimage (e.g. Mall of America) and visits to iconic tourist attractions such as historic battlefields (e.g. Culloden), Graceland and New Age Meccas such as Glastonbury and Sedona. Often the experience is not easy to characterize except by the individual stating that it was 'something special', standing out as a marker against which all such similar experiences will be measured during the course of one's lifetime.

The experience can be mediated through the intervention of a priest or similar, and can also be a validation for those on a spiritual quest that they are on the right path, but likewise it may be deeply distressing for the unprepared, who then are faced with trying to make sense of their experiences without an adequate belief system. By stating that one is on a pilgrimage is to create an expectation in the mind and heart of the pilgrim that he or she automatically will be rewarded with an encounter with the Other, but the literature seems to be silent on unmet pilgrim expectations particularly in today's consumer-driven marketplace. Does the pilgrim select another product from the spiritual supermarket shelf, or return at another time in the hope that the 'gods will smile on them' and that they will experience a transformative encounter with the Other? It should be noted that the history of medieval pilgrimage is full of references to individuals who endlessly move from shrine to shrine in search of miraculous cures, or today's New Agers who crave new spiritual experiences via what Aldred (2000) dubs 'plastic medicine people or gurus'. In Wilhelm's (1989: 215) doctorate focusing on the experiences of eight pilgrim-tourists during a two-week Catholic study tour to Ireland, there was an implicit but unstated view that the pilgrims expected the miraculous during their paid tour vacation. One of the main findings from this research was that they 'were searching for change in their lives through the experience of pilgrimage', but the message here is that commercial operators promoting pilgrimage tourism should be mindful of their marketing message so that they do not leave themselves open to charges of false or misleading advertising by creating unrealistic expectations in the minds of consumers.

The journey

To undertake a pilgrimage carries with it the implicit assumption of a physical journey, hence the oft-cited maxim by Turner and Turner (1978: 20) found in most discussions of pilgrimage: 'a tourist is half a pilgrim, if a pilgrim is half a tourist'. Certainly this is true of almost all pilgrimage experiences until the late twentieth century when travel was an essential element. However, the technological revolution (e.g. television, media and the Internet) has irrevocably changed this perception. Although the physical journey reminds pilgrims that they are engaging on a quest with the sacred, an essential element of any pilgrimage is also the inner journey that the pilgrim undergoes, namely the quest and search for meaning. Often the pilgrim's route is preordained through the annals of time so that to deviate in any way would be to undermine the authenticity of the experience. This is especially clear in the context of the *camino* to Santiago de Compostela in Spain, Shingon Buddhist pilgrimage on the Japanese island of Shikouko, and the *Hajj* to Mecca.

The journey can be taken individually or in groups, the mingling of pilgrims both on the journey and at the journey's end leading to Turnerian (1972) *communitas* where a temporary fellowship and comradeship develops between fellow pilgrims. *Communitas* combines two elements: an observable egalitarianism that sees pilgrims mingling freely, regardless of status, and a desire to be with like-minded people, sharing common interests and/or experiences. Eade and Sallnow (1991), on the contrary, did not find *communitas* in the case studies in their edited work on religious pilgrimage; however, Digance and Cusack (2001) found *communitas* at an alternative New Age event. From a more secular perspective, Moore (1980: 216) also found indications of *communitas* in Walt Disney's Magic Kingdom at Walt Disney World, with pilgrimage per se being 'susceptible to appropriation by commercial, secular interests'.

The search for meaning offers unlimited business opportunities for small and medium enterprises and multinationals alike; organized tours to pilgrimage sites, guided site tours, bed and breakfasts, self-help books, guide books, and a whole range of souvenirs are just some of the commercial opportunities associated with pilgrimage. In medieval times, souvenirs were sold at the major shrines and included badges (which the returning pilgrim often wore around his or her village to acquire additional prestige), relics of doubtful worth (the famous scallop shell associated with the *Camino de Santiago*), holy water, and items of a more secular nature, such as jewellery and fabric. One would expect today's pilgrims, given the opportunity, to act as their medieval counterparts did. However, it is suggested here that the collection of souvenirs is not only indicative of pilgrimage but is common for all tourists alike. The economic benefit accruing to long established sites and/or events has been well documented (cf. Fleischer 2000; Vukonić 1996), but some of the earnings from secular pilgrimage are equally as impressive. For instance, as far back as 1988, Americans were spending over $100 million annually on crystals (Aldred 2002 citing Chandler 1988), and Elvis Week is worth $50 million to the Memphis area (Warren 2002). Nonetheless, George (1994: 39) argues that the sacredness of the place and moment are carried back home not necessarily in the form of photographs and/or souvenirs, 'but in the continuing autographs of the inner person'.

Mass communication modes of television and radio allow vicarious pilgrimage in the safety and comfort of one's own living room so that 'armchair pilgrims' can savour the event without having to endure crowds or financial or physical hardships. Home entertainment systems with large plasma screens and surround sound deliver sanitized images of millions of Hindu pilgrims bathing in the Ganges during the Kumbha Mela with such dramatic realism that viewers can almost sense the water lapping around their feet. Also, essential to any discussion of pilgrimage is mention of Turnerian (1969) *liminality* whereby the pilgrims are in a transitory and initiatory 'betwixt and between' phase as they journey towards their hoped for encounter with the Other. This is also likened to a rite of passage (Van Gennep 1965), with ceremonies or rituals used to denote transitional moments in the pilgrim's journey. Historically, the pilgrimage journey has tended to emphasize notions of hardship, particularly in the journey to and from the site. Today, modern technological advances in transportation make it possible for most people to access their preferred pilgrimage site or event within a day or two, with few having the necessary resources of time and/or money to make lengthy pilgrimage trails by foot (Russell 1999). However, some religious pilgrimage routes base their claim to authenticity on foot trails, such as Santiago de Compostella in Spain (Spanish section alone totals 850km), Shingon Buddhism on Shikouko in Japan (one to two months needed to complete route in its entirety), and the one-day barefoot August pilgrimage to Croagh Patrick in Northern Ireland. Even today, pilgrims could expect to encounter some form of hardship on their journey, but compared with medieval times, such travails are usually minor and considered to be a quintessential part of the pilgrimage experience.

Issue 32 of the journal *Religion* (2002) is devoted, with one exception, to discussing the role of the Internet in disseminating religion and religious thought. Citing Leibovich (2000), MacWilliams (2002: 316) notes that 'over 25% of the 100 million Americans online use the Internet for religious purposes'. Cyberpilgrims can elect to visit sites of the institutionalized or mainstream religions, or alternative pagan and New Age sites (Karaflogka 2002). This also includes some tourist sites, such as Graceland, Holocaust museums, and battlefields, which are visited by secular pilgrims. Such websites act 'like religious hubs or syncretic sites . . . where people are exposed to a variety of traditions when they are searching for religion or spiritual information' (Helland 2002: 301). The Internet has become a virtual cyberspace supermarket as individuals use search engines, choose homepages and then follow links in their quest for the Other. Virtual pilgrimage may be the next best thing to being there. This often includes 360-degree pictures and mini-tours of holy places with online prayer circles, group meditations, articles, chat rooms, and guest books enabling one to 'do religion' online in the twenty-first century. The question remains, however, whether or not religious pilgrimage via the Internet is really pilgrimage at all, as the Internet and the World Wide Web are secular tools used to disseminate information – just because the site may contain religious material does not necessarily mean that those who access it are religious pilgrims. Perhaps they equate to Turner's (1969) *liminal personae*, and are sailing in unchartered waters as they hang-five on the Internet surfboard in their never-ending search for meaning!

Sacred place

Pilgrimage is usually characterized by a journey to a named place where an encounter with God or the divine figure(s) central to one's belief system or cosmology is the anticipated outcome. Here the profane becomes a sacred realm in any number of natural and human-made sites, such as temples, mountains, cathedrals, groves, secular sites and so on. While undertaking field studies in Ireland, the author came across a solo pilgrim meditating on the summit of the Hill of Tara, a sacred hilltop with a pagan past and background in Irish folklore, and somewhat tenuous Celtic links, located about an hour's drive north of Dublin. The Hill of Tara is generally described in the literature as a royal site where kings lived. According to Herity and Eogan (1996: 247), it offers 'good proof for a tribal society and kingship' and is near the three *Brú na Bóinne* cluster of mounded tombs, including Newgrange. Such sacred sites 'are venerated not for their own sake but their perceived facilitation of access to supernatural powers' (Jackson and Henrie 1983: 96). Medieval pilgrimage sites competed with each other in attracting pilgrims based upon the number of miracles that had been witnessed, notably at monasteries and cathedrals that were the repository of reliquary of saints. Expectation of the miraculous, events whose occurrence cannot be explained within nature, is also applicable to pilgrimage today, irrespective of whether it is religious or secular pilgrimage. Millions visit Lourdes, for example, in search of cures for their physical woes, while shopping pilgrims seek rare Beatles vinyls on eBay (a non-place); both could equally claim to be in search of the miraculous!

Place-centred sacredness owes much to the works of Eliade (1959) and Turner (1969; 1972), whereas Eade and Sallnow (1991) suggest that such an emphasis ignores where the sacredness vests in a holy person (e.g. Sai Baba or Mother Mary) or the notion of 'textual pilgrimage' (such as Roman Catholic pilgrimage to Jerusalem). Geo-piety, topophilia or place attachment (Tuan 1977) are also important when explaining pilgrimage whereby people develop, over time, a sense of belonging to specific geographical locales. Drawing on Turner's work, Cohen (1979: 190) develops the idea of 'Elective' spiritual centres that are 'external to the mainstream of his[/her] native society and culture'.

New pilgrimage sites are continuously evolving. Ground Zero in New York has developed as a secular pilgrimage centre since September 11, 2001 (Cusack and Digance 2003). Wallis (2002: 7) sees the opportunity for national transformation coming out of such horrific events: 'We don't visit holy sites just for the ritual, we go to be changed . . . since that day its been harder to find parking spaces outside our churches, synagogues and mosques.' Periodic pilgrimages to these elective centres provide spiritual sustenance that sustains the pilgrims upon return to their real world environment. With its supposed relicry of St Andrews, the cathedral in St Andrews was considered the second most important medieval pilgrimage site after Santiago de Compostella (Tobert 2001), but today the Old Course at the Royal & Ancient Golf Club in St Andrews is one of the world's more popular golf 'pilgrimage' sites. Zakus (2002: 850) argues that visiting the Old Course constitutes a pilgrimage journey, for 'as Muslims must make a pilgrimage to Mecca at least once in their lives, gazing upon or more importantly playing the Old Course must be understood in the same way for golfers and golf tourists'. In Australia there is potential for the 5.1-metre statue unveiled in July 2004 of homegrown rock 'n roll legend, Johnny O'Keefe, to develop into a secular pilgrimage site. Located on one of the main thoroughfares in Coolangatta, it attracts an eclectic visitor mix but it remains to be seen whether it becomes a pilgrimage site to O'Keefe and/or the Australian music industry in general or not. At iconic traditional religious pilgrimage sites, pilgrims, mass tourists and locals compete for use, and at some secular pilgrimage sites this also occurs. Nowhere has this sparked more angst than at prehistoric pilgrimage sites where the traditional indigenous inhabitants continue to use their sacred sites (Digance 2003). Jocks (1996) is extremely critical of the way in which the New Age has appropriated American Indian religious or spiritual systems, which see increased demand by non-indigenous groups to access sacred sites. At Mount Shasta, both Native Americans and the US Forest Service have become increasingly concerned about the "environmental effects of trampling and heavy use, and the 'lack of respect' for sacred sites they feel that some New Age activities indicate" (Huntsinger and Fernandez-Gimenez 2000: 538 citing Theodoratus and Evans 1991). Other New Age Meccas also report conflict between local inhabitants, mass tourists and New Age neo-tribes, such as at Glastonbury and Sedona (Ivakhiv 2001), and Byron Bay, Australia.

Consecration of the site

Transformation of profane space into sacred space may occur due to a hierophany, or its association with a myth, sacred object or event. Sacred space is also generally set apart with boundaries usually indicating where profane time and space make way for a sacred realm. Essential to effecting this transformation is the use of authentic ritual, either heavily formalized and ordained over time, or individualistic and/or universal by nature, which connect the pilgrim with the sanctity of the site and/or the event being conducted. For instance, to circumambulate a Tibetan Buddhist stupa in an anti-clockwise direction would be considered to be extremely inauspicious. Ritual is usually practised just prior to, or at the commencement of, use of the site. Events held at a consecrated site automatically tend to vest the site and/or event and its participants with sacredness. Once consecrated, the site is generally advertised to the world as a named sacred place, employing rules and procedures so that the possibility of visitors defiling and desecrating the site is avoided. This may need to be enforced by excluding pilgrims from certain parts of the site, such as Dome of the Rock in Jerusalem, and even employing security staff to enforce the exclusion. Ritual also certifies continuous links with centuries-old authentic practices and traditions, and reinforces the view that one has chosen the correct path in seeking to commune with the Other.

The foregoing is largely drawn from the religious pilgrimage tradition, of which there is much discussion in the literature. But, how does secular pilgrimage involve the use of ritual in denoting space as being sacred to a particular individual or group of individuals? One can only speculate on how individuals consecrate the time and place as being special; very often this is not reflected visibly but rather can represent a consecration by the mind, using visual imagery and/or meditation. Meditation is no more than pilgrimage of the mind, the inner journey and seeking the Other within. Chidester and Linenthal (1995) note that sacred space can be replicated anywhere one chooses by repeated ritualistic performances. Reporting on the twenty-fifth candlelight vigil outside Graceland, Warren (2002) reports on a devoted Elvis fan/secular pilgrim who has 'been first in line for the vigil 13 out of the 16 years he has attended Elvis Week'. The motive for coming is 'to pay back Elvis for what he's meant to me' and over the years a support group has developed to include his wife, son and 25 friends who stay in touch throughout the year. Cyberpilgrims too may routinely undertake certain rituals (e.g. making a cup of coffee, lighting a candle or incense, logging on only at a certain time of day, a prescribed order to visiting chat rooms, etc.) each time they use the Internet to 'do religion' online.

An essential part of consecrating the site by ritualistic actions is engaging in devotions and proffering votive offerings. Devotions can be organized, as with scheduled traditional religious events, or highly individual, such as the Hill of Tara example cited earlier or the writings on the wall at Graceland (Alderman 2002). In medieval times, these could include wax models of parts of the body cured by prayers and at St Léonard de Noblat near Périgueux in France, pilgrims released from prison would leave their chains and instruments of torture at the altar (Barber 1991). Nolan and Nolan (1989) discussed the many different types of offerings found in modern Christian pilgrimage in Western Europe, the most common being candles and fresh flowers. An unusual secular pilgrimage shrine visited by the author was a 'rag well' outside Inverness in the Scottish Highlands where the tying of rags on a tree adjacent to the well symbolizes a request for healing the pilgrim's bodily ills. Flowers laid on the ground at Avebury in England on summer solstice are also votive offerings, a reminder of ceremonies (possibly pagan in origin) carried out earlier that day.

Conclusion

As foreshadowed at the beginning of this chapter, the writer proposed that there are indeed many similarities between traditional religious and modern secular pilgrimages. However, one aspect that supposedly distinguishes one from the other is that the former pilgrimage is made as 'an act of faith'. As noted earlier, the world's major religions incorporate longstanding pilgrimage traditions to sites and/or events. To undertake a pilgrimage one expects there to be some linkage between attending an event/visiting a site and an individual's cosmology or belief system so that their motive could be seen as 'an act of faith'. Visiting Jerusalem, Mecca or Bodhgaya are examples of pilgrimage that respectively gives Christians, Muslims or Buddhists meaning to their belief systems. If the same criteria are applied to secular pilgrimage, then some difficulties are encountered when posing the same question. Asking two people, one tourist who is a practising Roman Catholic and one who is not, who visited the Vatican and St Peter's in Rome, one could expect the response to be different even though both described the visit as having been a special experience. What makes it special for the Roman Catholic tourist is that he or she was primarily motivated to visit the Holy See as a pilgrim, seeking a deeper connection with their belief system and higher being.

When speaking of secular pilgrimage, a grey area appears on two fronts: if an individual does not ascribe to a particular set of religious beliefs, and the amorphous New Age movement. To accept that those in the first category can experience exactly the same type of experience as the Vatican example noted earlier, both claiming to be pilgrims would be abhorrent to some because it cheapens and devalues the notion of pilgrimage by suggesting that the two are equal (hence using a discriminator of 'an act of faith'). But if an individual has eschewed all religion and instead is driven by accumulating economic wealth, can visiting the New York Stock Exchange and Wall Street be capable of fulfilling exactly the same criteria, so that the journey becomes a pilgrimage undertaken as 'an act of faith'? Many New Agers follow an eclectic mix of belief systems and ideas, more often than not being unable to be clearly defined as a discrete religious group. However a few distinct New Age groups, which undertake pilgrimage as part of their cosmology, namely certain longstanding pagan religions such as Wicca and Druidry, could lay claim to doing so as 'an act of faith'. Likewise, some extremely alternative New Age cyber community members occasionally come together in real-time to celebrate special events, citing their reason for attending as an act of faith (Digance and Cusack 2001).

There is no official external arbiter in these areas but the individual who owns both the inner and outer journey that are part of the essential pilgrimage experience. Perhaps at the end of the day, it does not really matter what motivates an individual to undertake a pilgrimage. By electing simply to call themselves 'pilgrims' denotes some people's journey as being special and set apart from the profane world; the heartfelt desire to enter a special time and place is no more than something that is part of humankind's existence. There are no 'pilgrim police' out there to ensure that only those who are indeed spiritually worthy take on the name of pilgrim. Has popular use of the word debased its once supposed meaning, or was the word not so inherently pure as once thought? Certainly, medieval historians suggest that this is the case and that journeys based solely upon a search for spiritual meaning were few and far between. Perhaps quantitative research soliciting pilgrims' motivations are irrelevant after all when all that really matters is that an individual took time to take a journey redolent with meaning in his or her desire to connect with the Other!

References

Alderman, D.H. (2002) 'Writing on the Graceland wall: on the importance of authorship in pilgrimage landscapes', *Tourism Recreation Research* 27(2): 27–34.

Aldred, L. (2000) 'Plastic shamans and astroturf sun dances', *The American Indian Quarterly* 24(3): 329–357.

Aldred, L. (2002) 'Money is just spiritual energy: incorporating the New Age', *Journal of Popular Culture* 35(4): 61–75.

Barber, R. (1991) *Pilgrimages*, Rochester: The Boydell Press.

Brainard, F.S. (1996) 'Defining "Mystical Experience"', *Journal of the American Academy of Religion* 64(2): 359–393.

Caplan, M. (2001) 'The fare and failings of contemporary spirituality', *ReVision* 24(2): 51–57.

Carlton, J. (2000) 'EBay CEO touts the company in a rare Wall Street pilgrimage', *Wall Street Journal* 19 May: C1.

Chandler, R. (1988) *Understanding the New Age*, Dallas: Word Publishing.

Chidester, D. and Linenthal, E.T. (1995) 'Introduction', in D. Chidester and E. Linenthal (eds) *American Sacred Space*, Bloomington: Indiana University Press.

Cohen, E. (1979) 'Phenomenology of tourist experiences', *Sociology* 13(2): 179–201.

Cusack, C. and Digance, J. (2003) 'Seeking spirituality in times of national trauma: a case study of September 11', *Australian Religion Studies Review* 16(2): 153–171.

Digance, J. (2000) 'Modern Pilgrims: Spiritual Warrior or Merely Mass Tourists', unpublished Doctorate of Philosophy, University of Sydney.

Digance, J. (2003) 'Pilgrimage at contested sites', *Annals of Tourism Research* 30 (1): 160–177.

Digance, J. and Cusack, C. (2001) 'Secular pilgrimage events: Druid Gorsedd and Stargate Alignments', in C.M. Cusack and P. Oldmeadow (eds) *The End of Religions? Religion in an Age of Globalization*, Sydney: the University of Sydney.

Eade, J. and Sallnow, M.J. (eds) (1991) *Contesting the Sacred: The Anthropology of Christian Pilgrimage*, London: Routledge.

Eliade, M. (1959) *The Sacred and the Profane: The Nature of Religion*, New York: Harvest.

Fahey, F. (2003) 'Pilgrimage as opposition in Latin American women's literature', *Mosaic: A Journal for the Interdisciplinary Study of Literature* 36(4): 33–42.

Fleischer, A. (2000) 'The tourist behind the pilgrim in the Holy Land', *International Journal of Hospitality Management* 19(3): 311–326.

George, K.M. (1994) 'Pilgrims or tourists?', *Contours* 7(7–8): 39–42.

Heelas, P. (1996) *The New Age Movement: The Celebration of the Self and the Sacralization of Modernity*, Oxford: Blackwell.

Helland, C. (2002) 'Surfing for salvation', *Religion* 32: 293–302.

Herity, M. and Eogan, G. (1996) *Ireland in Prehistory*, London: Routledge.

Hollinshead, K. (1999) 'Tourism as public culture: Horne's ideological commentary on the legerdemain of tourism', *International Journal of Tourism Research* 1(4): 267–292.

Huntsinger, L. and Fernandez-Gimenez, M. (2000) 'Spiritual pilgrims at Mount Shasta, California', *Geographical Review* 90(4): 536–559.

Ivakhiv, A.J. (2001) *Claiming Sacred Ground: Pilgrims and Politics at Glastonbury and Sedona*, Bloomington: Indiana University Press.

Jackson, R.H. and Henrie, R. (1983) 'Perception of sacred space', *Journal of Cultural Geography* 3(2): 94–107.

Jocks, C.R. (1996) 'Spirituality for sale: sacred knowledge in the consumer age', *American Indian Quarterly* 20(3/4): 415–432.

Karaflogka, A. (2002) 'Religious discourse and cyberspace', *Religion* 32: 279–291.

Kilbride, D. (2003) 'Travel, ritual and national identity: planters on the European tour 1920–1860', *The Journal of Southern History* 69(3): 549–568.

Leibovich, L. (2000) 'That online religion with shopping too', *New York Times* 16 April: G1.

McColl, C. (1989) *Living Spirituality: Contemporary Australians Search for the Meaning of Life*, Elwood: Greenhouse Publications.

MacWilliams, M.W. (2002) 'Virtual pilgrimages on the Internet', *Religion* 32: 315–335.

Moore, A. (1980) 'Walt Disney World: bounded ritual space and the playful pilgrimage centre', *Anthropological Quarterly* 53(4): 207–218.

Morinis, E.A. (1992) 'Introduction: the territory of the anthropology of pilgrimage', in A. Morinis (ed.) *Sacred Journeys: The Anthropology of Pilgrimage*, Westport, CT: Greenwood.

Nolan, M.L. and Nolan, S. (1989) *Christian Pilgrimage in Modern Western Europe*, Chapel Hill: University of North Carolina Press.

Olsen, D.H. (2003) 'Heritage, tourism and the commodification of religion', *Tourism Recreation Research* 28(3): 99–104.

Russell, P. (1999) 'Religious travel in the new millennium', *Travel & Tourism Analyst* 5: 39–68.

Solomon, L.I. (1999) 'Are you spiritually promiscuous?', *New Age Journal* March/April: 66–69.

Soyer, D. (2000) 'Back to the future: American Jews visit the Soviet Union in the 1920s and 1930s', *Jewish Social Studies* 6(3): 124–145.

Sumption, J. (1975) *Pilgrimage: An Image of Medieval Religion*, London: Faber and Faber.

Theodoratus, D.J. and Evans, N.H. (1991) *Statement of Findings: Native American Interview and Data Collection Study of Mt. Shasta, California*, Redding, CA: Department of Agriculture, Shasta-Trinity National Forest.

Tobert, M. (2001) *Pilgrims in the Rough: St Andrews Beyond the 19th Hole*, Edinburgh: Luath Press.

Tuan, Y.-F. (1977) *Space and Place: The Perspective of Experience*, Minneapolis: University of Minnesota Press.

Turner, V. (1969) *The Ritual Process: Structure and Anti-Structure*, London: Routledge and Kegan Paul.

Turner, V. (1972) 'The centre out there: pilgrim's goal', *History of Religions* 12(3): 191–230.

Turner, V. and Turner, E. (1978) *Image and Pilgrimage in Christian Culture: Anthropological Perspectives*, New York: Columbia University Press.

Van Gennep, A. (1965) *The Rites of Passage*, London: Routledge and Kegan Paul.

Vukonić, B. (1996) *Tourism and Religion*, New York: Pergamon.

Wallis, J. (2002) 'Report from Ground Zero', *Sojourners Magazine* 31(1): 7–8.

Warren, D. (2002) 'Pilgrims make annual trek to honour Presley memory. Thousands light candles at Graceland', *Boston Globe* 16 August: A28.

Wilhelm, K. (1989) 'Journeys: the dynamics of speciality travel tourists to Ireland', unpublished Doctorate of Philosophy, University of Maryland.

Zakus, D.H. (2002) 'Understanding nostalgia and sport tourism: the Old Course as "Mecca" and a "Museum without walls"', in E. Thain (ed.) *Science and Golf IV*, London: Routledge.

4 Paradigms of travel

From medieval pilgrimage to the postmodern virtual tour

Lutz Kaelber

Religiously motivated travel to sacred sites is perhaps the oldest and most prevalent type of travel in human history. Sacred journeys were part and parcel of the ancient worlds of yesteryear and may go back to the beginnings of many of the world's religions (Casson 1974; Coleman and Elsner 1995; Westwood 1997; Tomasi 2002). In the historical literature on tourism, the medieval pilgrim often emerges as a precursor to the modern tourist and traveler (MacCannell 1999; Bauman 1996). The pilgrim is portrayed as a person who ostensibly sought out a place of sacredness for reasons of personal piety and conceived of his journey there and back in terms of penitence, expiation, salvation, and liminality (Turner and Turner 1978; Feifer 1986: ch. 2). In the words of Thomas Kuhn (1970), this view of the pilgrim expresses a paradigm, predicated on the assumption that religious elements are at the core of medieval pilgrimage. Subsequently, as part of "normal science," subsequent research has proceeded on this assumption.

Yet while devotion and religious faith arguably loomed large in the repertoire of motives of medieval religious pilgrims, the following analysis will point to discrepancies between this still dominant paradigm and the results of newer research. The analysis then turns to cultural travel and mass tourism, the successors to pilgrimage in the early modern age and the modern era up to the present. As the societal framework of meaning for such travel shifted from a normative paradigm to secular models of travel, McDonaldization, Disneyfication, and postmodernism provide alternative (and complementary) approaches to explain the course of such a historical transformation and the shifting boundaries between the modern tourist, the post-tourist traveler, and postmodern pilgrim. Thematizing some possibilities for travel in the (near) future, the chapter returns to the topic of religious travel in the concluding section and address the virtual turn in travel in the context of pilgrimages in hyperreality.

Medieval religious pilgrimage

Even in the Middle Ages the experience of travel labeled religious pilgrimage was not immune from forms of commodification and criticism commonly ascribed to much later periods and more secularized forms of travel. Moreover, in religious travel, ludic elements and political issues sometimes intermingled with religious ones. The intermingling of secular and religious motives in medieval pilgrimage could be readily observed in the travels of royalty and other upper strata of medieval society. As Nicholas Vincent (2002: 15) shows for the pilgrimages of the Angevin kings of England in the twelfth and thirteenth centuries, their royal itinerary consisted of abundant journeying from one dominion to the next and included a "near ceaseless round of campaigning, hunting expeditions, crown-wearings, solemn entries and local visitations." It was thus difficult in this context to distinguish religious travel for pious purposes from recreational journeys. There were political elements present in these journeys as well, as royal pilgrimages were orchestrated to portray the Angevin kings as intermediaries between the profane and the sacred on their visits to shrines and churches, which not only consolidated their societal position but also spawned new pilgrimages to seek their favors, political and otherwise.

Ludic elements of travel were sometimes ascribed to pilgrim travelers from other strata as well, especially in what can be considered the forebear of the modern Baedeker, the literary genre of medieval travel writings for instructive purposes. In such writings, the German friar Felix Fabri was scandalized by being rushed around from place to place during his visit to Jerusalem (Kaelber 2002: 59–60); his words echo in Italian courtier Santo Brasco's admonition to visit holy places in a spirit of veneration rather than "with the intention of seeing the world or from ambition and pride to say, 'Been there! Seen that!'" (quoted in Morris 2002: 146). This type of "curiosity" was considered spiritually dangerous or at the very least useless because it expressed itself in a sensual experience and thus could lead people to temptation (Zacher 1976; Stagl 1995). For those who could not afford travel to far-away places or considered it too cumbersome, spiritual theme parks such as the *sacri monti* in Italy were established in the later Middle Ages, to bring the foreign spiritual experience closer to home. In fact, the emerging concept of a "spiritual pilgrimage," for which travel and guide books provided instruction and guidance, created the possibility of seeking an entirely stationary way of traveling to holy places. Such armchair pilgrimages required no physical exercise but could spiritually be just as effective. Moreover, similar to conscripts to armies at various points in military history, who could avoid their designated duties by hiring substitutes, penitents could avoid prescribed religious travel by hiring professional pilgrim-travelers to carry out a "pilgrimage by proxy" (Morris 2002; Swanson 1995: 168–169; Webb 2002: 50–57).

Though one of the most demonstrative forms of medieval piety, pilgrimage as a foremost religious undertaking is thus primarily a *normative* construct. Created and upheld by ecclesiastical authorities and only partly rooted in popular culture, it was nevertheless firmly entrenched as a shared societal expectation. Yet much of the literature composed by non-medievalists has failed to recognize the variation in the motives and experiences of medieval pilgrims, substituting (often unknowingly) a depiction of *what should have been* for a depiction of *what was*. Some of this literature has also yet to heed John Eade and Michael Sallnow's (2000) advice to take stock of *alternative* or *competing discourses* centering on pilgrimage, which in this case occurred both within dominant religion and outside of it. A prevailing concern about abuse and misuse of pilgrimage existed in orthodoxy, but critical voices were also loud in heterodoxy long before the Reformation – and would have been much louder yet had their members not been violently persecuted as heretics and silenced (Constable 1976; Webb 1999: 235–254, 2002: 71–77; Kaelber 2002: 60–66). Objections among Cathars, Waldensians, and Hussites were based on both a principled skepticism of pilgrimage's spiritual value as well as a critical eye towards actual pilgrim practices. These movements rejected the spiritual accounting reflected in indulgences, which were granted to relieve time in purgatory. Indulgences were commonly tied to pilgrimages, mostly local, but also granted wholesale to crusaders and those who visited the major shrines of Christianity, or participated in a jubilee. Penitentiary pilgrimage not only emphasized the obligatory element in pilgrimage, but boosted its commercial one as well, for pilgrims were travelers and had to be put up, fed, and guided on all but the most local pilgrimages. Pilgrim hostels along the major pilgrimage routes were a major contributor to the expansion of the commercial hospitality and catering business, and the production, marketing, and trade of pilgrim badges, which were physical proof of having been at a shrine, were a significant industry (Schmugge 1984; Haasis-Berner 2002). Moreover, the more affluent pilgrims could rely on a pre-modern version of the package tour: for the Holy Land, "inclusive packages of varying degrees of elaboration, depending on the depth of the pilgrim's purse, were obtainable by negotiation with licensed shipmasters at Venice" (Webb 2002: 28). Opportunities to make a profit from long-distance travel were a concomitant result of such voyages and led to the emergence of the merchant-pilgrim. Thus, toward the end of the Middle Ages the commercialization of pilgrimage had become a more common and salient feature than most modern commentators realize. If costs were still prohibitive to allow people of lesser means to go on a long-distance pilgrimage, and women were similarly circumscribed on an ideological basis, mass travel did still occur in the form of pilgrimages to local shrines (Webb 2002: 89–98, 100–113). John Urry's (2002: 5) view that "before the nineteenth century few people outside the upper classes traveled anywhere to see objects for reasons that were unconnected with work or business" would surely have mystified the medieval peasant or woman on such a pilgrimage. In the subsequent development of religious pilgrimage there is no evidence, in spite of some initial Protestant disquietude, that it fell out of favor among the populace of Europe in the early modern era, even though ludic, political, and other elements of pilgrimage continued to intermingle with spiritual ones. The penitential forms of religious pilgrimage saw their unique salutory value decrease in Catholicism, and they never acquired such a value in Protestantism (Kaelber 2002: 64–65).

Educative travel as a secular complement to pilgrimage

The early modern era ushered in, first, a pluralization of the forms of travel governed by a normative framework, and, second, a weakening of that framework on a global scale. Religious pilgrimage now had a secular equivalent: the educative type of travel. It was exemplified by the Grand Tour, first a typically very lengthy enterprise and more confined to young male aristocrats, then shorter and more inclusive of women and middle-class professionals (Towner 1996). Guided by romantic notions of the past and inspired by the ideas of humanism and the Enlightenment, travel, "that foolish beginning and excellent sequel to education," as Harvard President Charles Elliott was wont to say in his 1869 inaugural address (quoted in Brodsky-Porges 1981: 172), was still subsumable under a sanctionable societal norm of what one *ought* to do. In this case, the imperative was to develop and exercise one's intellect and thereby realize one's self through the exploration of past achievements in other European countries. To do so was ennobling and enabling, in that "travel in this form was secularized pilgrimage undertaken to regenerate the soul and revive one's faith in culture through contact through its visible remains," and achieving such travel's goals promised to turn the traveler into "a modern cultural hero ... [who] embodied the Romantic striving toward the infinite goals of truth and self-cultivation, stood for freedom from staidness of European society, and expanded the bounds of knowledge and sympathy between peoples" (Lieberson 1986: 617–618, 621). Among the classes capable of such "heroism," the peril in not having gone on an educative journey was to be considered untraveled. Being untraveled became synonymous with being uneducated, unenlightened, and, ultimately, uncultured – a stigma that seems to have existed across virtually all European cultures (Krasnobaev *et al.* 1980). But journeying itself was also not without perils, and one of them was more specific to women. Even though travel offered them otherwise unattainable opportunities to flee abusive relationships and achieve some literary prominence in travel writing, the dangers of losing control in self-indulgent, aimless traveling, and the duty not to indulge, were seen as more pertinent to them (Chard 1999; Dolan 2002). Another peril lay in cultural confrontation, even before the advent of mass tourism, between travelers as guests who did not want to be tourists and hosts who saw them as such, or worse. Scholars have reconstructed these types of clashes, for example, between natives of Italy and Victorian and Edwardian visitors whose puritanical social practices and religious views precluded a cultural understanding of the Catholic south (Pemble 1987). They also occurred between the French and traveling Americans, usually members of upper social strata. Americans' mythical admiration for French culture did not prevent them from attempting to import their institutional racism into France as late as the interwar years of the twentieth century, while, ironically, the abhorred French upper class as tourists employed their own racialized gaze on their lesser subjects in the French colonies (Levenstein 1998; Furlough 2002). In these and in other cases, anti-tourist literature and rhetoric pilloried the cultural prejudices and pretensions of the tourist, which questioned the method and purpose of this "way to culture" (Buzard 1992) – not unlike the ways in which anti-pilgrimage discourse in the Middle Ages had questioned the value of the dominant societal paradigm of travel. At the end of this historical development, the normatively exalted status of travel, be it as a salutary exercise (medieval penitential pilgrimage) or heroic cultural achievement (aristocratic educative travel), had waned. Societal norms no longer provided a strong and salient prescriptive impetus to travel.

Modern tourism: McDonaldized, Disneyfied, postmodernized

With the differentiation of societal spheres in industrial modernity and the expansion of tourism as a form of "travel-capitalism" (Böröcz 1992), a vast new array of motives for, experiences in, and objects of travel emerged. Expanding economic interconnectedness, innovations in transportation and communication technologies, rising incomes of the middle and lower classes, and an increase in leisure time, rang in a multitude of novel forms and functions of travel, whose vicissitudes now more than ever could be described, in Judith Adler's (1989: 1372) words, "as a history of coexisting and competitive ... styles whose temporal boundaries inevitably blur." The style to emerge most saliently in the modern industrialized age was, and in some ways continues to be, the well-chartered types of mass travel in the form of leisure tourism. Thomas Cook became the pioneer for highly packaged tours in the nineteenth century, and tourism, as recreational travel for larger segments of society, became an industry. American Express's ingenious invention of the traveler's check facilitated the taking of currency abroad and was the first step in a process toward a global credit card society (Brendon 1991; Ritzer 1995). First railroads, then cars, then airplanes democratized travel. The rising incomes of the middle classes in the twentieth century, from the postwar years to the mid-1970s, together with a considerably shortened work week – a trend that has since reversed itself in the US but not in most industrializing Asian and in European countries – as well as increased socioeconomic interpenetration allowed people to take advantage of these technological and organizational changes (Shaw and Williams 2002: ch. 9; Urry 2002: ch. 2). Moreover, the world was becoming smaller, in that travelers from industrialized nations found it possible and even convenient to explore the natural habitats of what at the time was still called the "Third World." For example, by the early 1970s, safari holidays had become Kenya's major export; its capital, Nairobi, was the seat of sixty-five safari companies, all but one foreign owned. Buses took tourists to sites of raw nature and savage beasts, with the indigenous population serving as drivers, guides, and servants (Feifer 1986: 231). Some ten thousand miles away the budding industry of sex tourism was about to take off in the Far East. Now was the time of leisure tourism, a time, to use McCrick's (1989) phrase, to enjoy "sun, sex, sights, savings, and servility."

Mass tourism stood and stands as a paradigm of the McDonaldization of travel and recreation, a term associated with George Ritzer. This perspective is broadly associated with what one might call a "modernist" point of view, conceiving of travel as being subjected to the imperatives of the industrial age. This includes a clear time–space order, the mass production of travel experiences, and the availability of certain meta-narratives to make these experiences meaningful. Ritzer (2004) sees modernity, and tourism as part of it, as guided by efficiency, calculability, predictability, and (non-human) control. Predictability in recreational travel means few surprises, as expressed by a Hawaiian tourist official: "The kids are safe here, there's low crime, you can drink the water, and you can speak the language" (Ritzer and Liska 1997: 99). In other words, McDonaldized tourists want the things they are familiar with on a daily basis, which, in an American cultural context, means CNN for television, McDonald's for food, and Mastercard for the moments that aren't priceless. Efficiency relates to tourists' expectations to have little or no unnecessary slack and avoid inefficiencies. To get tourists to as many tour sites as possible was, after all, one of the rationales for Thomas Cook's package tour. Calculability, the third element of McDonaldization, entails having as few surprises – at least unpleasant ones – as possible. Tourists, in this perspective, want to know exactly where they will go, how long they will be away, and how much it is going to cost. Ritzer mentions the popularity of cruise ship tours and other highly orchestrated journeys in this context. Finally, the fourth element is control over humans, which pertains, for example, to the use of scripts to deal with visitors in Walt Disney theme parks, or the ordering of day schedules through routines on cruise ship travel (Ritzer and Liska 1997: 99–100).

The literature on the "Disneyfication" of culture provides, in part, an alternative to this perspective, and, in part, complements it. Especially in the work of Alan Bryman (1999, 2004), this literature shifts the focus of analysis from productive to cultural elements of travel, specifically a type of consumption of time and space that leaves little to the imagination. Applied to travel, Disneyfication (which Bryman calls Disneyization) refers to four processes. The first process is theming, which refers to changes to unfamiliar space through a culturally shared frame of reference.

Disney's theme parks and Las Vegas's themed hotels and other built environments establish coherence and permanency in a tourist's transient and murky encounter with a new or different physical setting. The second process is the dedifferentiation of consumption, defined as a development by which forms of consumption associated with different institutional spheres become interlocked and are increasingly difficult to distinguish. Many airports, for example, have become mini-malls that tend to carry globalized products. Merchandising is the third process, which Bryman defines as the promoting of goods in recreational settings, which refers to the proffering of retail goods and other merchandise that may then serve as souvenirs. Fourth, there is a trend toward emphasizing emotional labor, in the way first and perhaps still best described by Arlie Hochschild (1985), who, incidentally, studied airline flight attendants as her main subjects. "Feeling rules," which in Hochschild's work refers to expectations about what the airlines' employees are to *feel* in social interaction, set a normative framework for how the tourist is to experience the cultural other emotionally, and the production and consumption of such an experience involves the commercialization of emotion.

Postmodernist scholarship on culture in general, and on travel in particular, has taken the cultural turn reflected in the Disneyfication thesis several steps further. Time–space compression, the dedifferentiation of social spheres, an erosion of boundaries between signifier and signified, the dominance of surface images and *simulacra*, and the ensuing ephemeral character and volatility of social practices, as well as the fragmentation of identity, are the familiar diagnoses for a global consumer society run amok (Harvey 1989; Baudrillard 1983; Jameson 1992; Lyotard 1984). Meta-narratives, it is argued, are no longer meaningful, which implies that skepticism toward any overarching explanatory scheme, such as McDonaldization. Instead, a plurality of narratives exists that refract tourist experiences and motives differently. Some of the experiences, increasingly mediated and image-driven, may indeed be orchestrated for cursory consumption through the "tourist gaze," so masterfully described by Urry (2002), or provide "ways of escape" from monotony and daily routines by intensifying the routines and extending or inverting them through spectacle, as Chris Rojek (1993) has argued. At the same time, the fragmented nature of postmodern existence has its consumptive equivalents in tourism and is compatible with an array of tourism styles.

Thus, the McDonaldization and Disneyfication perspectives and the postmodernist approach to travel place different emphases in analyzing travel at the beginning of the twenty-first century and come to different conclusions about it. Each view has its particular strengths and problems. The McDonaldization thesis essentially focuses on the production (or delivery) of tourist experiences as a standardized consumer good. It focuses on factors that have enabled tourism to become one of the largest industries in the world through the fast-food and factory-like supply of leisure travel. It does not concern itself with the ways in which tourists appropriate these goods, and the reasons they have for doing so. It also presupposes an at least somewhat gullible consumer, who, just as people who eat at McDonald's, may not know much – or care – about issues of quality and long-term adverse effects on the environment. Leaving aside the issue that this depiction of consumer motivation and behavior is simplistic and runs counter to scholarship that shows a fragmented consumer culture (Gabriel and Lang 1995), the question still remains: What makes the consumption of mass-produced tourism possible, or even perhaps desirable?

Unlike McDonaldization, Disneyfication is capable of addressing the shaping of consumer preferences and behavior in tourism. It relates tourism experiences to the formation (or, as some would say, manipulation) of demand factors. However, the Disneyfication thesis, too, presents the tourist largely as a passive and reactive subject, and it leaves two main questions unanswered: Why would tourists want to consume Disneyfied environments, which extend far beyond the realm of travel (Gottdiener 2001), in the first place? Moreover, both McDonaldization and Disneyfication imply increasing standardization and homogenization of tourism experiences – how does this mesh with the reality of a bewildering array of tourist choices and practices in the new millennium, from political tourism to adventure tourism, ecotourism, heritage tourism, etc., when at the same time "traditional" mass tourism is alive and well?

Postmodernism provides an answer to these questions by relating travel to a fundamental transformation of industrialized societies toward image-based societies. However, the perspective of postmodernism tends to over-identify the modern traveler with the nomadic sightseer and all-too-readily-duped consumer of Boorstinian pseudo- and non-events (Boorstin 1964). The most prominent critic of the image of the tourist merely gazing at surface events is Dean MacCannell, who in his magisterial *The Tourist* emphatically argued that a "quest for authenticity" (1999: 105) is at the core of modern touristic attitudes. Tourists are reflective of the fact that tourism providers stage touristic events, so they demand a deeper, more real experience, a process that the providers themselves are aware of and anticipate, setting up an almost infinite loop of reflexivity. While MacCannell (1999: 194), in the third and most recent edition of his book, has acknowledged that since the book's original publication in the mid-1970s "an aggressive invasion of the touristic field by corporate entertainment interests" has occurred, he nevertheless steadfastly continues to argue that tourists actively construct their own experiences and, in doing so, demand authenticity, or, in other words, employ a "second gaze" piercing the veil of surface images (MacCannell 2001).

Yet while MacCannell is correct in pointing to tourist agency, he bases his argument on doubtful empirical evidence, never properly disclosing his methods beyond vaguely hinting at his ethnographic studies in California and Paris – evidence that is at least thirty years old by now anyway. His view applies perhaps best to the "post-tourist" (Feifer 1986) who emphasizes a type of Habermasian communicative rationality in travel over McDonaldized and Disneyfied tourism. The post-tourist travel style, currently best reflected in strands of volunteer tourism and ecotourism, is tellingly captured in this description of a tour in a (recycled-paper) travel brochure: travelers to the Caribbean are warned of "ants, mosquitoes, and cockroaches ... erratic water and electricity supplies" and a flight "that can be tedious as can protracted entry formalities" upon arrival (quoted in Munt 1994: 103). One might also suspect the "post-tourist" to be behind discourse critical of mass travel and supportive of organized movements that advocate de-Disneyfication and de-McDonaldization (Ritzer 2004: 201–212; Gottdiener 2001: 162–167).

But not everyone wants to (or can) be a post-tourist. While MacCannell locates much of modern tourism on the "sustainable authenticity" end of the continuum of tourism motives and experiences, the fact is that main-stream tourist brochures tend to foster little interaction between tourist and native, and tend to portray the latter in a stereotypical fashion (Dann 1996). They are much closer to the other end of the continuum, that of "contrived post-modernity" (see Cohen 1995). MacCannell's tourist is also not typic-ally found on the ever more popular cruise ships. On some lines the deck chairs are set up to face inward, away from the sea, and passengers are seemingly oblivious to the significant marine pollution that the ships are responsible for and that threatens to destroy the very beauty these types of boats were initially meant to view. Nor are they aware of the repro-duction of older colonial-type class hierarchies in the ships' staff, the lower rungs of which purposely remain hidden from them (Wood 2000, 2004). MacCannell's (2001: 26) tourist is also, so it seems, the proper audience of "tourist bubbles," as he himself acknowledges. In the United States, such bubbles manifest themselves as pacified and class-segregated shop-ping islands in formerly downsized and unsafe neighborhoods, which are reborn as a requisite mix of entertainment malls, posh retail stores, large convention centers, atrium-lobbied hotels, restored historical neighbor-hoods, sport stadiums, formulaic restaurants, and casino gambling facilities (Judd 1999).

Thus, different perspectives account for different facets of an eclectic mix of tourism motives, experiences, and styles. In the final, concluding section I will turn to the future of tourism by coming back to a topic explored in the beginning sections of this chapter, religious travel, to dis-cuss the recent expansion of tourism into multimedia generated hyperreality and present some foundational arguments for future research.

Conclusion: twenty-first-century pilgrimages – toward hyperreality?

Researchers of tourism are well advised to take note of empirical studies on contemporary religious pilgrimages, for they have ramifications for the study of tourism at large. Religious pilgrimage is alive and well, not only because many world religions have reasserted themselves in the public sphere (Casanova 1994), but also because its disconnection from other forms of religious tourism and other travel are increasingly fuzzy. For religious travel, Rinschede (1992: 57) estimated that in the early 1990s 200 million people annually engaged in international, national, and supra-regional pilgrimage journeys. Ten to fifteen years later, these figures seem far too low, as, for example, press reports estimated the number of attendees at the 2001 Maha Kumbha Mela festival, which involves a ritual bathing at the confluence of the Ganges and Yamuna rivers in India, at some 70 million Hindus, and more than 20 million pilgrim-travelers are believed to have visited Rome in the jubilee of 2000 (Cipriani 2000; Cipolla and Cipriani 2002; Rinschede 1999: 197–221).

Empirical studies of such jubilee pilgrims and of travelers to other modern pilgrimage shrines and routes have repeatedly shown that religious convictions are at the root of the journey. A core portion of religious pilgrims is indeed motivated by genuinely religious concerns and seeks "authentic" corresponding experiences. Motives are hardly static, however, as some develop religious motives only after starting on such journeys (Cipolla and Cipriani 2002; Dubisch 1995; Frey 1998). These studies also show that most pilgrims are not solely focused on the sacred. A plurality of motives, religious as well as secular, guide them, and factors such as gender, politics, and leisure orientation play a role in shaping their experiences.

At the same time, transcendent purposes are evident in other types of travel that have become popular and, while per se less intimately tied to sacrality, have become very much infused with sacred meaning. Recent work on dark tourism, also called necro- or thanatourism, to war graves, cemeteries, and other "dark spots" associated with disaster, tragedy, or atrocity more often than not paint a serene picture of those who come to commemorate such events. Many such sites have indeed acquired a Durkheimian sacred quality (Foote 2003; Lennon and Foley 2000; Rojek 1993: ch. 4; Seaton 1999, 2002; Walter 1993). John Sears (1989) has shown that the same can be said of secular sites that became redefined as sacred places in nineteenth-century American tourist landscapes, where Americans in search of a cultural identity described their visits to tourist attractions such as Mount Rushmore as pilgrimage and their travels as journeys to manifestations of a divine presence. Landscapes and cultural icons could be consumed in the spirit of patriotism and a nationalist affirmation of what it meant be American. This trend continues to this day. Juan Campo's (1998) research on the "American pilgrimage landscape" evinces a plethora of popular quests to national shrines. Icons of worship, from Jefferson to Elvis, and their sites (Alderman 2002; Kiely 1999) are firmly woven into the American ritual tapestry (Deegan 1998) by ritual dramas that are increasingly orchestrated in commodified settings. Sacralization also extends to malls as sacred centers and other "cathedrals of consumption" that promise an "enchanted" yet also McDonaldized shopping experience (Ritzer 2002; Timothy 2005; Zepp 1997).

Many of these new cathedrals now extend into cyberspace, with its virtual shopping malls à la Amazon.com, where opportunities for combining commodified McDonaldization with themed Disneyfication abound. The Disneyfied virtualization of pilgrimage promises to add yet another layer of "travel in hyperreality" to a decentered landscape (Eco 1986) where time and space boundaries collapse. Already "virtual cemeteries" exist to cyber-congregate and commemorate the dead (Schwibbe and Spieker 1999), and, as I have argued elsewhere, the "virtual gaze" is increasingly part of the consumption of postmodern religion in cyberspace. This involves such events as "virtual guided tours" to the Holy Land, the possibility to create a medieval pilgrimage by becoming a "virtual pilgrim," and a form of virtual penance and cleansing in the Ganges River. In the latter case, a web site supported by the government of Uttar Pradesh, India, invites visitors to fill out a survey form concerning their caste, gender, and physical attributes and attach their picture. They also choose a date at which time they can see their picture superimposed as a head on a virtual body (of the type they previously indicated on the form), to be absolved in an animation depicting their virtual cleansing in the river, and printouts of their photographs will actually be dipped into the Ganges (Kaelber 2001; Government of Uttar Pradesh 2001). Another form of cyber-tourism is hinted at in the following discussion of advances in virtual animation and the future of dark tourism:

> [M]useum cyberguides and curators will take their virtual tourists on real time tours of active detention camps, killing fields, death rows, and execution chambers. . . . [L]ongstanding psychological distinctions between real and virtual, here and there, subject and object may them-selves loosen. If so, then the dark cybertourist may not in fact sense a substantial difference between walking and browsing through Auschwitz.
>
> (Miles 2002: 1177)

Disneyfication by means of cyber-theming Nazi *Lebensraum* ideology as "death space" (Aldor 1940)? – Ominous words indeed.

References

Adler, J. (1989) "Travel as performed art," *American Journal of Sociology* 89(6): 1366–1391.

Alderman, D.H. (2002) "Writing on the Graceland wall: on the importance of authorship in pilgrimage landscapes," *Tourism Recreation Research* 27(2): 27–34.

Aldor, F. (1940) *Germany's Death Space: The Polish Tragedy*, London: Francis Aldor.

Baudrillard, J. (1983) *Simulations*, New York: Semiotext(e).

Bauman, Z. (1996) "From pilgrim to tourist – or a short history of identity," in S. Hall and P. Du Gay (eds) *Questions of Cultural Identity*, Thousand Oaks, CA: Sage.

Boorstin, D.J. (1964) *The Image: A Guide to Pseudo-Events in America*, New York: Harper and Row.

Böröcz, J. (1992) "Travel-capitalism: the structure of Europe and the advent of the tourist," *Comparative Studies in Society and History* 34: 708–741.

Brendon, P. (1991) *Thomas Cook: 150 Years of Popular Tourism*, London: Secker and Warburg.

Brodsky-Porges, E. (1981) "The Grand Tour: travel as an educational device, 1600–1800," *Annals of Tourism Research* 8: 171–186.

Bryman, A. (1999) "The Disneyization of society," *Sociological Review* 47: 25–47.

Bryman, A. (2004) *The Disneyization of Society*, Thousand Oaks, CA: Sage.

Buzard, J. (1992) *The Beaten Track: European Tourism, Literature, and the Ways to Culture, 1800–1918*, Oxford: Clarendon Press.

Campo, J.E. (1998) "American pilgrimage landscapes," *The Annals of the American Academy of Political and Social Sciences* 558: 40–56.

Casanova, J. (1994) *Public Religions in the Modern World*, Chicago: University of Chicago Press.

Casson, L. (1974) *Travel in the Ancient World*, London: Allen and Unwin.

Chard, C. (1999) *Pleasure and Guilt on the Grand Tour: Travel Writing and Imaginative Geography, 1600–1830*, Manchester: Manchester University Press.

Cipolla, C. and Cipriani, R. (2002) *Pellegrini del Giubileo*, Milan: Franco Angeli.

Cipriani, R. (2000) "The jubilee as a global event: research on Roman pilgrims," paper presented at the Annual Meeting of the Society for the Scientific Study of Religion, Houston, Texas.

Cohen, E. (1995) "Contemporary tourism, trends and challenges: sustainable authenticity or contrived post-modernity?," in R.W. Butler and D. Pearce (eds) *Change in Tourism: People, Places, Processes*, London: Routledge.

Coleman, S. and Elsner, J. (eds) (1995) *Pilgrimage: Past and Present in the World Religions*, Cambridge, MA: Harvard University Press.

Constable, G. (1976) "Opposition to pilgrimage in the Middle Ages," *Studia Gratiana* 19: 125–146.

Dann, G. (1996) "The people of tourist brochures," in T. Selwyn (ed.) *The Tourist Image: Myths and Myth Making in Tourism*, Chichester: Wiley.

Deegan, M. (ed.) (1998) *The American Ritual Tapestry: Social Rules and Cultural Meaning*, Westport, CT: Greenwood Press.

Dolan, B. (2002) *Ladies of the Grand Tour: British Women in Pursuit of Enlightenment and Adventure in Eighteenth-Century Europe*, New York: Harper Collins.

Dubisch, J. (1995) *In a Different Place: Pilgrimage, Gender, and Politics of a Greek Island Shrine*, Princeton, NJ: Princeton University Press.

Eade, J. and Sallnow, M.J. (2000) "Introduction," in J. Eade and M.J. Sallnow (eds) *Contesting the Sacred: The Anthropology of Christian Pilgrimage*, Urbana: University of Illinois Press.

Eco, U. (1986) *Travels in Hyperreality*, San Diego: Harcourt Brace Jovanovich.

Feifer, M. (1986) *Tourism in History: From Imperial Rome to the Present*, New York: Stein and Day.

Foote, K.E. (2003) *Shadowed Ground: America's Landscapes of Violence and Tragedy*, Austin: University of Texas Press.

Frey, N. (1998) *Pilgrim Stories: On and Off the Road to Santiago*, Berkeley: University of California Press.

Furlough, E. (2002) "Une leçon des choses: tourism, empire, and the nation in interwar France," *French Historical Studies* 25: 441–473.

Gabriel, Y. and Lang, T. (1995) *The Unmanageable Consumer: Contemporary Consumption and Its Fragmentation*, Thousand Oaks, CA: Sage.

Gottdiener, M. (2001) *The Theming of America: American Dreams, Media Fantasies, and Themed Environments* (2nd edn), Boulder, CO: Westview Press.

Government of Uttar Pradesh, Department of Information (2001) "Kumbh Mela 2001," available at www.webdunia.com/kumbhupinfo (accessed January 1, 2002).

Haasis-Berner, A. (2002) "Pilgerzeichenforschung: Forschungsstand und Perspektiven," in H. Kühne, W. Radtke, and G. Strohmaier-Wiederanders (eds) *Spätmittelalterliche Wallfahrt im mitteldeutschen Raum*, Berlin: Humboldt-Universität.

Harvey, D. (1989) *The Condition of Postmodernity: An Enquiry into the Origins of Cultural Change*, Oxford: Blackwell.

Hochschild, A. (1985) *The Managed Heart: Commercialization of Human Feeling*, Berkeley: University of California Press.

Jameson, F. (1992) *Postmodernism, Or, the Cultural Logic of Late Capitalism*, Durham, NC: Duke University Press.

Judd, D.R. (1999) "Constructing the tourist bubble," in D.R. Judd and S.S. Fainstein (eds) *The Tourist City*, New Haven: Yale University Press.

Kaelber, L. (2001) "The virtual gaze: religious tourism and the consumption of postmodern religion in cyberspace," paper presented at the Annual Meeting of the Society for the Scientific Study of Religion, Columbus, Ohio.

Kaelber, L. (2002) "The sociology of medieval pilgrimage: contested views and shifting boundaries," in W.H. Swatos Jr and L. Tomasi (eds) *From Medieval Pilgrimage to Religious Tourism: The Social and Cultural Economics of Piety*, Westport, CT: Praeger.

Kiely, R. (1999) "From Monticello to Graceland: Jefferson and Elvis as American icons," in M. Garber and R.L. Walkowitz (eds) *One Nation Under God?: Religion and American Culture*, London: Routledge.

Krasnobaev, B.I., Robel, G., and Zeman, H. (eds) (1980) *Reisen und Reisebeschreibungen im 18. und 19. Jahrhundert als Quellen der Kulturbeziehungsforschung*, Berlin: Camen.

Kuhn, T. (1970) *The Structure of Scientific Revolutions* (2nd edn), Chicago: University of Chicago Press.

Lennon, J. and Foley, M. (2000) *Dark Tourism*, London: Continuum.

Levenstein, H. (1998) *Seductive Journey: American Tourists in France from Jefferson to the Jazz Age*, Chicago: University of Chicago Press.

Lieberson, H. (1996) "Recent works on travel writing," *Journal of Modern History* 68: 617–628.

Lyotard, J.-F. (1984) *The Postmodern Condition: A Report on Knowledge*, Minneapolis: University of Minnesota Press.

MacCannell, D. (1999) *The Tourist: A New Theory of the Leisure Class* (3rd edn), Berkeley: University of California Press.

MacCannell, D. (2001) "Tourist Agency," *Tourist Studies* 1: 23–37.

McCrick, M. (1989) "Representations of international tourism in the social sciences: sun, sex, sights, savings, and servility," *Annual Review of Anthropology* 18: 307–344.

Miles, W.F.S. (2002) "Auschwitz: museum interpretation and darker tourism," *Annals of Tourism Research* 29: 1175–1178.

Morris, C. (2002) "Pilgrimage to Jerusalem in the late Middle Ages," in C. Morris and P. Roberts (eds) *Pilgrimage: The English Experience from Becket to Bunyan*, Cambridge: Cambridge University Press.

Munt, I. (1994) "The 'other' postmodern tourism: culture, travel and the new middle classes," *Theory, Culture and Society* 11: 101–123.

Pemble, J. (1987) *The Mediterranean Passion: Victorians and Edwardians in the South*, Oxford: Oxford University Press.

Rinschede, G. (1992) "Forms of religious tourism," *Annals of Tourism Research* 19: 51–67.

Rinschede, G. (1999) *Religionsgeographie*, Braunschweig: Westermann.

Ritzer, G. (1995) *Expressing America: A Critique of the Global Credit Card Society*, Thousand Oaks, CA: Pine Forge Press.

Ritzer, G. (2002) "Cathedrals of consumption: rationalization, enchantment, and disenchantment," in G. Ritzer (ed.) *McDonaldization: The Reader*, Thousand Oaks, CA: Pine Forge Press.

Ritzer, G. (2004) *The McDonaldization of Society* (rev. new century edn), Thousand Oaks, CA: Pine Forge Press.

Ritzer, G. and Liska, A. (1997) "'McDisneyization' and 'post-tourism': contemporary perspectives on contemporary tourism," in C. Rojek and J. Urry (eds) *Touring Cultures: Transformations of Travel and Theory*, London: Routledge.

Rojek, C. (1993) *Ways of Escape: Modern Transformations in Leisure and Travel*, Lanham, MD: Rowman and Littlefield.

Schmugge, L. (1984) "Die Anfänge des organisierten Pilgerverkehrs im Mittelalter," *Quellen und Forschungen aus italienischen Archiven und Bibliotheken* 64: 1–83.

Schwibbe, G. and Spieker, I. (1999) "Virtuelle Friedhöfe," *Zeitschrift Für Volkskunde* 95(2): 220–245.

Sears, J.F. (1989) *Sacred Places: American Tourist Attractions in the Nineteenth Century*, Oxford: Oxford University Press.

Seaton, A.V. (1999) "War and thanatourism: Waterloo, 1815–1914," *Annals of Tourism Research* 26: 130–158.

Seaton, A.V. (2002) "Thanatourism's final frontiers? Visits to cemeteries, churchyards and funerary sites as sacred and secular pilgrimage," *Tourism Recreation Research* 27(2): 73–82.

Shaw, G. and Williams, A.M. (eds) (2002) *Critical Issues in Tourism: A Geographical Perspective* (2nd edn), Oxford: Blackwell.

Stagl, J. (1995) *A History of Curiosity: The Theory of Travel, 1550–1800*, Chur: Harwood Academic Publishers.

Swanson, R.N. (1995) *Religion and Devotion in Europe, c. 1215–c.1515*, Cambridge: Cambridge University Press.

Timothy, D.J. (2005) *Shopping Tourism, Retailing and Leisure*, Clevedon: Channel View.

Tomasi, L. (2002) "*Homo Viator*: from pilgrimage to religious tourism via the journey," in W.H. Swatos, Jr and L. Tomasi (eds) *From Medieval Pilgrimage to Religious Tourism: The Social and Cultural Economics of Piety*, Westport, CT: Praeger.

Towner, J. (1996) *An Historical Geography of Recreation and Tourism in the Western World, 1540–1940*, Chichester: Wiley.

Turner, V. and Turner, E. (1978) *Image and Pilgrimage in Christian Culture*, New York: Columbia University Press.

Urry, J. (2002) *The Tourist Gaze* (2nd edn), London: Sage.

Vincent, N. (2002) "The Pilgrimages of the Angevin Kings of England, 1154–1272," in C. Morris and P. Roberts (eds) *Pilgrimage: The English Experience from Becket to Bunyan*, Cambridge: Cambridge University Press.

Walter, T. (1993) "War grave pilgrimage," in I. Reader and T. Walter (eds) *Pilgrimage in Popular Culture*, London: Macmillan.

Webb, D. (1999) *Pilgrims and Pilgrimage in the Medieval West*, New York: Tauris.

Webb, D. (2002) *Medieval European Pilgrimage, c.700–c.1500*, New York: Palgrave.

Westwood, J. (1997) *Sacred Journeys: An Illustrated Guide to Pilgrimages Around the World*, New York: Harry Holt.

Wood, R.E. (2000) "Carribbean cruise tourism: globalization at sea," *Annals of Tourism Research* 27(2): 345–370.

Wood, R.E. (2004) "Global currents: cruise ships in the Caribbean Sea," in D.T. Duval (ed.) *Tourism in the Caribbean: Trends, Development, Prospects*, London: Routledge.

Zacher, C.K. (1976) *Curiosity and Pilgrimage: The Literature of Discovery in Fourteenth-Century England*, Baltimore: Johns Hopkins University Press.

Zepp, I.G. (1997) *The New Religious Image of Urban America: The Shopping Mall as Ceremonial Center* (2nd edn), Niwot: University of Colorado Press.

5 Travel and journeying on the Sea of Faith

Perspectives from religious humanism

C. Michael Hall

And so we go on taking pilgrimages, in part because every discovery, however unlooked-for, is a step forward; but also, more deeply, because every one of us carries around, inside, a certain, unnamed homesickness, a longing for a place we left and don't know how to find again.

(Iyer 1999: xi)

The phrase 'Sea of Faith' refers to a line from the poem *On Dover Beach* by the nineteenth-century English poet Matthew Arnold. The phrase was used by English theologian Don Cupitt in a book and BBC television series of the same name to explore the changing role and nature of Christianity and religion in Western society (Cupitt 1984). Arnold's poem lamented the end of certainty of religious, and particularly Christian, teachings and beliefs in the face of modernity and scientism. Central to Cupitt's thinking, and related to some of the thinkers of radical Christian theology and religious humanism (e.g. Spong 1998; Geering 2002; Leaves 2004), is the notion that god is created by humankind. This approach is therefore reflective of the influence of postmodernity and the linguistic turn in much of contemporary thinking in the humanities and social sciences that creates a basis of celebrating the liberating effects of the twentieth century's revolutionary understanding of 'the word made flesh'.

Is such a reflection of any significance for understanding tourism and pilgrimage? Absolutely! One concern in the study of tourism is why people travel for non-work reasons. What drives people and sustains them and, just as importantly, what constrains many of them from travelling voluntarily? Such considerations are some of the fundamental questions of thinking about tourism and are embedded in our understandings of the history and development of tourism, as well as understanding motivations for travel (Hall 2005). Such theorizing is embedded in our understanding of structure and agency, of the individual's interaction with and location within society and its institutions. In the same way that one might search for the historical Jesus, or the historical Buddha, one might also be searching for the historical 'tourist'.

   An understanding of religion and the religious experience has to be central in understanding travel over time because of the role of religious institutions and structures in affecting the capacity of individuals to travel and cross borders not only in historical time but also in the present. Moreover, an understanding of the religious experience is arguably also significant to those who do not even adhere to a particular stated faith. Religion can be usefully defined as 'a total mode of interpreting and living of life' (Geering 2002: 147). It is a significant part of the structure that influences individual agency, thereby affecting a whole range of behaviours and actions. Such behaviours not only include whether one travels but also how one travels. What conscious ethical decisions, if any, are made with respect to where one travels, what one does, and what one consumes? Where does the ethical base of sustainable tourism, ecotourism or pro-poor tourism come from? Surely such issues of tourism and modernity have a metaphysical dimension which means that it is likely to connect with the varieties of religious experience and consideration of appropriate ethical behaviour. Yet such considerations are rarely dealt with in the study of tourism, although the notion of secular pilgrimage is clearly one area (perhaps the only area) where some of these connections have been made.

   In looking at modern tourism textbooks it often seems that there is a presentation of 'tourism as modernity' versus 'pilgrimage as premodernity' as if part of some seamless evolutionary development that has undergone a gradual and 'normal' transition. Unfortunately, the history of tourism as with the history of evolution is actually fraught with discontinuity and speciation in which different categories of being occupy and sometimes compete within the same space. As many of the chapters in this book attest, even in this so-called postmodern period, religious pilgrimage is alive and well, even though much of the contemplation of tourist movement is undertaken in a secular fashion. There is therefore, arguably, a disjoint between much of the actual nature of human movement as is experienced by many of the individuals undertaking it and what is being described in terms of voluntary mobility at a global scale. Much of this disjoint may occur because many examinations of tourism tend to focus on what happens at destinations or, to be more precise, at a specific point of time at a destination. Arguably, there are few studies in tourism that seek to explain tourism within the context of the overall lifecourses of individuals (Hall 2005). In other words, few studies have sought to provide a much more rounded view of people's mobility that is as interested in their actions, behaviours and influences at 'home' as those in the destination. Most observers merely place people in the simplistic category of 'tourist' as if what they do away from their home environment will be the best explanation for future travel behaviour without understanding all the other influences on an individual's path through life. Such a notion of understanding tourism is perhaps representative of positivist empiricism in the social sciences at its worst, rather than seeking to provide a perspective of the tourist as a whole person. Clearly, in terms of understanding the totality of someone's life and mobility such a perspective is important, as while someone may be categorized as a tourist and seemingly not being engaged in any overt form of religious behaviour, the experiences of such journeys will likely inform religious perspectives, concepts of being, and relations with others (Coleman and Elsner 1995; Garabedian 1999).

A more gestalt perspective of travelling would also suggest that what happens in the religious life and experiences of people in their home environment must also inform their travel behaviours, often perhaps even in unconscious ways. From such a perspective, of course, notions of religion and the secular, or sacred and the secular, should not be seen as absolute opposites but are rather at least highly permeable concepts if not better recognized as lying at two ends of a continuum. Moreover, a lifecourse perspective on the religious experience in tourism also provides a platform from which to better examine the religious experience of humanism, a point that will now be addressed.

Religious humanism

The term 'religious humanism' may seem to many readers at first glance to be a contradiction. It is not, however. As noted earlier, religion is a way of interpreting, experiencing and living life. The notion of a religious experience can embrace many different notions of god and the spiritual, including that of those who do not believe in the existence of a god or divine being independent of themselves. For example, Cupitt (1984: 269), one of the leading exponents of Christian religious humanism, defines god as 'the sum of our values, representing to us their ideal unity, their claims upon us and their creative power'. From such a humanist position, the notion of god as objectively occurring independently of ourselves is a mythological idea that human beings could do without. For some, the scientific understandings of the twenty-first century are at odds with the occult notion of god, 'out there' in another world. A notion of an independent god does little to explain the way

> he functions as *our* God, chosen by us, our religious ideal, our life-aim and the inner meaning of our identity. Just as you should not think of justice and truth as independent beings, so you should not think of God as an objectively existing superperson.
>
> (Cupitt 1984: 270)

Cupitt (2002: xi) defines his non-realist Christianity in a 'nutshell':

> Suppose we become acutely aware of our own limits: we realize that we are always inside human language, and only ever see the world

through our human eyes. All that is ever accessible to us is the relative god, my god. As I see this, metaphysics dies and I am left knowing only my god, my guiding religious idea.

Such a position is obviously immensely challenging to many people who cling to fundamentalist and obsolete dogmas that are at odds with contemporary understandings of the world. Yet with historical change, people change, and hence their ideas about god also change. This is witnessed in the constant reformations and reinterpretations of the institutions of religion, as well as in their understandings of god itself (see the theses of Bishop John Spong (1998) with respect to a new reformation of the Christian Church from a Christian humanist perspective (Table 5.1)). That god is human-made is only startling if one is unaware that everything is. god is therefore our religious ideal. Such a stance is extremely challenging, as it means that rather than relying on someone else for one's particular ethical and moral stance and the whole way of life, one has to work it all out and be responsible for himself or herself. Individuals can no longer bathe themselves in the waters and wash their sins away because someone else can cleanse them of their irreconciled thoughts, behaviours and actions. According to this line of thinking, only individuals can do this as they, and they alone, are responsible for finding unity within themselves and with others. However, such a task is tragic, in the same way that Christ's life is also tragic, in that while people strive to find ultimate harmony and unity of moral and spiritual values they can never succeed, even though perhaps they seem to experience transcendental moments of wholeness. Of course, this also means that not only must people assume internal conflicts within the achievement of their internal notion of god – that is their set of values and ideas and the means by which they examine those ideas – so it is also the case that such conflicts also become external, because

> the personal God is tied rigorously to the study of a particular set of scriptures, particular religious practices, and life as a member of a particular religious community . . . There is a Jewish God, a Muslim God, a Russian God and so on.
>
> (Cupitt 1982: 71)

> When we have fully accepted these ideas and have freed ourselves from nostalgia for a comic Father Christmas, then our faith can at last become fully human, existential, voluntary, pure, and free from superstition. To reach this goal is Christianity's destiny, now approaching.
>
> (Cupitt 1984: 271)

However, as Cupitt (1984) rightly observed such a view of the world and of god is traumatic. For many people it removes the certainty that they

Table 5.1 Bishop John Spong's theses for a new reformation

1	Theism, as a way of defining God, is dead. So most theological God-talk is today meaningless. A new way to speak of God must be found.
2	Since God can no longer be conceived in theistic terms, it becomes nonsensical to seek to understand Jesus as the incarnation of the theistic deity. So the Christology of the ages is bankrupt.
3	The biblical story of the perfect and finished creation from which human beings fell into sin is pre-Darwinian mythology and post-Darwinian nonsense.
4	The virgin birth, understood as literal biology, makes Christ's divinity, as traditionally understood, impossible.
5	The miracle stories of the New Testament can no longer be interpreted in a post-Newtonian world as supernatural events performed by an incarnate deity.
6	The view of the cross as the sacrifice for the sins of the world is a barbarian idea based on primitive concepts of God and must be dismissed.
7	Resurrection is an action of God. Jesus was raised into the meaning of God. It therefore cannot be a physical resuscitation occurring inside human history.
8	The story of the Ascension assumed a three-tiered universe and is therefore not capable of being translated into the concepts of a post-Copernican space age.
9	There is no external, objective, revealed standard writ in scripture or on tablets of stone that will govern our ethical behaviour for all time.
10	Prayer cannot be a request made to a theistic deity to act in human history in a particular way.
11	The hope for life after death must be separated forever from the behaviour control mentality of reward and punishment. The Church must abandon, therefore, its reliance on guilt as a motivator of behaviour.
12	All human beings bear God's image and must be respected for what each person is. Therefore, no external description of one's being, whether based on race, ethnicity, gender or sexual orientation, can properly be used as the basis for either rejection or discrimination.

Source: Spong (n.d.); the theses are also discussed in Spong (1998) and various other Christian and religious websites.

may have previously had in their lives in terms of dependence on an external Other. Indeed, the present author would note that parallels and relationships exist with the loss of certainty that occurs to many students as they work through key concepts and understandings of the making of society, culture and individual values in the social sciences and humanities as they go through university. Moreover, such a position is also traumatic because of the way in which it challenges the authority of religious institutions.

> When critical analysis thus begins to expose the power-interests that have shaped our basic concepts, concepts such as those of nature, culture and sacred authority, we become more and more uncertain as to where and how we are to lay the foundations for a reconstruction of the moral order.
>
> (Cupitt 1982: 159)

Nevertheless, within the notion of 'Christianity without God' (Geering 2002) there is still a place for Jesus of Nazareth, the man who was a teacher of great wisdom though not a place for the traditional figure of Jesus Christ the saviour.

> Of relevance to us is not the Jesus who is elevated into mythical heaven but Jesus the fully human person who shared the tensions, enigmas and uncertainties that we experience. It is Jesus who told stories which shocked people out of their traditional ways of thinking and behaving, who can free us from mind-sets in which we have become imprisoned. The Jesus most relevant to us is he who provided no ready-made answers but by his tantalizing stories prompted people to work out their own most appropriate answers to the problems of life.
>
> (Geering 2002: 145)

Of course, although the religious humanist still embraces religion – in the sense that it is embracing the historic task of religion, which is embodying, enriching, conserving and witnessing values in symbolic and actual form in relation to human life – it still raises the question of why someone would choose one religious life over another. To Cupitt (1982, 1984), Kierkegaard provides an answer for the modern apologetic. His method is to expound each particular form of life from within, like a novelist. He does not need to test it against any supposed metaphysical realities external to it. It is sufficient to explore its values, its inner logic, and the life-possibilities that it opens up. By this method we may be able to show purely from within that a particular way of life eventually runs into difficulties that can be solved only by making a transition to another one. This is what is meant by Kierkegaard's allegedly irrational 'leap of faith'.

As Cowdell (1988: 34) has commented, Cupitt's post-theist view of religion that rejects the notion that Christianity depends for its life and spiritual worth on the existence of an invisible, world-transcending individual called god 'is certainly not a comfortable theology, nor one for the faint-hearted'. However, for many people Cupitt's theological and philosophical analysis does provide a sense of the religious for the implications of postmodernity. Moreover, Cupitt takes a different stance on the more usual view of the wowserism of Christian morality, which stifles human drives, and replaces it with a notion that the ethical 'has become that which makes us most alive' (Cowdell 1988: 35). Instead, Cupitt, and other commentators such as Geering (2002), seek to affirm human desire without giving way to unbridled hedonism. The resultant 'theology of the cessation of desire' is an attempt to demonstrate that with effort and attention, one can live out an ethic that balances freedom and constraint. Table 5.2 presents a potential outline for such an ethic of religious humanism as indicated by Cupitt (n.d.) and Bishop John Spong's (1998) theses. Instead, the problem that is to be surmounted is that of vicarious living, whereby cultural pressure, and the media in particular, creates unrealistic expectations. As Cowdell (1988: 35) notes: 'This overspill of desire must be curbed initially, so that thereafter it may be freely channeled in more productive and realistic directions.' In other words, 'one must discipline oneself to live one's own life, within one's own limitations. Only, thus can we be fully alive'. And, for Cupitt, this is the primary purpose of ascetical religion.

Table 5.2 Don Cupitt's ten-point 'sketch of modern philosophy'

• Until about two centuries ago human life was seen as being lived on a fixed stage, and as ruled by eternal norms of truth and value. (This old-world picture may nowadays be called 'realism', 'platonism' or 'metaphysics'.)
• But now everything is contingent: that is, humanly postulated, mediated by language and historically evolving. There is nothing but the flux.
• There is no Eternal Order of Reason above us that fixes all meanings and truths and values. Language is unanchored.
• Modern society then no longer has any overarching and authoritative myth. Modern people are 'homeless' and feel threatened by nihilism.
• We no longer have any ready-made or 'dogmatic' truth and we have no access to any 'certainties' or 'absolutes' that exist independently of us.
• We are, and we have to be, democrats and pragmatists who must go along with a current consensus world-view.
• Our firmest ground and starting-point is the vocabulary and world-view of ordinary language and everyday life, as expressed for example in such typically modern media as the novel and the newspaper.
• The special vocabularies and world-views of science and religion should be seen as extensions or supplements built out of the life-world, and checked back against it.
• Science furthers the purposes of life by differentiating the life-world, developing causal theories, establishing mathematical relationships and inventing technologies.
• Religion seeks to overcome nihilism, and give value to life. In religion we seek to develop shared meanings, purposes, narratives. Religion's last concern is with eternal happiness in the face of death.

Source: Cupitt (n.d.).

Journeying on the Sea of Faith

The development of religious humanism pre-dates the work of Cupitt, but the public attention given to his work did provide the basis for the formation of an association of like-minded people, known as the Sea of Faith network, who argue that 'we must make our own meaning, create our own purpose, find ways of working out our own salvation' (Boulton 1996: n.p.). The network, which has thousands of members in Britain, New Zealand and the United States, is atheistic in the sense that it does not recognize the existence of a real metaphysical god that exists outside of human consciousness but does see a role for god as metaphor and symbol. In this there is a substantial relationship with the Buddhist faith. Therefore, the Sea of Faith network has 'become a place where humanism and radical religion meet and overlap, and "humanist" has taken its place alongside "radical", "post-Christian", "non-literalist" and "non-realist" as alternative descriptions ... [used] ... within the Network' (Boulton 1996: n.p.). Interestingly, the religious humanist perspective is primarily associated with reformation and a reformulation of the Christian tradition, although Buddhism also plays a significant role. Cupitt (2001) argues that Christianity, unlike Islam, has opened itself up to the possibilities of humanism. There is no equivalent of the reformation in Islam in which the legitimacy of independent critical thinking about cultural facts is accepted.

> Islam has never undergone such a change. It has never reconciled itself to critical thinking, or to the idea that the individual thinker may be right against the world. It cannot accept the idea that religion needs continual self-criticism and reform in order to develop aright. It does not accept the idea of an autonomous, secular sphere of life that can and should function independently of religious control.
>
> (Cupitt 2001: n.p.)

Nevertheless, even if it is not possible for something analogous to the Protestant Reformation to occur in Islam, it must also be noted that much of Christian doctrine in the West is unreformed and surrounded by dogma and religious thinking that is not yet free. As witnessed, for example, by debates over the teaching of evolutionary theory and creationism in many American schools, as well as the issue of gay marriage.

> Protestantism has largely decayed into fundamentalism. If we are still not able fully to accept our own principles, we can scarcely expect Islam to embrace them. Perhaps none of us yet understands the magnitude of the religious and cultural revolution the world now needs.
>
> (Cupitt 2001: n.p.)

Within the context of religious humanism the notion of pilgrimage is therefore something that can be regarded as life affirming and enriching and relates to spiritual ideas of pilgrimage even though it is undertaken without a belief in 'god'. Indeed, clearly, from the religious humanist perspective all human life is a pilgrimage. However, rather than the pilgrimage map being laid out as part of some cosmic plan, the route is cast by individuals themselves. As Chu (2004) comments: 'The pilgrim is no ordinary traveler. His map is in the heart.'

Rather than turning one's back on faith and religion, such an approach is instead regarded as an opportunity to embrace life. Therefore, within this context the idea of journeying is extremely important. As Iyer (1999: vii) observes,

> every journey is a question of sorts, and the best journeys for me are the ones in which every answer opens onto deeper and more searching questions. Every traveler is on a quest of sorts, but the pilgrim stands out because his every step is a leap of faith.

However, journeying may take several forms. Ideally, the journey makes us more experienced people, if not better people, as it is a process of learning, of building social relations and creating conscious connections with our surrounds that serve to encourage reflection on life, values and ethics. Indeed, the very notion of pilgrimage is apt for those who leave the security of traditional monotheistic religions that have a god independent of man for the word 'pilgrim' comes from the Latin term, *per agrum* – one who leaves the security of village and road and walks 'through the fields' (Cheer 2000). This is not to say that all travel is continually consciously seen as a spiritual pilgrimage. Even the most ardent religious pilgrims have difficultly in concentrating on the spiritual rather than the secular and the banal over the entire course of their journey, while different pilgrims are also seeking different things from the same route or sacred space (Pfaffenberger 1979; van der Veer 1992; Hayes 1999; Reynolds 2001). For example, the tension between the purely sacred and necessarily secular aspects of pilgrimages (e.g. provision of inns and hospices, supplies, souvenirs) was a theme that ran through Davies and Davies' (1982) study of pilgrimage to Compostela. Instead, it is perhaps reflective of an overall attitude towards how people understand their life experiences and how they travel through life in an effort to bear witness to leading 'a good life'. Such a non-specific approach to the notion of pilgrimage may not be particularly apt for a volume that addresses pilgrimage tourism as a specificity but arguably does shed greater light on the need to link personal ethics, and the spiritual and religious experience as a source for such ethics, to wider attitudes towards where and how one travels.

> In a secular age, the spiritual impulse is more likely to manifest itself in a cycling or mountain-climbing adventure, or the quiet contemplation of an English garden. Mass worship may take place in a football

stadium; and wine is served, with all the reverence of communion, in the caves of Loire Valley châteaus. Many of today's most secular pilgrimages have a ritualistic quality that makes them part of the ancient tradition.

(Chu 2004)

In terms of more specific forms of pilgrimage it is the case that humanists can engage in pilgrimage in which meaning is intended to be derived from both the process of travel and the object, or destination, of such travel. Elements of such an approach towards pilgrimage can be witnessed in the experience of people who travel to find their family history or significant places, people and events in their lives. With respect to family history, the desire to find out from whence one came is regarded as a response by some people to the seeming chaos of modernity and issues of identity and place (e.g. see Delaney 1990; Baldassar 2001; Ioannides and Cohen Ioannides 2002; Stephenson 2002; Coles and Timothy 2004).

The notion of secular pilgrimage has also been applied to significant historical sites that may contribute to national and cultural identity. For example, Hall (2002) describes the pilgrimage significance of Anzac Day (April 25), a national day of remembrance in Australia and New Zealand that commemorates the 1915 landing of Australian and New Zealand army corps on the Turkish Gallipoli peninsula. The pilgrimage associated with Anzac Day is not a religious pilgrimage, although Anzac commemorative services do have religious overtones, nor is one of a migrant returning to his or her ancestral home to locate a cultural identity. Instead, Anzac is related to a secular pilgrimage of national identity in which the myths of nationhood and cultural self are paramount.

The location of meaning in secular pilgrimage can also be understood in terms of travel to sporting grounds. For example, for many Australians, the Melbourne Cricket Ground is often described as a 'holy of holies' because of its association with Australian rules football and cricket. This is even the case with respect to entertainers, such as Elvis Presley (Alderman 2002; Doss 1999; Rinschede and Bhardwaj 1990) and the Beatles (Kruse 2003), or even television programmes such as Star Trek (Porter and McLaren 1999). Regarding a visit to Graceland or Strawberry Fields as a pilgrimage, and therefore placing such places in a category with that of Jerusalem, Mecca or Rome, may seem like blasphemy to some people. But why should it? If one takes the religious humanist position that humans have created god, if a visit to Liverpool has a deep personal meaning that helps people explain their lives, then it is of equivalent value to a Christian's journey to Jerusalem. Indeed, Kruse (2003) in locating Strawberry Fields as a place of pilgrimage, accurately notes that in world religions, pilgrimage is the physical/exteriorized quest for the holy. A pilgrim leaves home and the familiar world behind on a journey to a sacred centre(s) 'out there' often through extremely proscribed routes and forms (Campo 1991).

But pilgrimage is also a journey 'in there', a spiritual interior quest within the heart of those who feel something lacking in their lives – a sense of mystery and wonder, power, health, meaning and connection with others, an observation long recognized in studies of the pilgrimage experience (Turner 1973; Coleman and Elsner 1995). Indeed, as the Very Reverend Robert Willis, Dean of Canterbury, observed, pilgrimage is the search for 'blessing and enrichment. In pilgrimage, body, mind and spirit come together in an individual quest ... Jesus was always walking, walking, walking – all the way to Calvary'. Today's quests need not be so momentous. 'Any journey that adds a mini-jigsaw piece to the puzzle of you can be a mini-pilgrimage' (quoted in Chu 2004). Similarly, Iyer (1999: vii) commented:

> A pilgrim does not have to be moving towards something holy ... so much as toward whatever resides in the deepest part of him: it could be a poet who gave wings to his soul, or a lover who broke his heart open.

The overt search for meaning by religious humanists may even occupy the same pilgrimage spaces as those from more traditional theistic faiths (i.e. Christianity, Judaism and Islam). However, as suggested earlier, the liminality and *communitas* of religious pilgrimage is not the sole domain of the traditional religions (Turner 1973, 1982, 1984; Bleie 2003). For example, Chu (2004) notes:

> Part of the joy of pilgrimage is a spirit of community that comes from identifying with something bigger than oneself. The pilgrim who sets out solo shares a bond with others who journey on the same path: the aches, the pains and the triumphs.

Yet, as Chu (2004) went on to note, 'each pilgrim is utterly alone, because a pilgrimage is a trip not just to a physical place but also into a person's soul'. Such a theme was picked up in Bouldrey's (1999) collection of essays and short stories of contemporary pilgrimage entitled *Travelling Souls*. The majority of stories are observations of the spiritual dimensions of secular pilgrimage and quest. Someone who goes on a pilgrimage is usually quite clear about the route and the destination. A pilgrimage is, in one sense, a very regular sort of journey, often taken amidst a big crowd of other people all with their eyes on the same goal. Nevertheless, as Bouldrey (1999: xvi) comments: 'Every person might take the same pilgrimage and bring home another story.' By contrast, people who go on a quest may know what they are seeking, but they do not necessarily know the route. Nonetheless, the journey is still worthwhile for it is the journey itself that is important. Regardless, from the perspective of religious humanism the shrine lies within.

Conclusion

For the religious humanist then, the broad understanding is that all life is a pilgrimage and a journey in which people are 'essentially traveling deeper into faith and doubt at the same time: deeper into complexity' (Iyer 1999: viii–ix). No step on a trip is ever wasted and everything that happens, however hard and difficult, is good. Yet the religious humanist is still open to the possibility of the religious experience, although it must be emphasized that such experiences are understood as coming from within, not from without. Conscious pilgrimage, in that a particular route or path to a place will be chosen, is also still open to the secular pilgrim. However, the end place is no longer regarded as an institutionally defined location, although the beauty and aesthetics of such locations may also be open to those who find meaning in religious humanism. Indeed, arguably much of what is described as religious tourism belongs to the travels of the secular who are curious or just interested rather than only the realm of the faithful. Such journeys can be undertaken by foot and sometimes vicariously. Yet the important thing is that the journey is taken.

Acknowledgements

Several people have contributed to this chapter although they may not necessarily realize it. In particular the author would like to thank Jackson Browne, Julie Pitcher, Sarah Kristina Pollmann, Madelaine Mattson, David Press, Anna-Dora Saetorsdottir, Chrissy Schriber and Sandra Wall.

References

Alderman, D.H. (2002) 'Writing on the Graceland wall: on the importance of authorship in pilgrimage landscapes', *Tourism Recreation Research* 27(2): 27–34.

Baldassar, L. (2001) *Visits Home: Migration Experiences Between Italy and Australia*, Melbourne: Melbourne University Press.

Bleie, T. (2003) 'Pilgrim tourism in the Central Himalayas', *Mountain Research and Development* 23(2): 177–184.

Bouldrey, B. (ed.) (1999) *Travelling Souls: Contemporary Pilgrimage Stories*, San Francisco: Whereabouts Press.

Boulton, D. (1996) *A Reasonable Faith: Introducing the Sea of Faith Network*, available at www.sof.wellington.net.nz/ukfaq.htm (accessed 25 January 2005).

Campo, J.E. (1991) 'Authority, ritual, and spatial order in Islam: the pilgrimage to Mecca', *Journal of Ritual Studies* 5(1): 65–91.

Cheer, N. (2000) 'Pilgrimage: linear and circular', *Sea of Faith Network N.Z., SOF Newsletter* 36, May, available at www.sof.wellington.net.nz/sofnnl36.htm#PLC (accessed 25 January 2005).

Chu, J. (2004) 'The roads now taken', *Time Europe* 164(1): 5 July.

Coleman, S. and Elsner, J. (1995) *Pilgrimage Past and Present: Sacred Travel and Sacred Space in the World Religions*, Cambridge, MA: Harvard University Press.

Coles, T.E. and Timothy, D.J. (eds) (2004) *Tourism, Diasporas and Space*, London: Routledge.

Cowdell, S. (1988) 'The recent adventures of Don Cupitt', *St Mark's Review* Winter: 32–35.

Cupitt, D. (1982) *The World to Come*, London: SCM Press.

Cupitt, D. (1984) *The Sea of Faith: Christianity in Change*, London: BBC Books.

Cupitt, D. (2001) 'Comparative religions', *Guardian* 27 October, available at www.sofn.org.uk/The_Collection/comparative_religions.htm (accessed 25 January 2005).

Cupitt, D. (2002) *Is Nothing Sacred? The Non-Realist Philosophy of Religion*, New York: Fordham University Press.

Cupitt, D. (n.d.) *A Democratic Philosophy of Life, UK Sea of Faith Magazine*, available at www.sofn.org.uk/The_Collection/Theses/cuplist.html (accessed 25 January 2005).

Davies, H. and Davies, M. (1982) *Holy Days and Holidays: The Medieval Pilgrimage to Compostela*, Lewisburg: Bucknell University.

Delaney, C. (1990) 'The Hajj: sacred and secular', *American Ethnologist* 17: 513–530.

Doss, E. (1999) *Elvis Culture: Fans, Faith, and Image*, Kansas City: University Press of Kansas.

Garabedian, M. (1999) 'Kerouac's pursuit of possibility: the logistics of secular pilgrimage in "On the Road"', paper presented at Writing the Journey: A Conference on American, British, and Anglophone Travel Writers and Writing, 10–13 June 1999, University of Pennsylvania, Philadelphia.

Geering, L. (2002) *Christianity Without God*, Wellington: Bridget Williams Books.

Hall, C.M. (2002) 'ANZAC Day and secular pilgrimage', *Tourism Recreation Research* 27(2): 87–91.

Hall, C.M. (2005) *Tourism: Rethinking the Social Science of Mobility*, Harlow: Prentice-Hall.

Hayes, D.M. (1999) 'Mundane uses of sacred places in the central and later Middle Ages, with a focus on Chartres Cathedral', *Comitatus* 30: 11–36.

Ioannides, D. and Cohen Ioannides, M.W. (2002) 'Pilgrimages of nostalgia: patterns of Jewish travel in the United States', *Tourism Recreation Research* 27(2): 17–25.

Iyer, P. (1999) 'Foreword: a journey into candlelight', in B. Bouldrey (ed.) *Travelling Souls: Contemporary Pilgrimage Stories*, San Francisco: Whereabouts Press.

Kruse, R.J. (2003) 'Imagining Strawberry Fields as a place of pilgrimage', *Area* 53(2): 154–162.

Leaves, N. (2004) *Odyssey on the Sea of Faith – The Life and Writings of Don Cupitt*, Santa Rosa, CA: Polebridge Press.

Pfaffenberger, B. (1979) 'The Kataragama pilgrimage: Hindu-Buddhist interaction and its significance in Sri Lanka's polyethnic social system', *Journal of Asian Studies* 38(2): 253–270.

Porter, J.E. and McLaren, D.L. (eds) (1999) *Star Trek and Sacred Ground: Explorations of Star Trek, Religion, and American Culture*, Albany: State University of New York Press.

Reynolds, J.M. (2001) 'Ise Shrine and a modernist construction of Japanese tradition', *Art Bulletin* 83(2): 316–341.

Rinschede, G. and Bhardwaj, S.M. (eds) (1990) *Pilgrimage in the United States*, Berlin: Dietrich Reimer Verlag.

Spong, J. (1998) *Why Christianity Must Change or Die: A Bishop Speaks to Believers in Exile*, San Francisco: Harper.

Spong, J. (n.d.) *A Call for a New Reformation*, available at www.dioceseofnewark.org/jsspong/reform.html (accessed 25 January 2005).

Stephenson, M. (2002) 'Travelling to the ancestral homelands: the aspirations and experiences of a UK Caribbean community', *Current Issues in Tourism* 5(5): 378–425.

Turner, V. (1973) 'The center out there: pilgrim's goal', *History of Religions* 12(3): 191–230.

Turner, V. (ed.) (1982) *Celebration: Studies in Festivity and Ritual*, Washington, DC: Smithsonian Institution Press.

Turner, V. (1984) 'Liminality and the performative genres', in J.J. MacAloon (ed.) *Rite, Drama, Festival, Spectacle: Rehearsals Toward a Theory of Cultural Performance*, Philadelphia: Institute for the Study of Human Issues.

van der Veer, P. (1992) 'Playing or praying: a Sufi saint's day in Surat', *Journal of Asian Studies* 51(3): 545–564.

6 Religious tourism as an educational experience

Erik H. Cohen

This chapter explores a phenomenon at the crossroads of three fields: religion, tourism and education. To understand how religious tourism functions as an educational experience, changes that have occurred in each of the three sub-fields are considered, including their cognitive, affective and instrumental impacts. The relationships between the various pairings of these three concepts (religious education, religious tourism and educational tourism) will be discussed in each of their respective fields. In the post-modern age, they have been melded into a concept increasingly important in all three fields. Travelers seeking knowledge and spirituality they feel cannot be found at home have given rise to a growing phenomenon: the educational pilgrimage.

In the course of an in-depth empirical study of youth tours to Israel, the inseparability of the three concepts has become apparent. The tour would not be complete if it lacked the religious or the educational dimension, and the same experience could not be gained in the participants' home countries. The "Israel Experience" is the longest running, most thoroughly researched and well-documented example of educational religious tourism. Half a million participants have joined such tours since 1947, and a wealth of data on their motivations, impressions and reactions exists. Much can be learned from this example and applied to other populations and case studies.

Building on previously developed typologies (e.g. Apostolopoulos *et al.* 1996; Collins-Kreiner and Kliot 2000; Cohen 1979; Cohen 2003; Jafari 1987; MacCannell 1976; Smith 1992), tourism and pilgrimage are here interpreted neither as opposed concepts or even as opposite ends of a spectrum, but rather as the two axes in a theoretical model of pilgrimage and tourism. The popularization of religion, institutionalized religious education, the development of heritage tourism, and the importance of informal education in identity formation have all influenced the emergence of educational-religious tourism for religious reasons but also for educational purposes.

MacCannell (1973, 1976, 1992) introduced the concept of tourism as a type of modern pilgrimage. Through travel, particularly to places perceived as pristine and untouched, many individuals in modern (or post-modern) societies hope to experience a more authentic and holistic reality in which their fragmented world is reunited (Allcock 1988; Cohen 1986; Cohen 1988). Their feelings of alienation and fragmentation of relationships may be described as a *diaspora of consciousness* – an intellectual or spiritual, rather than physical, dispersion in which people are unable to exist fully in the here and now; they are torn between here and there, between now and another time, typically a romanticized past and idealized future (Cohen 1986). Travelers also may be seen as embarking on internal journeys of self-change, a search for "personal authenticity" (Desforges 1998; Noy 2004). Like traditional pilgrims, the traveler expects to be transformed in a significant way. According to MacCannell's (1976, 1992) view, tourism is essentially a cognitive activity, but the impacts may also be emotional or behavioral.

Therefore, in Jewish tradition and Zionist ideology, the Land of Israel is the locus of authenticity and the most "natural" place for Jews to live. By traveling there, Jews (particularly adolescents) reunify their fractured identity by encountering different aspects of Jewishness (spiritual, religious, political, social, etc.) and participate in ritually important acts expressing national identity (Zerubavel 1995; Ben-David 1997; Almog 2000; Noy 2004). Changes sought among visitors/participants may be cognitive (increased knowledge, changed attitudes), affective (enhanced feelings of attachment or commitment) or behavioral (practice of religious rituals, involvement with local Jewish community, return trips to Israel). This interpretation of fragmented identities being reunified through travel may be applied to other cases of heritage or religious travel in the postmodern world such as African-American travel to Africa or descendents of immigrants visiting the homes of their ancestors (Bruner 1991; Jackson and Cothran 2003; Lanfant *et al.* 1995; Timothy and Teye 2004). However, the process of fragmentation and the "healing" expected to take place in Israel has not been homogenous across the Jewish Diaspora. A study of pre-trip images of Israel held by participants in Israel Experience tours found that American Jews had a romanticized image of Israel as the Holy Land, while French Jews held a more pragmatic and balanced image of the modern State of Israel. These images are related to the Jewish educational systems in the respective countries. The Americans seek in Israel a sense of history they lack in the United States, while the French students seek the community they are unable to form in the political culture of France (Cohen 2001a).

For individuals in most of the Western world, religion must be reconciled with a rationalist worldview. Pilgrims expect not so much a supernatural encounter with the Divine, but an intellectually and spiritually fulfilling exploration into their cultural-religious roots, which will help them establish or enhance their religious identity within the context of the larger society to which they will return at the end of the journey.

Any pilgrimage, classic or modern, is a journey to an *axis mundi* (a Center), which represents a peak experience (a special moment in Time), where the sacred and profane meet (Eliade 1968). Pilgrimages are also made to centers of civil religion (Shils 1975). Understanding the nature of the Center is dependent upon cultural-religious education and/or inter-pretation by a teacher or guide. In the post-modern age, the Center has been replaced by multiple centers, impacting the nature of pilgrimage. What once was a journey to the axis of the world may now be only one stop among many in a tour, which combines elements of pleasure, spir-ituality and learning about history and culture (Apostolopoulos *et al.* 1996; Boniface and Fowler 1993).

Additionally, for the same individual or group, different Centers (phys-ical and symbolic) may exist simultaneously, corresponding to various needs or aspects of identity: religious, cultural, political or familial. Each of these Centers may have its own set of values, norms and attitudes associ-ated with it. The educational pilgrimage to each various type of Center would include familiarizing visitors with the protocol appropriate to it (dress, customs, language, core beliefs and rituals).

The experience cycle outlined in Kelly's (1963) theory of "man as scientist" may be used to understand and evaluate travel and educational pilgrimage. The cycle consists of: anticipation, investment in the event, encounter, confirmation/disconfirmation, and constructive revision. Whether or not a visit to a religious site is an educational experience depends largely on the perception of the traveler. Does he or she anticipate being changed in some way? What kind of investment (emotional, time, monetary, phys-ical exertion) is made? Is the experience approached with mindfulness? Who does the traveler expect to be in control of the experience (self, guide, no-one)? How is the experience evaluated and integrated into the person's life upon return home?

In this chapter, various educational facets of religious tourism are examined, and some of the main concepts that link them together will be discussed. These are: quest for understanding, informal education, identity development, the role of the guide, and formation of community.

Quest for understanding

Travel, religion and education all share as one of their possible goals a quest for understanding. Some travel, of course, may be for purely recre-ational reasons. Fundamentalist religious movements may be opposed to learning certain subjects or being exposed to certain ideas. In general, however, travelers, spiritual seekers and students all hope to gain deeper and new understandings of the world and oneself. "The desire to travel in order to satisfy the need to know both mundane reality and celestial mystery is an impulse that has constantly driven humankind" (Tomasi 2002: 1). In educational tourism and pilgrimage, places are interpreted so as to enrich the visitors' understanding of the world and history, as opposed to a type of tourism in which places are products to be consumed (Wood 2001).

In religion: Though a quest for understanding has always been an integral part of pilgrimage, the emphasis on acquiring knowledge as a motivation has waxed and waned. During medieval Christian pilgrimage, expiating sins, demonstrating faith or the hope to be healed eclipsed the desire to learn by visiting new places. In contrast, travel, religion and education were linked in the earliest days of Islam (Gellens 1990) as recorded in the prophet Muhammed's dictate to "Seek knowledge even unto China" (translation in Sardar and Malik 1994: 35). As medieval pilgrimage evolved into modern religious tourism, the emphasis on gaining knowledge as a motivation for undertaking the journey increased (Swatos and Tomasi 2002). Visits to religious sites as part of the Grand Tours of Europe were considered cultural education (Brodsky-Porges 1981; Kilbride 2003). An anthropological interest in the exotic Other and in one's own religious roots similarly increased as a motivation for travel to sites of religious significance (Van Den Berghe 1994; Hollinshead 1998; Galbraith 2000). Today, opportunities for learning are often emphasized by organizers of and participants in religious tours.

In travel: Learning as a motivation for non-religious travel similarly increased. In some cases, study is an explicitly stated and predominant motivation for the tour, such as in study-abroad programs, or tours offering opportunities to participate in scientific, artistic or historical studies. To meet the needs of today's travelers, many destinations have started telling their stories in more sophisticated ways (Wood 2001). Vacations often include trips to museums or historical sites where interpretation and explanation by guides or written information is provided. Travel expands one's understanding of others and the world in a global era, and the desire to learn about other cultures through travel is a popular reason for travel, though the extent to which tourists actually learn about the cultures they visit varies greatly with the length and type of tour (Cohen and Cohen 2000; Sales 1998).

Informal education

Kahane (1997) stressed the growing role of informal educational tools in the post-modern age. Informal education not only transmits information, but also is concerned with group dynamics and interpersonal relations, identity formation, values and beliefs (Jeffs and Smith 1990).

In religion: In the past, almost all religious and values education took place informally, in homes and community settings. In recent decades, given the dissolution of extended families and communities, institutions have taken over much of the task of religious, cultural and values education. However, formal school settings have not been found to be sufficient, at least not alone, and informal educational programs have become an important supplement. Youth groups, informal study circles, and mentoring programs, to name only a few, have proliferated. In many cases, the key goals are to instill feelings of group solidarity, impart values, and allow participants to explore a variety of roles. Field trips and travel are often part of such informal educational programs.

In travel: The emotional impact of physical presence in places to be learned about is paramount. The informal learning opportunities of study-abroad programs are a basis of their appeal, even if they are organized through formal educational institutions. Overnight summer camps are popular and well-established forms of informal education, often expected to improve participants' self-confidence and independence. A national leader of Jewish formal education in South Africa concluded that the most effective setting for educating and motivating youth to pray is not in school but in a short-term camp (Cohen 1992). If the camp experience takes place in a "sacred" site, the impact should be even greater.

Identity development

In religion: The establishment and enhancement of "religious identity" has becomes a goal of religious educational institutions. In the past, the concept of religious identity did not exist because it was unquestioned. In homogenous or segregated societies, members' identities are assumed. As societies become more open and boundaries between groups more permeable, identities are increasingly called into question. Berger (1979) called this the "heretical imperative." In reaction to this social phenom-enon, religious institutions and movements have launched campaigns to encourage members, or potential members, to include religious affiliation as part of their personal and social identities (Herman 1977; Lakeland 1997; Cohen and Horencyzk 1999; Khan 2000; Limage 2000).

The expanded possibilities for choosing identities are inherently bound up with the increased mobility of people. This includes migration as well as tourism. Choice, economic necessity and political conflicts have scattered co-religionists across the globe. Now, however, technological advances in transportation and communication allow them to maintain con-nections with each other. This scenario necessarily makes a more conscious type of identification. Tourism to the "homeland," spiritual centers and holy sites are an increasingly important aspect of the formation of reli-gious identity. However, the anticipated climatic experience at a pilgrimage site may be difficult for post-modern travelers to achieve. For example, it is not uncommon for participants in Israel Experience tours seeing the Western Wall in Jerusalem for the first time to describe feelings of dis-appointment or emptiness (Cohen 1994; Goldberg 1995; Kelner 2001), although at the end of the trip, the visit to this site consistently receives the highest evaluations in terms of emotional impact. During the course of the trip, the context for understanding the experience is established.

In the case of civil religion, specific sites may be among the "realms of memory" through which national identity and self-definition are formulated (Nora 1984). These sites are often considered "musts" in the itinerary of tours owing to the insight they offer into local culture.

In travel: Personal identity development is a key motivator among certain types of travelers (MacCannell 1973, 1976; Cohen 1979; Urry 1990). Encountering the Other as a way to understand the Self, exploration of one's heritage (Timothy 1997), exploring roles not available at home, and gaining a wider view of the world are all aspects of identity development found in travel. Additionally, being a traveler, in and of itself, may be a part of one's identity. This is especially prevalent among young people who generally think of themselves not as tourists but as explorers, backpackers, or travelers (Desforges 1998; Richards and Wilson 2003). Upon returning home, a person who has undertaken a significant journey may have a different self-image and image in the eyes of others. This is true of secular travel, but may be especially relevant in the case of pilgrimage. A graphic example would be the pilgrimage to Mecca. Those who have completed this religious obligation are afterwards known as *al-Hajj* or *al-Hajji*, and receive special respect. Expectations for transformation of the Self through travel, whether a pilgrimage, adventure or honeymoon (Michie 2001), are motivations for undertaking the journey. Telling others about adventures and personal transformations experienced is a critical and culminating aspect of travel and pilgrimage (Noy 2004).

In education: Personal and social identity development is also an important outcome of the educational experience, both in school and informal settings. In the case of informal education, identity development may in fact be the primary goal of the program. While it is beyond the scope of this chapter to explore the relationship between education and identity development, the literature on this subject is rich and broad (e.g. Durkheim 1957; Jeffs and Smith 1990, 1999; Kahane 1997; Cohen and Cohen 2000; Horowitz 2000; Flores-Gonzàlez 2002; Sadowski 2003).

Education through informal settings may be seen as following a process of identification followed by experimentation and concluding with commitment (Cohen 2001b). Each of these steps may occur in relation to the group, the activities or the content (philosophy) of the educational program. In other words, participants first identify themselves with a group (a religious youth group), an activity (visiting a religious site) and/or with the content (religious faith). Throughout the course of the educational program, participants may experiment with a number of roles: leader, rebel, passive observer, etc. At the end of the process, a sense of commitment may be achieved. This follows Erikson's (1968, 1974) process of identity formation, in which role confusion and experimentation in the adolescent years gradually are resolved and a personal/social identity confirmed. Travel offers opportunities for experimentation.

One must particularly examine educational and identity-forming aspects of tours to places offering few recreational opportunities. Tourism to sites of tragedies can only be understood if the visit has motivations other than pleasure. For example, in recent years Jewish tourism to sites of the European Holocaust has increased. These visits are conceptualized as pilgrimages to help Jews understand their identity and to learn about the Holocaust in a way not possible from books (Kugelmass 1994; Greenblum 1995; Ashworth 1996, 2003). The March of the Living (Helmreich 1992, 1994) and a similar program sponsored by the Israeli Ministry of Education bring Jewish teens from the Diaspora and Israel to sites of the Holocaust in Europe. These may be understood as a pilgrimage of civil religion, as it reinforces one of the underlying reasons for the existence of the State of Israel, a refuge for Jews in a hostile world (Feldman 2001a, 2001b, 2002). Similarly, African-Americans visit historical sites on both sides of the Atlantic related to the slave trade (Bruner 1996; Finley 2001; Timothy and Teye 2004). If the goal of an educational act is the transformation of the learner into something more than he or she was beforehand, then pilgrimages to sites of tragedy and conflict may be understood as a type of informal education.

There may also be the intention of transforming the site itself into something more positive: a shrine to martyrs, a lesson for future generations, a museum or educational center. For example, millions of pilgrims have visited Medjugorje in Bosnia and Herzegovina (where the Virgin Mary is said to have appeared) to pray for peace:

> It is clear that Medjugorje and the surrounding region are weighted down with ethnic and religious conflict. And yet pilgrims have made a great effort to come to terms with the evils that have been perpetrated by intolerance and even to turn them to good.
>
> (Jurkovich and Gesler 1997: 450)

The role of the guide

Guides and teachers have a number of simultaneous functions. They frame experiences, select what will be seen, direct attention, provide interpretation and act as role models. Though teacher-guides may not have total freedom or control in these areas (a curriculum or itinerary may be developed by a larger organization), they have the most personal contact with the student-travelers and therefore the greatest impact. Changes in religious tourism have been accompanied by changes in the style and role of the guide.

In religion: Spiritual guides often also function as community leaders and educators. Expectations for religious guides to embody the values they teach are generally higher than for other types of leaders or teachers. Failure to do so may negatively affect the faith of followers, as seen in the current crisis in the Roman Catholic Church (Thigpen 2002) involving church educators and mentors. At the same time, many religious leaders today must try to adapt traditional values to changed social situations, and failure to do so may lead to alienation of followers, particularly in the case of youth or migrant populations. Religious guides are not limited to professional clergy. Taking on such a role (as in the case of parents) forces the guide to clarify his or her own beliefs (Fay 1993; Dollahite 1998).

In travel: In medieval pilgrimages, religious leaders often functioned as guides and the Bible or other religious texts as guidebooks. In modern religious tourism, professional guides most often accompany groups, and they almost universally give information about history, art, architecture, archeological findings and/or local culture related to a site. In educational pilgrimages, particularly those for adolescents, guides act as informal educators and role models and are therefore expected to personify the traits they are encouraging participants to develop (e.g. self confidence, ethnic pride, and tolerance) (Cohen *et al.* 2002).

The itinerary of an educational tour is analogous to a curriculum. Since religious sites are heavily laden with meanings and symbolisms, sites are chosen to raise particular emotions in visitors. The guides then provide a context for the sites. Professional guides may try to offer a "neutral" (for example historical and artistic rather than spiritual) interpretation. Guides accompanying groups of pilgrims are often members or teachers within the religious organization and interpret sites according to the orientation of the group. One need only scan the multitude of tours available to the Holy Land to see that itineraries are constructed to impart very different perceptions (religious, political and historical) of the same region. Within the Israel Experience program, various groups tailor their itineraries to emphasize certain aspects of the destination (e.g. religious or nationalist) by selecting from a menu of possible sites (Cohen and Cohen 2000). Counselor-guides are usually recruited from within the ranks of the sponsoring organization (e.g. youth movement and community center).

In pilgrimages to sites of civil religion, interpretation provided by the guide is similarly important and dependent on the guide and the type of group. For example, for African-American tourists, visiting cells in a building where slaves were held before being shipped to the United States is part of a heritage tour laden with meaning for the formation of their personal identities as descendents of those slaves (Austin 1999; Timothy and Teye 2004). Informal educational techniques, such as dramatizations, are used by some agencies offering tours to this site (Bruner 1996). Tours to the same site for different populations emphasize the building's other historical uses. Pilgrimages to sites of civil religion most often uphold and reinforce the values of the general society and demonstrate patriotic beliefs, or homage to national heroes, though they may critique traditional views of a site or offer alternative histories, as in the case of African-American Heritage Tours (Kolemaine 1994). Combined with visits to sites of an organized religion, such civil-pilgrimages can be used to link societal values/assumptions with those of the religion. For example, a tour of Washington, DC organized by Christian Heritage Tours (2004: n.p.) encourages participants to: "See how God has stamped His indelible mark on the landmarks, museums, monuments and memorials of our Nation's Capital." The Israel Experience tours include visits to both sites of civil religion, such as the Kenesset building and the grave of Yitzhak Rabin, and sites of the Jewish religion such as ancient synagogues and the Western Wall. The guide, acting as an informal instructor, provides the interpretation and infuses the visit with a particular meaning (Cohen 1985; Fine and Speer 1985; Katz 1985; Cohen *et al.* 2002).

In education: As schools are one of the major socializing institutions of children, teachers transmit values and ethics in addition to knowledge and information. This may become controversial when issues related to religion or social roles are broached. Even in classroom settings, teachers impact children's attitudes about group dynamics and self-image (Bourdieu 1984; Bourdieu and Passeron 1977, 1979). Though expectations that teachers "practice what they preach" may be less apparent than in the case of religious leaders, teachers' personal attitudes toward the subject affects students (Powell-Brown 2004). This was addressed by a program to revitalize Jewish studies in Israeli public schools through a course to improve teachers' personal attitudes toward the Jewish religion and texts (Cohen 2002). In the case of informal education, the guides' personal values take on even greater importance. In the process of clarifying their own values to be effective guides, informal educators are essentially trained to be future community leaders (Cohen *et al.* 2002). Counselors of religious informal educational tours may be close in age to the participants, providing a bridge between traditional religious leadership and contemporary youth culture (Cohen and Cohen 2000).

Formation of community

In religion: Group pilgrimages may be a bonding experience, giving simultaneously a social element to a spiritual experience and a spiritual element to a social experience, as Turner (1973) noted with the idea of *communitas*. As Jeter (1995: 161) noted in observing Jewish-Moroccan pilgrimages to gravesites: "The family reunion occurs in sacred space and time." Travel to participate in family reunions on religious holidays may be considered a type of pilgrimage.

Within the immediate group of travelers, special feelings of connection develop through the course of the trip. An emotional connection to the larger community of co-religionists may also develop and may, in fact, be one of the motivating factors for the journey. The interpersonal bonds between pilgrims, especially those who travel together by foot for long distances, may be as important as the eventual goal (Galbraith 2000). The pilgrim group becomes a kind of temporary community outside of the norms of the general society, which allows members to experiment with new roles (Turner and Turner 1978). Some view the pilgrimage group as a microcosm of society, not a separate entity (Salnow and Salnow 1987; Sered 1999; Juschka 2003). Whether the pilgrim community is a reflection of the larger society or a withdrawal from it, an analysis of various types of religious journeys as educational endeavors found the journey of the pilgrim to be identified with the community of fellow travelers (Senn 2002).

Travel to sites of religious significance may be part of a larger religious education program in the home community, and visitors are likely to travel as part of a school or adult study group, rather than as individual pilgrims. In this way, the pilgrimage is indeed a continuation of the home community, rather than a complete break from it. The intense group experience serves to bond members, and may serve as a catalyst for increased involvement in the home community (Grosso 1996; Frey 1998; Cohen 2004).

In travel: Most travel includes visits to pre-determined sites of religious, historic and/or cultural importance. The experience at these sites is connected to the group experience and its designation as a tourist destination. The impact of famous sites is affected by the number of people who have been there before. It is a collective spiritual experience. Tourists also form bonds with others in the traveling group, a type of temporary community. The micro-society of the tour group is often insulated from the society they are visiting.

In education: Forming communities in classrooms may be a conscious goal (Ellis *et al.* 2001; Levine 2003) or a by-product of the vast amount of time children and adolescents spend in school (Lynd and Lynd 1929; Mirel 1991). In informal settings, the creation of a community among participants may be one of the primary goals, and the activities (e.g. sports and camping) may be used as a means to achieve group solidarity (Jeffs and Smith 1990).

Conclusion

The desire for an in-gathering of the Diaspora in religious consciousness may cause travelers to seek a certain type of tour, influencing the nature of tourism to sites of religious importance. In the same vein, widespread participation in tourism and the development of educational and heritage tourism may influence the way people view and relate to religion in their home communities. Education, both during the tour and at home, mediates the dialog between these two realms. The educational implications of travel to religious sites are related to changes in how religion is taught, expectations for travel, and priorities in education. This chapter has explored some of the many issues related to educational pilgrimages.

Continued evolution of all three fields may be expected to influence the direction this new type of pilgrimage will take over the course of the next few decades. Globalization, the reemergence of religious fundamentalism and religious nationalist movements, "virtual" travel, and the importance of the Internet in education all will have significant impacts. One of the hallmarks of the post-modern age is the increased rate of social change, making predictions risky at best. Nevertheless, the human desire to undertake pilgrimages to spiritual centers and to gain a deeper understanding of themselves and the world through such experiences has a long history and deep roots in the human psyche. The themes explored here as linking religion, travel and education (quest for knowledge, informal education, identity development, formation of community, informal education and the role of the guide) may serve as guidelines for evaluating the changes that will continue to shape this phenomenon.

Acknowledgments

Thanks to Allison Ofanansky for her valuable contribution in the preparation of this chapter.

References

Allcock, J.B. (1988) "Tourism as a sacred journey," *Loisir et Société* 11(1): 33–48.
Almog, O. (2000) *Sabra: The Creation of a New Jew*, Berkeley: University of California Press.
Apostolopoulos, Y., Leivadi, S. and Yiannakis, A. (eds) (1996) *The Sociology of Tourism*, London: Routledge.
Ashworth, G.J. (1996) "Holocaust tourism and Jewish culture: the lessons of Krakow-Kazimierz," in M. Robinson, N. Evans and P. Callaghan (eds) *Tourism and Culture: Towards the Twenty-first Century*, London: Athenaeum.
Ashworth, G.J. (2003) "Heritage, identity and places: for tourists and host communities," in S. Singh, D.J. Timothy and R.K. Dowling (eds) *Tourism in Destination Communities*, Wallingford: CABI.
Austin, N.K. (1999) "Tourism and the transatlantic slave trade: some issues and reflections," in P.U.C. Dieke (ed.) *The Political Economy of Tourism Development in Africa*, New York: Cognizant Communications.
Ben-David, O. (1997) "*Tiyul* (hike) as an act of consecration of space," in E. Ben-Ari and Y. Bilu (eds) *Grasping Land: Space and Place in Contemporary Israeli Discourse and Experience*, New York: State University of New York Press.
Berger, P. (1979) *The Heretical Imperative*, New York: Anchor Press/Doubleday.
Boniface, P. and Fowler, P. (1993) *Heritage and Tourism in the Global Village*, London: Routledge.
Bourdieu, P. (1984) *Distinction: A Social Critique of the Judgment of Taste*, Cambridge, MA: Harvard University Press.
Bourdieu, P. and Passeron, J.C. (1977) *Reproduction in Education, Society and Culture*, London: Sage (Studies in social and educational change).

Bourdieu, P. and Passeron, J.C. (1979) *The Inheritors: French Students and their Relation to Culture*, Chicago: University of Chicago Press.

Brodsky-Porges, E. (1981) "The Grand Tour: travel as an educational device, 600–1800," *Annals of Tourism Research* 8: 171–186.

Bruner, E. (1991) "Transformation of self in tourism," *Annals of Tourism Research* 18: 238–250.

Bruner, E. (1996) "Tourism in Ghana: the representation of slavery and the return of the black diaspora," *American Anthropologist* 98: 290–304.

Christian Heritage Tours (2004) Viewed online February 4, 2004, available at www.christianheritagetours.org/tours.htm

Cohen, E. (1979) "A phenomenology of tourist experiences," *Sociology* 13: 179–201.

Cohen, E. (1985) "The tourist guide: the origins, structure and dynamics of a role," *Annals of Tourism Research* 12: 5–29.

Cohen, E. (1988) "Authenticity and commoditization in tourism," *Annals of Tourism Research* 15: 371–398.

Cohen, E.H. (1986) "Tourisme et identité," *Pardès* 4: 84–97.

Cohen, E.H. (1992) *The World of Informal Jewish Education, Executive Summary*, Jerusalem: The Joint Authority for Jewish Zionist Education.

Cohen, E.H. (1994) *A Compilation of Direct Quotes of Participants' Personal Comments and Evaluations* (13 volumes), Jerusalem: Israel Experience Survey & Evaluation.

Cohen, E.H. (2001a) "Images of Israel: diaspora Jewish youth's preconceived notions about the Jewish State," in D. Elizur (ed.) *Facet Theory: Integrating Theory Construction with Data Analysis*, Prague: Matfyz Press.

Cohen, E.H. (2001b) "A structural analysis of the R. Kahane Code of Informality: elements toward a theory of informal education," *Sociological Inquiry* 71(3): 357–380.

Cohen, E.H. (2002) *The Hartman Institute's Community Program for Jewish Studies Teachers in State Schools 1995–2001: A First Overall Evaluation*, Jerusalem: Research & Evaluation, Avi Chai Foundation.

Cohen, E.H. (2003) "Tourism and religion, a case study: visiting students in Israel," *Journal of Travel Research* 42: 36–47.

Cohen, E.H. (2004) *From Four Corners of the World: Executive summary of the Survey of Birthright Israel Participants from Argentina, Australia/New Zealand, Brazil and France 2002–2003*, Jerusalem: Research & Evaluation, Birthright Israel Foundation.

Cohen, E.H. and Cohen, E. (2000) *The Israel Experience Program: A Policy Analysis*, Jerusalem: Jerusalem Institute for the Study of Israel.

Cohen, E.H., Ifergan, M. and Cohen, E. (2002) "A new paradigm in guiding: the *madrich* as a role model," *Annals of Tourism Research* 29: 919–932.

Cohen, S. and Horencyzk, G. (eds) (1999) *National Variations in Jewish Identity: Implications for Jewish Education*, New York: State University of New York Press.

Collins-Kreiner, N. and Kliot, N. (2000) "Pilgrimage tourism in the Holy Land: the behavioural characteristics of Christian pilgrims," *GeoJournal* 50: 55–67.

Desforges, L. (1998) "Checking out the planet: global representations/local identities and youth travel," in T. Skelton and G. Valentine (eds) *Cool Places: Geographies of Youth Cultures*, London: Routledge.

Dollahite, D. (1998) "Fathering, faith and spirituality," *The Journal of Men's Studies* 7: 3–15.

Durkheim, E. (1957) *Education and Sociology*, Glencoe, IL: Free Press.

Eliade, M. (1968) *The Sacred and Profane: The Nature of Religion*, New York: Harvest Books.

Ellis, J., Small-McGinley, J. and De Fabrizio, L. (2001) *Caring for Kids in Communities: Using Mentorship, Peer Support, and Student Leadership Programs in Schools*, New York: Peter Lang.

Erikson, E.H. (1968) *Identity: Youth and Crisis*, New York: Norton.

Erikson, E.H. (1974) *Dimensions of a New Identity*, New York: Norton.

Fay, M. (1993) *Do Children Need Religion? How Parents Today Are Thinking About the Big Questions*, New York: Pantheon Books.

Feldman, J. (2001a) "Roots in destruction: the Jewish past as portrayed in Israeli youth voyages to Poland," in H. Goldberg (ed.) *The Life of Judaism*, Berkeley: University of California Press.

Feldman, J. (2001b) "In the footsteps of the Israeli Holocaust survivor: Israeli youth pilgrimages to Poland, Shoah memory and national identity," in P. Daly, K. Filser, A. Goldschläger and N. Kramer (eds) *Building History: The Shoah in Art, Memory, and Myth*, New York: Peter Lang.

Feldman, J. (2002) "Marking the boundaries of the enclave: defining the Israeli collective through the Poland Experience," *Israel Studies* 7(2): 84–114.

Fine, E. and Speer, J. (1985) "Tour guide performances as sight sacralization," *Annals of Tourism Research* 12: 73–95.

Finley, C. (2001) "The door of (no) return: identity politics and cultural heritage tourism in Ghana," *Common Place* 1(4), online journal viewed October 9, 2005 at http://www.common-place.org/vol-01/no-04/finley/

Flores-Gonzàlez, N. (2002) *School Kids/Street Kids: Identity Development in Latino Students*, New York: New York Teachers College Press (Sociology of Education series).

Frey, N. (1998) *Pilgrim Stories: On and Off the Road to Santiago*, Berkeley: University of California Press.

Galbraith, M. (2000) "On the road to Częstochowa: rhetoric and experience on a Polish pilgrimage," *Anthropological Quarterly* 73(2): 61–73.

Gellens, S. (1990) "The search for knowledge in medieval Muslim societies: a comparative approach," in D.E. Eickelman and J. Piscatori (eds) *Muslim Travelers*, Berkeley: University of California Press.

Goldberg, H. (1995) *A Summer on a NFTY Safari 1994: An Ethnographic Perspective*, Jerusalem: CRB Foundation.

Greenblum, J. (1995) "A pilgrimage to Germany," *Judaism* 44(4): 478–485.

Grosso, S. (1996) *Experiencing Lourdes: An Intimate View of the Miraculous Shrine and its Pilgrims*, Ann Arbor, MI: Servant Publications.

Helmreich, W. (1992) *Against All Odds, Holocaust Survivors and the Successful Lives They Made in America*, New York: Simon and Schuster.

Helmreich, W. (1994) "The March of the Living: a follow-up study of its long-range impact and effects," unpublished manuscript, Jerusalem.

Herman, S.N. (1977) *Jewish Identity: A Social Psychological Perspective*, New York: Sage.

Hollinshead, K. (1998) "Tourism, hybridity and ambiguity: the relevance of Bhabha's 'third space' cultures," *Journal of Leisure Research* 30(1): 121–157.

Horowitz, B. (2000) *Connections and Journeys: Assessing Critical Opportunities for Enhancing Jewish Identity*, New York: UJA-Federation of New York.

Jackson, J. and Cothran, M. (2003) "Black versus black: the relationships among African, African American, and African Caribbean persons," *Journal of Black Studies* 33(5): 576–604.

Jafari, J. (1987) "Tourism models: the sociocultural aspects," *Tourism Management* 8: 151–159.

Jeffs, T. and Smith, M. (eds) (1990) *Using Informal Education*, Buckingham: Open University Press.

Jeffs, T. and Smith, M. (1999) *Informal Education: Conversation, Democracy, and Learning*, Ticknall: Education Now.

Jeter, K. (1995) "Pilgrimage: an extension of the social work agenda; one response in social artistry by families to modernity," *Marriage & Family Review* 21(1/2): 157–180.

Jurkovich, J. and Gesler, W. (1997) "Medjugorje: finding peace at the heart of conflict," *Geographical Review* 87(4): 447–468.

Juschka, D. (2003) "Whose turn is it to cook? Communitas and pilgrimage questioned," *Mosaic* 36(4): 189–205.

Kahane, R. (1997) *The Origins of Postmodern Youth*, Berlin and New York: Walter de Gruyter.

Katz, S. (1985) "The Israeli teacher-guide: the emergence and perpetuation of a role," *Annals of Tourism Research* 12: 49–72.

Kelly, G. (1963) *A Theory of Personality: The Psychology of Personal Constructs*, New York: W.W. Norton & Company.

Kelner, S.J. (2001) "Authentic sights and authentic narratives on Taglit," paper presented at the 33rd Annual Meeting of the Association for Jewish Studies, Washington, DC.

Khan, S. (2000) *Muslim Women: Crafting a North American Identity*, Gainesville: University Press of Florida.

Kilbride, D. (2003) "Travel, ritual, and national identity: planters on the European tour, 1820–1860," *The Journal of Southern History* 69(3): 549.

Kolemaine, R. (1994) "Black operators in heritage touring," *American Visions* 9(2): 52–56.

Kugelmass, J. (1994) "Why we go to Poland: Holocaust tourism as secular ritual," in J. Young (ed.) *The Art of Memory Holocaust Memorials in History*, Washington, DC: Prestel.

Lakeland, P. (1997) *Postmodernity: Christian Identity in a Fragmented Age*, Minneapolis, MN: Fortress (Guides to Theological Inquiry).

Lanfant, M., Allcock, J. and Bruner, E. (eds) (1995) *International Tourism: Identity and Change*, London: Sage.

Levine, D.A. (2003) *Building Classroom Communities: Strategies for Developing a Culture of Caring*, Bloomington, IN: National Educational Service.

Limage, L. (2000) "Education and Muslim identity: the case of France," *Comparative Education* 36(1): 73–95.

Lynd, R.S. and Lynd, H.M. (1929) *Middletown: A Study in Modern American Culture*, New York: Harcourt Brace Jovanovitch.

MacCannell, D. (1973) "Staged authenticity: arrangements of social space in tourist settings," *American Journal of Sociology* 79(3): 589–603.

MacCannell, D. (1976) *The Tourist: A New Theory of the Leisure Class*, New York: Schocken Books.

MacCannell, D. (1992) *Empty Meeting Grounds*, London: Routledge.

Michie, H. (2001) "Victorian honeymoons: sexual reorientations and the 'sights' of Europe," *Victorian Studies* 43(2): 229–251.

Mirel, J.E. (1991) "Adolescence in twentieth century America," in R. Lerner, A.C. Petersen and J. Brooks-Gunn (eds) *Encyclopedia of Adolescence, Vol. 2*, New York: Garland Publishing.

Nora, P. (ed.) (1984) *Realms of Memory: The Construction of the French Past*, New York: Columbia University Press.

Noy, C. (2004) "This trip has really changed me: backpackers' narratives of self-change," *Annals of Tourism Research* 31: 78–102.

Powell-Brown, A. (2004) "Can you be a teacher of literacy if you don't love to read?," *Journal of Adolescent & Adult Literacy* 47(4): 284–286.

Richards, G. and Wilson, J. (2003) *New Horizons in Independent Youth and Student Travel: Summary Report*, Arnhem, Netherlands: International Student Travel Confederation (ISTC) and the Association of Tourism and Leisure Education (ATLAS).

Sadowski, M. (ed.) (2003) *Adolescents at School: Perspectives on Youth, Identity, and Education*, Cambridge, MA: Harvard Education Press.

Sales, A. (1998) *Israel Experience – Is Length of Time a Critical Factor?*, Waltham, MA: Cohen Center for Modern Jewish Studies.

Salnow, J.E. and Salnow M. (1987) *Pilgrims of the Andes: Regional Cults in Cusco*, Washington, DC: Smithsonian Institution Press.

Sardar, Z. and Malik, Z.A. (1994) *Muhammad for Beginners*, Cambridge: Icon Books.

Senn, C. (2002) "Journeying as religious education: the shaman, the hero, the pilgrim and the labyrinth walker," *Religious Education* 97(2): 124–140.

Sered, S. (1999) "Women pilgrims and women saints: gendered icons and the iconization of gender at Israeli shrines," *NWSA Journal* 11(2): 48.

Shils, E. (1975) *Center and Periphery: Essays in Macrosociology*, Chicago: University of Chicago Press.

Smith, V.L. (1992) "Introduction: the quest in guest," *Annals of Tourism Research* 19: 1–17.

Swatos Jr, W. and Tomasi, L. (2002) *From Medieval Pilgrimage to Religious Tourism: The Social and Cultural Economics of Piety*, Westport, CT: Praeger.

Thigpen, T. (ed.) (2002) *Shaken by Scandals: Catholics Speak Out About Priests' Sexual Abuse*, Ann Arbor: Charis Books.

Timothy, D.J. (1997) "Tourism and the personal heritage experience," *Annals of Tourism Research* 34: 751–754.

Timothy, D.J. and Teye, V.B. (2004) "American children of the African diaspora: journeys to the motherland," in T. Coles and D.J. Timothy (eds) *Tourism, Diasporas and Space*, London: Routledge.

Tomasi, L. (2002) "*Homo Viator*: from pilgrimage to religious tourism via the journey," in W.H. Swatos Jr. and L. Tomasi (eds) *From Medieval Pilgrimage to Religious Tourism: The Social and Cultural Economics of Piety*, Westport, CT: Praeger.

Turner, V. (1973) "The center out there: pilgrim's goal," *History of Religion* 12(3): 191–230.

Turner, V. and Turner, E. (1978) *Images and Pilgrimage in Christian Culture: Anthropological Perspectives*, New York: Columbia University Press.

Urry, J. (1990) *The Tourist Gaze*, London: Sage.

Van den Berghe, P. (1994) *The Quest for the Other: Ethnic Tourism in San Cristobal, Mexico*, Seattle: University of Washington Press.

Wood, C. (2001) "Educational tourism," in N. Douglas, N. Douglas and R. Derrett (eds) *Special Interest Tourism: Context and Cases*, Milton: Wiley Australia.

Zerubavel, Y. (1995) *Recovered Roots: Collective Memory and the Making of Israeli National Tradition*, Chicago: University of Chicago Press.

7 Empty bottles at sacred sites

Religious retailing at Ireland's National Shrine

Myra Shackley

This chapter examines aspects of merchandising at Ireland's National Shrine at Knock, in the context of souvenir retailing at sacred sites. Knock is one of the many Christian pilgrimage sites in Europe that have developed over the last 150 years into popular pilgrimage and tourism destinations. In particular, major Marian shrines, such as Fatima and Lourdes, have grown in popularity and create huge volumes of tourist business in the cities themselves and the surrounding localities. For example, after the death of the Capuchin Friar Padre Pio in 1968, the mountain-top friary where he spent his life in the remote town of San Giovanni Rotondo became the number one pilgrimage destination in Europe, attracting between 7 and 8 million visitors each year for several years to its newly built basilica (Margry 2002). Although Knock typically receives only about 1.5 million visitors per year, it is of immense economic significance to a relatively undeveloped area of western Ireland and a major element in the Catholic pilgrimage network of Europe.

The economics of religious tourism

Pilgrimage has long been a common mode of travel. On a global scale, pilgrimage probably attracts some 250 million people a year (Jackowski 2000), although this is very difficult to measure because most countries (with the exception of Saudi Arabia) do not differentiate between religiously motivated visitors and other tourists in their data collection efforts (Russell 1999). Regardless of the debates between the definitions and classifications of pilgrimage and tourism, the classifying of religious sites, the travel patterns of religious tourists, or the differentiation between pilgrims and tourists (Boorstin 1964; MacCannell 1973; Cohen 1979; Olsen and Timothy 1999; Rinschede 1992; Smith 1992; Sizer 1999; Fleischer 2000), what is known is that religious travel has long been big business, particularly at both pilgrimage destinations and the communities located along the route(s) leading to the site, with the development of a pilgrimage infrastructure consisting of hospices, monasteries and other such services. Currently, religious tourism is seen by many government and tourism officials as a way to either diversify or save struggling economies (Olsen 2003). Jackowski and Smith (1992) note that religious tourism could become an important source of income and employment in Poland if the infrastructure issues that stem back to the Second World War were to be resolved. In Spain, an emphasis on tourism to the El Rocio shrine has led to increased employment and local revenues (Crain 1996).

Another way that religious tourism can have an economic impact is through the sale of religion-based souvenirs. Bywater (1994) and Fleischer (2000) suggest that sales from religious souvenirs currently reach hundreds of millions of dollars a year. While the sale of official religious items and relics has existed for centuries, enterprising entrepreneurs and vendors also produce devotional items and other various articles that could be considered *kitsch*, taking away a degree of authenticity from the souvenir or religious item. While some religious groups and managers of sacred sites try to limit this type of commercialization owing to the perceived negative image of making a profit from visitors, others embrace the creation of religion-based souvenirs. In many cases the economic benefits of tourism to sites of religious importance outweigh the negative impacts associated with tourism, particularly when religious site managers are responsible for maintaining the site without a steady income (Shackley 2001). It is in the context of religious souvenir retailing and merchandising that the discussion turns to the Shrine of Our Lady at Knock.

The Shrine of Our Lady at Knock

Knock was a small, obscure bog-side village in County Mayo where, on 21 August 1879, a vision of Our Lady accompanied by St Joseph and St John appeared, surrounded by a soft brilliant light. The apparition lasted for about two hours and had twenty witnesses, with the light visible for miles around. Validated by a Commission of Enquiry, it generated a great wave of religious enthusiasm in Ireland. Huge crowds thronged to Knock for healing, with many cures being claimed. For twenty-five years afterwards the shrine enjoyed the momentum of the apparition, although the second twenty-five years saw some decline (Walsh 2000). Official church approval was first given in 1929, followed by fifty years of pilgrimage and the development of pilgrimage and shrine facilities. The Vatican agreed that there were special spiritual favours to pilgrims who prayed at Knock, and four popes visited. Today, Knock remains a great centre of prayer and pilgrimage, typically attracting around 1.5 million visitors a year, although precise visitor data are unavailable. It is primarily a location for organized group tours, many requiring facilities for the disabled. A huge new circular church with a capacity of 10,000 people was completed in 1976, accompanied by landscaping of the extensive shrine grounds, the building of the Knock Folk Museum and assorted oratories, churches, confessional halls, assembly and processional areas. This large area is necessary to support large numbers of people – in 1993 more than 40,000 people attended a single mass on the visit of Mother Theresa of Calcutta.

Mayo County Council has reconstructed the small village of Knock by widening streets and roads, building a new shopping centre, and constructing a huge car park. The shrine is served by its own airport, Knock International Airport, some 11 miles north, opened in 1985 and developed by the Connaught Regional Airport Company, with the mission to provide a gateway to open up this disadvantaged region of Ireland and help the shrine achieve its full pilgrimage and tourism potential. Currently, Knock is linked to different Irish and English airports by scheduled and charter flights, including budget airlines such as Ryanair. Tourism to Knock is seasonal with little (including the tourist office) open outside the main tourism season (May–October). Knock is not just a pilgrimage destination but also a major attraction for non-pilgrim tourists. Its religious visitors come mainly as day-trippers, motivated by a desire to see where the apparition happened (and with the faint hope that it might happen again), and the wish to share in the sanctity of a holy place. Some wish to pray for healing, others use a visit as a focus for family or community life, and some are simply curious. The shrine itself generates low levels of income (as donations) from its visitors with most money being made by private-sector service businesses. No precise visitor data are available, as the shrine has open access, but it is likely that the percentage of non-Irish and international visitors is increasing, partly as a result of easier air access and partly as a result of the decline in Catholic religious observance in Ireland.

Retailing and visitor services at Knock

Visitor services required at a Catholic shrine are much the same as those at any other cultural tourism attraction, being dominated by transport, parking, information, catering, accommodation and merchandising. However, shrines and other religious sites do provide extra visitor facilities, including places for prayer, worship and meditation, and gardens for meetings and discussion. Most of the 'religious' infrastructure is included within the shrine grounds (e.g. churches, basilica, confessionals, shrine office, convents and monasteries), although some may be found outside the shrine boundaries including ancillary churches, pilgrimage venues, houses of religion, grottoes and hospices. Within the grounds, facilities are provided and managed directly by the religious authorities who manage the shrine, although, in common with other heritage attractions, these may include partnership or franchising arrangements with bookshops and catering outlets. However, the facilities inside the shrine grounds are limited in number and always exceeded by the number of private-sector firms operating outside and around the shrine to cater to the needs of visitors.

The shrine at Knock has no 'official' catering outlets except those within hospices and care centres within the shrine enclosure, although this is relatively unusual. Despite the small size of Knock village (only 400 people) and a limited population base within the surrounding catchment area, more than seventy local businesses have developed in the village, which primarily cater to pilgrims and other visitors. Most of these are involved with different aspects of merchandising and catering, often in some form of combination, as well as the provision of ancillary services such as a post office, flower shop, general stores and numerous bars. Few of the businesses in Knock cater specifically to locals, with the exceptions being the garage and newsagent. Bearing in mind the large numbers of visitors received by the shrine it is perhaps surprising that the immediate area of Knock can offer no more than 250 accommodation bed-spaces, although more are available in hospices and houses of religion, as well as in nearby camping and caravan sites. As noted previously, Knock is primarily a day-trip destination, with the result that visitors to the shrine may have come some considerable distance during the day, some from as far as Dublin, with many staying overnight 50–100 miles distant in the Galway/Westport area. This has the added benefit of disseminating the economic benefits of the shrine more widely.

In Knock, as at many other shrines, the location and characteristics of retail outlets form spatially distinct retailing zones around the shrine. There are primary retail outlets within the shrine itself selling 'authorized' material, which when purchased has religious significance for the customer. Typically, their stock is dominated by tasteful and expensive items including books, posters, icons and music. Outside the shrine grounds, but located as near to the shrine as possible, a secondary retail zone develops which is characterized by well-established, privately owned shops selling religious souvenirs and memorabilia. These offer an eclectic mixture of wares with retailing and other service functions often muddled together, so that a coffee shop may also be a souvenir stand and a flower shop. In the case of Knock this development is both planned and themed, having been carried out as part of the official re-modelling of the village, although in the case of older shrines it is often more haphazard. Shop layouts are informal, considerable duplication of merchandise exists and many stores are very small (12–50m^2) with the smallest being located furthest away from the shrine. Shops are crowded, disorganized and generally self-serviced. These retail outlets sell four major categories of material:

1   Irish souvenirs (clothing and miscellaneous items in green or patterned with shamrocks or leprechauns). The range is tasteless and includes everything from tea towels to underpants and mugs.
2   Souvenirs that are specific to Knock (books, cards and photographs about the shrine, statues and guidebooks).
3   Items associated with death and funerals (grave markers, urns, plaques).

4   Worship aids (prayer books, rosaries) plus other things taken as religious souvenirs and gifts (religious medals and scapulars). This includes the full range of Catholic *kitsch*.

As at most Catholic religious sites there is an extensive range of pictures and medals of the saints. These are subject to the whims of fashion – at present the iconography is dominated by Mother Theresa of Calcutta and Padre Pio. Religious retailers have an annual trade fair to help them cater to these trends. It is noticeable that although the general Irish souvenirs and material from Knock are available widely throughout the surrounding catchment area of the shrine, religious souvenirs can only be bought at the shrine itself. All retailers in Knock report very low levels of visitor spending, as many pilgrims are disabled and others on low incomes.

A tertiary retailing zone has developed at Knock located near the car and coach parks (which have spaces for 500 cars and 50 coaches) in informal and often poorly maintained units. These shops closely resemble seaside stalls and sell low-cost merchandise, dominated by *kitsch*. These small units have low overheads, no marketing or promotion costs and are located in areas with very high footfall. Within the developing world these stalls/shops would themselves be surrounded by unregulated street traders and hawkers, although this is not the case at Knock. Knock merchandise may also be purchased at towns and transport hubs up to 50km from the shrine, particularly at Knock and Galway airports and in Galway town itself. The differentiation of these kinds of retail outlets is interesting and occurs both spatially and by visitor spending. Although it was impossible to gain precise figures, projections from other shrine retailing experiences observed by the author suggests that the per capita spend in bookshops within the shrine grounds is likely to be approximately £7–£10. In small retail stalls near the car park it may only reach 20 per cent of that amount.

The spatial differentiation of retail outlets at Knock is interesting. It is well known that demands for services from tourists and pilgrims can change both the cultural landscape and urban land use patterns at pilgrimage centres. The landscape at Lourdes, for example, is divided between a profane/commercialized zone (where all the restaurants, shops and hotels may be found) and a sacred zone (of pilgrimage shrine and churches). The urban morphology of Lourdes has also changed as a result of the intensive agglomeration of souvenir shops and restaurants along roads leading from town to the pilgrimage area (Rinschede 1986).These phenomena are also seen, although to a smaller and less developed extent, in the zoning of retailing outlets around the shrine at Knock.

Knock's visitation levels have been important enough to warrant the construction of an international airport. Tourism to Knock is changing from being primarily domestic to being more mixed domestic and international. Irish tourism (both domestic and international) is flourishing; visitor numbers and revenue grew by approximately 4 per cent in 2003, with

Ireland gaining significantly in its main markets. However, because of the large volume of short-break business, city tourism grew to a greater extent than in other parts of the country. Currently, the west of Ireland receives some 2.5 million tourists annually, with 60 per cent of these visiting Knock. This, alone, emphasizes the community's economic importance. Even in a time of decline in levels of traditional Catholic worship, Knock is favourably positioned to profit from high levels of private-sector investment in nearby secular urban destinations, including Galway and Westport. It can look to a supplementary market related to urban short breaks, supplementing the existing primary markets of group pilgrimage. In this case it will be interesting to see whether the character of the shrine retailing alters and, as one might expect, the percentage of religious material declines in favour of 'Irish' souvenirs.

Retailing at sacred sites

This level of commercialization is a common phenomenon at most visitor attractions, and sacred sites are no exception (Shackley 2001). For example, annual sales of religious souvenirs in Italy were estimated over a decade by Bywater (1994) to be some US$255,200. Shops in Lourdes, and indeed the entire community, are dependent on sales of religious objects. Most of the studies associated with commercialization at sacred sites concentrate on either the commercialization process (Vukonić 1996; 2002) or its socio-cultural impacts (Greenwood 1989). However, it remains uncertain to what extent this commercialization affects the quality of the visitor experience and how this is related to the different types of shrine visitors and their motivations (Nash 1966) as well as their perceptions of levels of authenticity and enthrallment by the visit. At sacred sites the quality of experience is directly related to maintenance of the 'spirit of place' (Shackley 2001: 24) but it seems probable that the provision of these varied merchandising opportunities at Knock, at varying levels of expense and catering for all tastes, meet the expectations of most of the visitors in one way or another.

However uncertain it is that the provision of commoditized merchandise adversely affects the pilgrims' view of the authenticity of the site, Cohen (1992) would argue that one of the tests of authenticity is an absence of commoditization. The present writer would suggest that it is perfectly possible for the visitor/pilgrim to have an authentic religious experience at the shrine by visiting the place of the apparition, attending Mass and prayers and rounding off the visit by the purchase of religious souvenirs that may appear to be tasteless *kitsch*. It is a fashionable intellectual pose to disapprove of commoditization at sacred sites (Greenwood 1989; Vassallo 1979; Houlihan 2000), but such disapproval is usually articulated by people who are unaware of the emotions stirred up in the pilgrim by the act of visiting such a site. Moreover, Catholic visitors to a sacred site are likely to be experienced shrine visitors whose levels of expectation are precise, including the ready availability of souvenirs that they consider appropriate. Taste, after all, is in the eye of the purchaser not the beholder.

Nolan and Nolan (1992) noted that secular pilgrims and Protestants who may have a rather more puritanical view of pilgrimage would be more likely to be shocked by perceived over-commercialization of a sacred site than Catholic pilgrims. Eade (1992) noted that there are, as one might expect, considerable disagreements within groups of pilgrims over the value of souvenirs. Some ignore and ridicule them, while others purchase a wide range of goods that for them have sacred associations because of the purchase location. What counts as authentic depends on the cultural lens of the seeker, which in turn guides the direction from which authenticity is sought (Littrell *et al.* 1993; Spooner 1986). A souvenir can be defined both as a remembrance of some place (Anderson and Littrell 1995) and a mnemonic device around which to tell stories (Hitchcock and Teague 2000). Many souvenirs are bought to give away to friends or relatives reminding the recipient of the donor and occasion. At many sacred sites throughout the world, mementos can be purchased that have deep spiritual and religious meanings for the consumers (Shenhav-Keller 1993). At shrines, these will include cards issued to prove that a Mass has been said for a specific individual. In Rome it can include a certificate of papal indulgencies. There is some evidence to suggest that tourists make more planned purchases for others than for themselves (Anderson and Littrell 1995; Kim and Littrell 2001; Timothy 2005), which accounts for the wide range of souvenirs available at Knock. Pilgrims who come for a once in a lifetime experience are willing to spend a significant amount of money on shopping for a tangible memory of their spiritual experience (Fleischer 2000), although at many Marian shrines the large proportion of disabled and elderly pilgrims tends to militate against this.

Reflections and concluding remarks

Pearce and Moscardo (1986) argue that people's perceptions of a situation play an important role in determining its judged authenticity, but that people's needs or demands for authenticity also vary. Fine and Speer (1985) suggest that an authentic experience involves participation in a collective ritual, where strangers get together in a cultural production to share a feeling of closeness or solidarity. Very large numbers of fellow tourists are supposed to inauthenticate this experience, but this does not seem to be true of religious sites. Indeed, the nature of the experience of most pilgrims (as opposed to non-pilgrim tourists) visiting religious sites is greatly enhanced by the presence of large numbers of fellow worshippers, a fact which can be seen clearly by any participant in an open air papal mass at St Peter's Square in Rome, for example, or any communal worship activity engaged in by shrine visitors. *Communitas* is the essence of pilgrimage, in all religious traditions.

It is undeniable that many of the souvenirs being offered for sale at Knock do not pass the conventional tests of authenticity as articulated, for example, by Littrell *et al.* (1993), which include uniqueness, workmanship, aesthetics and use, cultural and historical integrity and genuineness. But the present writer argues that the real meaning of authenticity in the context of shrine retailing may be found in the most commonly purchased item, which at Knock is an empty plastic bottle, used for the collection of holy water from the holy water fonts in the shrine precinct. Since very few pilgrims come equipped with an empty bottle, the vast majority must buy one at the site where, as at all major Catholic shrines and churches, the bottles are customized with the name of the site and/or an image of the Virgin or saints. Furthermore, the bottles are typically available in an infinite variety of shapes and sizes. Such bottles are priced from 50c upwards, but they are merely being used as containers for the real souvenir, the holy water, which is available free of charge. The pilgrims are in this fashion taking a tangible part of the site home with them. Pilgrims often bring back a token of the place both as proof that the journey has been completed and as a physical manifestation of the charisma of a sacred centre (Coleman and Elsner 1991). The most powerful and significant souvenirs are those that include a physical fragment of the site itself – a practice which has led to vandalism and looting at religious sites from all traditions as pilgrims try to remove a fragment of the site for veneration (Shackley 2001). Bottling and retailing of water is not confined to Knock. At Lourdes, for example, a commercial firm will visit the grotto on your behalf and send Lourdes water delivered to your doorstep in a matter of days (www.lourdesdirect.com). Likewise, there are a number of firms in Israel that will sell a bottle of water from the Jordan River, a box containing genuine earth from the Mount of Olives, or even a can which, when opened, releases 'genuine' Holy Land air. This phenomenon reinforces both the authenticity and the intensity of the pilgrim's experience, which is centred and focused around the need to submerge him/herself in the location where a miraculous event occurred. Grace, in the Christian sense, may be obtained even by just visiting the site, and more so by attending some type of communal prayer or worship. Being physically present at the site enables the pilgrim to share in the experience of those who actually witnessed the original apparition; being able to take a physical part of the site away enables this experience to be prolonged indefinitely. The phenomenon is not far from the medieval practice of preserving the body parts of the saints or items associated with Jesus himself as holy relics, which, when touched, transferred part of their holiness to the visitor. Even being in the presence of such a relic without physically being able to touch it can have the same effect – as is witnessed by the outpourings of devotion at the periodic expositions of the Turin shroud, considered by many to be the winding-sheet of Christ. Here, at Knock, the most sacred and authentic souvenir is water from the site itself with the plastic container acquiring a measure of sanctity from its contents.

Another phenomenon is also evident, and that is the relationship between the perceived sanctity of the object and the relationship between the location at which it is purchased and the shrine itself. Items purchased within the shrine are often regarded as the most holy and will be bought by visitors at an inflated price in preference to identical items being offered for sale outside the shrine, precisely for that reason. The further one gets from the shrine, the less holy the object, hence the fact that shops at airports or towns near Knock sell only routine local souvenirs, not items intrinsically connected with the shrine. In this context authenticity may be found not in the nature of the object itself but also in the purchase location, with the most authentic objects, and thus the most sacred, being those bought nearest the site of the original apparition. However, for most visitors, the item of greatest spiritual value is that which has no material value at all: just plain water.

References

Anderson, L.F. and Littrell, M.A. (1995) 'Souvenir purchase behaviour of women tourists', *Annals of Tourism Research* 22: 328–348.

Boorstin D.J. (1964) *The Image: A Guide to Pseudo-events in America*, New York: Harper and Row.

Bywater, M. (1994) 'Religious travel in Europe', *Travel & Tourism Analyst* 2: 39–52.

Cohen, E. (1979) 'A phenomenology of tourist experiences', *Sociology* 13: 179–202.

Cohen, E. (1992) 'Pilgrimage and tourism: convergence and divergence', in A. Morinis (ed.) *Sacred Journeys: The Anthropology of Pilgrimage*, Westport, CT: Greenwood.

Coleman, S. and Elsner, J. (1991) 'Contested pilgrimage: current views and future directions', *Cambridge Anthropology* 15(3): 63–73.

Crain, M.M. (1996) 'Contested territories: the politics of touristic development at the Shrine of El Rocío in Southwestern Andalusia', in J. Boissevain (ed.) *Coping with Tourists: European Reactions to Mass Tourism*, Providence, RI: Berghahn Books.

Eade, J. (1992) 'Pilgrimage and tourism at Lourdes, France', *Annals of Tourism Research* 19: 18–32.

Fine, E. and Speer, J. (1985) 'Tour guide performance as site sacralization', *Annals of Tourism Research* 12: 780–797.

Fleischer, A. (2000) 'The tourist behind the pilgrim in the Holy Land', *International Journal of Hospitality Management* 19: 311–326.

Greenwood, D.J. (1989) 'Culture by the pound', in V. Smith (ed.) *Hosts and Guests: The Anthropology of Tourism* (2nd edn), Philadelphia: University of Pennsylvania Press.

Hitchcock, M. and Teague, K. (eds) (2000) *Souvenirs: The Material Culture of Tourism*, Aldershot: Ashgate.

Houlihan, M. (2000) 'Souvenirs with soul: 800 years of pilgrimage to Santiago de Compostela', in M. Hitchcock and K. Teague (eds) *Souvenirs: The Material Culture of Tourism*, Aldershot: Ashgate.

Jackowski, A. (2000) 'Religious tourism – problems with terminology', in A. Jackowski (ed.) *Peregrinus Cracoviensis*, Cracow: Publishing Unit, Institute of Geography, Jagiellonian University.

Jackowski, A. and Smith, V.L. (1992) 'Polish pilgrim-tourists', *Annals of Tourism Research* 19: 92–106.

Kim, S. and Littrell, M.A. (2001) 'Souvenir buying intentions for self versus others', *Annals of Tourism Research* 28: 638–657.

Littrell, M., Anderson, L. and Brown, P. (1993) 'What makes a souvenir authentic?', *Annals of Tourism Research* 20: 197–215.

MacCannell, D. (1973) 'Staged authenticity: arrangements of social space in tourist settings', *American Journal of Sociology* 79(3): 589–603.

Margry, P.J. (2002) 'Merchandising and sanctity: the invasive cult of Padre Pio', *Journal of Modern Italian Studies* 7(1): 88–115.

Nash, D. (1966) *The Anthropology of Tourism*, Oxford: Elsevier.

Nolan, M.L. and Nolan, S. (1992) 'Religious sites as tourism attractions in Europe', *Annals of Tourism Research* 19: 68–78.

Olsen, D.H. (2003) 'Heritage, tourism, and the commodification of religion', *Tourism Recreation Research* 28(3): 99–104.

Olsen, D.H. and Timothy, D.J. (1999) 'Tourism 2000: selling the Millennium', *Tourism Management* 20: 389–392.

Pearce, P.L. and Moscardo, G. (1986) 'The concept of authenticity in tourist experiences', *Australian and New Zealand Journal of Sociology* 22: 121–132.

Rinschede, G. (1986) 'The pilgrimage town of Lourdes', *Journal of Cultural Geography* 7(1): 21–34.

Rinschede, G. (1992) 'Forms of religious tourism', *Annals of Tourism Research* 19: 51–67.

Russell, P. (1999) 'Religious travel in the new millennium', *Travel & Tourism Analyst* 5: 39–68.

Shackley, M. (2001) *Managing Sacred Sites*, London: Continuum.

Shenhav-Keller, S. (1993) 'The Israeli souvenir: its text and context', *Annals of Tourism Research* 20: 182–196.

Sizer, S.R. (1999) 'The ethical challenges of managing pilgrimages to the Holy Land', *International Journal of Contemporary Hospitality Management* 11(2/3): 85–90.

Smith, V.L. (1992) 'Introduction: the quest in guest', *Annals of Tourism Research* 19: 1–17.

Spooner, B. (1986) 'Weavers and dealers: the authenticity of an oriental carpet', in A. Appadurai (ed.) *The Social Life of Things: Commodities in Cultural Perspective*, Cambridge: Cambridge University Press.

Timothy, D.J. (2005) *Shopping Tourism, Retailing, and Leisure*, Clevedon: Channel View.

Vassallo, M. (1979) *From Lordship to Stewardship: Religion and Social Change in Malta*, The Hague: Mouton.

Vukonić, B. (2002) 'Religion, tourism and economics: a convenient symbiosis', *Tourism Recreation Research* 27(2): 59–64.

Vukonić, B. (1996) *Tourism and Religion*, Oxford: Elsevier.

Walsh, M. (2000) *The Glory of Knock* (8th edition), Knock: Custodians of Knock Shrine.

8   Management issues for religious heritage attractions

Daniel H. Olsen

People have long traveled to sites they deem as sacred, special, or set apart from the mundane, everyday world (Eliade 1961). Today, travel to major pilgrimage destinations has seen a rapid increase, in part because of the coinciding growth of both religious pilgrimage and other forms of tourism (Lloyd 1998). In countries around the world, religion and its associated sites, ritual, festivals, and landscapes are seen by many government officials and tourism industry promoters as a form of heritage (Timothy and Boyd 2003), as a resource that can be transformed and commodified for tourist consumption, which in turn encourages the growth of leisure and other activities. This is most evident through the images and meanings assigned to sacred sites and adopted in various local, regional, and national tourism promotional literature.

The end result of this touristification of religious sites is an overlapping of religious and tourist space, creating, in the words of Bremer (2001a: 3), a "duality of place." This convergence of religious/sacred and tourist/ secular space, however, adds a complexity to traditional management practices at sacred sites where the focus has been the needs of pilgrims and worshippers rather than other visitors. As well, this convergence raises questions pertaining to the maintenance, interpretation, and meaning of sacred sites, particularly where they have become multi-use in nature and function, acting as places of recreation, education, and leisure rather than strictly for religious instruction and ritual. For example, at many religious sites there is conflict over decisions to charge all visitors entrance fees, leaving those who have come to worship questioning why they have to "pay to pray" (Shackley 2001a; Willis 1994). In some instances religious rituals and festival processions have been altered to appease the gaze of leisure-oriented tourists (Shackley 1999a). As well, the sheer numbers of visitors often degrade the aesthetic qualities of sacred place, particularly when noisy and insensitive tourists disrupt those who have come to worship (Cohen 1998; Shackley 2002). In addition to dealing with internal management issues, religious site managers increasingly have to negotiate and communicate with various tourism stakeholders at local and regional levels who are increasingly interested in using sacred places as part of their tourism product – in essence having to mediate between religious goals and the operation of the tourism industry.

Even though all religious sites are part of a heritage environment, not all heritage sites are religious sites. Many of the challenges religious site managers deal with on a daily basis are akin to the challenges facing managers at other types of tourist attractions. However, many religious site custodians rely on organizational structures largely unaffected by modern management trends. In fact, management situations may exist in a management vacuum, where things are done by custom rather than focusing on specific goals or targets (Shackley 2003). Some religious groups, however, are not as traditional and have developed full management plans for their sites. Regardless of whether a site is managed with a detailed management plan or not, for the most part managers at religious sites are not trained tourism professionals. Rather, they tend to be professional or volunteer clergy who have been appointed by religious leaders to run a particular site (Shackley 2001a). Some management professionals and government officials fear that these "amateur" volunteers, with their lack of formal training in tourism and sustainable site management, including issues of interpretation, finances, and visitor impacts, will be unable to maintain religious sites adequately in the face of mass tourism. Stevens (1988: 44) argues, "there is an urgent need to equip those responsible; they are unlikely to be able to survive the pressures of tourism on the basis of prayer alone." In addition, Millar (1999) notes that ecclesiastical sites, or places operated by ecclesiastical leaders, differ from other tourism attractions because they have different organization goals rather than any physical differences. The *core business* of religious sites is the provision of a "focus and facility for those who wish to worship, pray or meditate" versus other tourism attractions which focus on entertainment and leisure (Shackley 2001a: 7).

Owing to the uniqueness of management challenges at religious heritage sites, it is surprising that with the exception of Shackley (1996, 1998, 1999a, 1999b, 2001a, 2001b, 2002, 2003) and a handful of others (O'Guinn and Belk 1989; Hobbs 1992; Jackson and Hudman 1994; Langley 1999; Bremer 2001a, 2001b, 2004; McGettigan and Burns 2001; Olsen and Timothy 2002; Digance 2003), very few researchers have examined the complex management issues endemic to important places where tourism and religion coincide. This is probably the case because religious heritage sites, ceremonies, festivals, and landscapes are typically considered part of a destination's cultural heritage product and fall within discussions of heritage tourism planning and management in general (Shaw and Jones 1997; Graham *et al.* 2000).

Against this background, the purpose of this chapter is to discuss some of the management challenges religious heritage site managers face when trying to service both the faithful and other tourists – challenges that become heightened in cases where the majority of visitors are non-pilgrim tourists who outnumber pilgrims (Vukonić 1996; Shackley 2001a), or where conflict over religious and secular forces create inter-faith contention (Kong 2001; Shackley 2002). This chapter is divided into two sections, highlighting the two major areas on which religious site managers need to focus. The first section deals specifically with the internal management issues with an emphasis on maintaining the tangible (physical) and intangible qualities of the site, including upholding a religious sense of place (Shackley 2002). The second section draws attention to external or off-site management pressures site caretakers face when negotiating with various tourism stakeholders who have a vested interest in the site.

Internal management issues

The main task of any site manager is to supervise and mediate the interactions between people and both the natural and built environment (Mitchell 1994). Like managers of other tourist attractions, religious site managers also have impacts to deal with, including visitor pressures, flows and experiences, site marketing and interpretation, accomplishing specific organizational goals, and planning special events. These concerns revolve around one important issue: the maintenance of a sense of place (Shackley 2001b, 2002). This sense of place is critical to providing an atmosphere of worship and meditation for those who wish to communicate with the divine. Though the meeting-house type of worship space is adequate for daily or weekly worship, many people travel to sites where a miracle, vision, or special event in the life of a religious leader occurred. Such places are sometimes considered higher places of worship where there is a greater likelihood of an encounter with deity (Marshall 1994). As a result, many religious sites attract committed visitors who travel long distances to experience something – whether spiritual or religious in nature – that they cannot experience closer to home (Stevens 1988: 3). Therefore, the main focus of religious site custodians is to preserve the emotive qualities of the place as a way of creating and maintaining an atmosphere conducive to worship and contemplation. They must also give attention to aesthetics and the nature and quality of instruction given to people attending worship services. This is important, as the nature of the experience desired by these religiously motivated travelers, as Shackley (2002: 345) notes, "is highly complex; being both intangible and including elements such as nostalgia, a closeness to God, 'atmosphere,' and the gaining of spiritual merit."

However, religiously motivated travelers and pilgrims are not the only people visiting religious heritage places. Shackley (2001a: 5) suggests that in addition to visitors whose primary purpose is to gain a religious experience (pilgrims), there are other visitors whose major motivation is either to visit an element of their international, national, local, or personal religious heritage or to be educated about a particular site or cultural group. However, this dichotomy may be too simplistic, as visitor motivations and their knowledge of the characteristics and purposes of a particular site vary considerably (Smith 1992; Fish and Fish 1993). Thus, visitors to religious sites cannot simply be labeled as pilgrims or tourists, as their motivations and previous experiences are not uniform. Even in instances where visitors are united by particular religious motivations or a common spiritual world-view, they come from "widely dispersed locations and thus do not belong to a single cultural constituency or interpretive community" (Coleman and Elsner 1994: 74), and thus come with a variety of expectations and needs. It is likely that only a small number of people traveling for religious purposes can satisfy the stereotype of the medieval pilgrim who was motivated by purely religious desires, seeking deep, meaningful experiences. In general, however, visitors to religious locations exhibit different motivations, expectations, and behavioral patterns (Graham and Murray 1997: 401).

This means that site managers must deal with a multitude of visitor motivations and expectations, which increases the frequency and difficulty of management challenges and issues, for when religious sites become tourism attractions they change from being a religious space for worship and ritual into profane tourism space, or tourism in religious space and time (Santos 2003). One immediate issue is increased visitation to religious sites as a result of tourism. If this increase in visitation were to deal only with religious travelers and pilgrims, managers might only need to make a few adjustments to maintain their sense of place and an atmosphere conducive for prayer and meditation. However, at most religious sites, non-worshipping visitors greatly outnumber those who visit for religious reasons (Vukonić 1996). Because of the different cultural and travel backgrounds of these varied visitors, it can be difficult to "convey a concept of sacredness across cultural boundaries" (Shackley 2001a: 9). While it used to be that people entering a cathedral or temple were quiet and respectful, many visitors today, even if they view a site as sacred, are unaware of the accepted codes of behavior in hallowed settings (Shackley 2001a: 8) and may not realise that loud talking, photographing pilgrims who are praying or meditating, and acting disruptively distracts worshippers and dilutes the sense of sanctity and calmness. As Shackley (2002: 348–349) notes:

> The perception of sanctity is central to the idea as sacred space exists only for those who know its characteristics and the reason for its delineation. This is at the root of many problems associated with the management of cathedrals. Although the premises are recognized as sacred by the worshipping community, who behave accordingly, tourists may not perceive them as sacred and behave in an inappropriate manner, creating tensions that provide interesting dilemmas for site managers. Christian cathedrals, for example, traditionally impose a basic dress code that bans the wearing of revealing tops, beachwear and shorts (at least for women). Many contemporary visitors are

unaware of such unwritten rules and feel that, if they are not believers, they should not be subject to the same rules.

This is particularly acute when coupled with overcrowding, which can violate a site's sanctity, as in the case of Mount Sinai, Egypt (Hobbs 1992) and Buddhist monasteries in Tibet and Nepal (Shackley 1999a). Over-crowding, whether consisting of mainly tourists or pilgrims, can have a detrimental effect on the built and natural environment through vandalism/ graffiti, theft, accidental damage, general wear and tear, microclimatic change, graffiti, and litter. Each of these physical impacts diminishes the experience of religiously motivated and other visitors, and in some extreme cases can potentially lead to the site being unable to retain its original functions, becoming in essence a dedicated tourist attraction (Shackley 2001a: 37), or closing down altogether.

Because of negative physical impacts there is often considerable tension between religious sites and tourism. Many managers would prefer to give access only to a small number of believers because of the physical impacts tourism causes. The tension between religious sites and tourism is not as great as it might seem, however, as many places are open to non-worshipping tourists, albeit cautiously in many cases. Because shrines or other sacred structures are the destination of travelers who wish to worship and commune with the numinous or sacred, managers do not typically want to deny access. At the same time, religious sites are often required to be financially self-sustaining, in that caretakers are responsible for maintaining architectural quality, staffing, and meeting utility costs.

Many religious sites encourage donations from visitors to help with site maintenance and support managers who are usually unpaid priests or other religious officials. Most sites do not ask for a specific donation amount (Nolan and Nolan 1992), relying solely on the charity of visitors, for according to one study, there is very little increase in the amount of donations per visitor when a specific amount is requested (Willis 1994). However, because donations alone do not cover maintenance and opera-tional costs, more and more religious sites are charging an entrance fee and allowing greater access to visitors in order to meet these expenses (Stevens 1988). As well, entrance fees can be used as a deterrent for people who are not seriously interested in visiting (more leisure-oriented visitors) or to limit access to some of the more physically sensitive areas (Shackley 2001a, 2002).

However, the idea of charging entrance fees is anathema to some who question whether charging all visitors is good practice, particularly when believers are required to "pay to pray" (Shackley 2001a, 2002). Local people who visit a religious site on a frequent basis might question why they must pay the same price to use *their* facilities as non-believing or long-distance visitors. This has resulted in some religious sites instituting a dual fee system based on discounts or free access to local people or setting up a pay perimeter located beyond the area where worship and prayer takes place (Shackley 2002). A loss of extra income from local visitors, however, can be a problem for non-profit religious sites that have little capacity to generate income in the first place (Shackley 2003). As well, while religious groups have long commodified themselves through the sale of holy items, relics, and souvenirs (Olsen 2003), some managers become concerned about the loss of meaning when religious rituals and artifacts are commoditized for tourists, and therefore try to distance themselves from any trace of commercialization on site (Vukonić 1996, 2002).

In addition to the direct economies of exchange that take place at religious heritage sites, there are also indirect economies of exchange (Bremer 2001b; Koskansky 2002), where "instead of money being passed between parties, religious teachings and feelings are exchanged" (Olsen 2003: 101). As such, religious site caretakers, depending on their organizational goals, may wish to engage in proselytizing efforts, viewing all non-believing visitors as potential converts, or engage in pastoral care or outreach programs with visitors (Olsen 2006).

However, because of the expenses that come with day-to-day operational costs, many sites have no choice but to open their doors to tourists. Therefore, managers have turned their attention to site conservation, in which emphasis is on combining site preservation and visitor interaction in a sustainable manner. Indeed, most managers of functioning religious heritage sites are faced with issues of how to manage the increasing volumes of visitors while "balancing the need to conserve the cathedral fabric with the provision of a high quality experience for the visitor" (Shackley 2002: 347). Managers therefore face major challenges in maintaining a sense of place while catering to the needs and expectations of both pilgrims and non-pilgrim tourists and preserving the site's physical integrity.

Site custodians from various religious traditions have implemented some management strategies to cope with the difficulties of balancing locational sense of place, the expectations and goals of tourists and pilgrims, conservation, and increased tourist arrivals. For example, where meditation and prayer are difficult as a result of overcrowding, lack of visitor controls, or the lack of signage encouraging reverence and respect while religious rites or services are in progress, sanctity may be maintained in part at least by allowing non-adherents to enter only at specific times by hardening the site (Schouten 1995), where specific and well-marked access routes are used to usher visitors through the area, both minimizing human impact and controlling access to sensitive areas. This can help stop visitors from entering restricted areas reserved for staff or administration or for sacred purposes (Shackley 2002). The use of signs can also encourage proper behavior and help with visitor flows. Some sites move non-believers around places designated for worship, leaving pilgrims room to pray in a reverent environment.

Grimwade and Carter (2000) suggest that heritage sites are best cared for when a well-planned preservation and conservation plan is combined with an effective interpretive program. This then allows pragmatic management strategies that foster community pride and considers the specific needs of the site and of visitors and pro-active presentation to preserve its cultural meaning and values. As such, many religious sites are actively interpreted to visitors, attempting to connect the meanings and messages of place "to something within the personality or experience of the visitors" (Rimmawi and Abdelmoneim 1992: 9). This has been done through the creation of visitor centers and the training of formal tour guides, tools that both educate visitors about the site and enhance their experiences (Hall and McArthur 1996).

To make the site more reverent and peaceful, some managers have turned to what can be called a "Ministry of Welcome," in which there are three focal points (Askew 1997) when actively presenting history and religious beliefs to visitors. First, there should be an emphasis on the "welcome," where visitors feel comfortable and not intimidated or wary of the environment or the people with whom they are interacting. Second, basic information about the faith should be available, whether through written or spoken word, showing how the site embodies theology in fixed space through the combination of art, architecture, and liturgy (Coleman and Elsner 1994: 78). Lastly, visitors should be encouraged to pray or meditate, as many tourists are searching for an authentic experience, which may manifest itself at a given sacred location. In so doing, tourists may become pilgrims (Askew 1997: 13).

Staff can also play a critical role in maintaining the religious sense of place and interpreting the importance of a site. Coleman and Elsner (1994: 78), writing about pilgrimage to Mt Sinai, note that:

> from the pilgrims' standpoint [the monks] functioned not only as interpreters of and guides to the holy, but also as enablers who could make a site comprehensible to believers of any nationality or provenance. While the network of paths, monastic chapels and prayer niches marked the ascent of Sinai, the monks provided an interpretive and liturgical form of pilgrim-experiences.

Indeed, while specialized guides might lead a tour group around a destination, it is religious site staff that generally lead tour groups and explain the history and significance of the location.

Prabhu (1993: 66) suggests that to enhance the experience(s) of visitors fully, several goals need to be met, including a quality welcome; a well-maintained physical structure; authentic ceremonies; an atmosphere of silence and recollection which facilitates worship; and the exclusion of merchants whose main aim is to turn a profit. However, this is not easily balanced and managed. For example, visitor discontent can occur if the wait to get into a site or to gain access to certain areas is too long. This leads to restlessness and irritation, and increases the possibility of disruption of the spiritual atmosphere. The lack of suitable infrastructure (e.g. toilets and food services) may also cause discontent and have a negative effect on visitor experiences. In addition, caretakers rarely understand the religious motivations of people who travel, and few empirical studies have been done by site managers or researchers. This may be because of the implicit assumption that people who visit religious sites are driven purely by religious motives. This gives credence to the view that researchers and religious site custodians "have theories about visitors' interests and motivations" which "appear to be based on subjective impressions rather than empirical research" (Heintzman and Mannell 2003: 56). Understanding the needs and expected outcomes of visitors is key, and like any other tourist attraction, religious sites have to compete for people's time against other attractions. Thus, religious site managers "need to understand their visitors, understand what draws them to attractions in the first instance, and understand what satisfies their thirst for fun, education, or whatever it is they are searching for" (de Sousa 1993: 334).

External management issues

In addition to internal management issues, there are several external or off-site issues that can influence management strategies and decisions. Various stakeholders have different interests in the way religious sites are conserved, managed, and consumed by tourists. Maybe the most important stakeholders are the religious leaders who operate the site. While beliefs and practices are interwoven with culture throughout the world, most religious organizations manifest themselves spatially through the development of stationary ecclesiastical institutions (e.g. parishes, temple complexes) and non-ecclesiastical institutions (e.g. ecumenical coalitions, etc.). With regard to built structures, these places are considered sacred space, as they are mystico-religious places where supernatural events have occurred, places where special or significant historical happenings occurred (Jackson and Henrie 1983), or religious administrative centers. Other sacred spaces, however, may be taboo in the sense that a site is so sacred that only certain people can enter. This idea of sacred space being taboo can extend to denying outsiders access to spaces and places (e.g. Mecca), as the right to enter may only belong to believers of the faith. Views of sacred space tie directly into how religious officials view tourism as a social phenomenon. Pious views of tourism can either be explicit in nature such as in the case of the Roman Catholic Church whose leaders have published various documents with their views on tourism, or implicit, where even though there have been no formal statements about tourism, particular attitudes can be drawn out by examining how the management of religious tourist locations becomes an expression of theology.

Not only do these views of sacred space and tourism influence pre-scriptions and proscriptions related to permissions to travel to destinations within a religious faith, but they also influence management practices. If officials of a particular faith view tourism in a positive manner, their associ-ated sacred sites may be used to engage in pastoral or missionary activities or as a place of outreach for believers and non-believers, using tourism as a tool to reach goals. Positive views of tourism also determine whether or not place custodians will engage in dialog or cooperate with outside stake-holders in the area. If religious authorities view tourism in a negative light, tourism may be prohibited or restricted in some way (Olsen 2006).

Caretakers of religious heritage sites can be constrained by the views and belief structures of their religion, depending on its administrative organization. For example, hierarchical religions (e.g. most Christian sects) tend to be well defined in terms of organizational structure, exercising tight control between and among groups in various geographical areas. While there might be very little variation between congregations and subsequently religious heritage sites, there tends to be more support from religious leaders and more management expertise available from experienced site directors because the sites historically have been clerically dominated for many years. Autonomous or decentralized religions are self-sufficient in the sense that the organizational structure involves a loose cooperation between believing communities (e.g. Islam and Hinduism). This adminis-trative structure allows for a more flexible agenda in terms of religious heritage site management.

Government organizations, urban planners, local and regional tourism managers, regional economic development agencies, and state-sponsored historical societies, are also major stakeholders in terms of influencing how religious sites function. Governments, with their economic mandate to increase tourism revenues, commodify physical and socio-cultural resources to create tourism products, transforming them from their original use in a manner that makes them suitable for tourism development (Dietvorst and Ashworth 1995). Tourism promotion is critical in changing religious sites into tourist places, as the symbolic meaning of place can be trans-formed into sites to be gazed at (Urry 1990) rather than sites of worship and contemplation "through the meanings ascribed to them by visitors and promotional agencies" (Young 1999: 374). Religious sites are commonly used in tourism promotional literature as cultural resources to be consumed by tourists. Because of the economic potential of religious tourism, some government organizations also take holy histories and events and trans-form them into commercial enterprises. For example, in 1998 the British Tourist Authority established a Wesley Trail that commemorated the two hundred and fiftieth anniversary of the founding of the Methodist Church by John and Charles Wesley. Hughes (1998: 17) notes that "the objectives for this project did not address the religious sensibilities of Methodists or conceive of the trail as a mark of respect for the Wesleys or Methodism. Rather, the project was addressed in exclusively commercial terms." In other instances, the spiritual purposes of religious sites are mitigated by crowds of hawkers, salespersons, and beggars (Orland and Bellafiore 1990; Shackley 2001a). This causes friction with religious site managers and leaders who view their sites as places of prayer and contemplation rather than centers of profiteering. This may lead some religious site managers to refuse cooperation with outside authorities when it comes to utilizing their site as a tourist attraction.

Governments not only influence the management of religious heritage sites through marketing and commodification but also through specific policies related to maintenance and interpretation, which range from suppression to subsidization. While suppression of religious groups rarely manifests itself in an overt fashion in western societies owing to the greater division between church and state, in other areas of the world many cases exist where government policies influence visitor travel patterns and inter-pretation of religious sites. A classic example of this is in Burma, where Buddhist shrines have been taken over by the reigning government, which in turn reinterprets the shrines for tourists in a sanitized manner, focusing more on reinforcing political and economic claims than on presenting the Buddhist views of site sacrality (Philip and Mercer 1999). This, however, is not new, for even as far back as the seventeenth-century secular author-ities in English cathedral cities attempted to appropriate cathedrals to claim legal and popular legitimacy in the eyes of the public (Estabrook 2002). Some religious heritage site custodians accept government funds to pay their bills and maintain current levels of conservation. However, when government money is accepted, it often means there will be concessions in terms of interpretive content and how tourism functions at the site.

Other interest groups include pilgrims and tourists with their own sets of motivations, expectations, and needs; local interest groups; and tourism-related businesses. Of note here is Post *et al.*'s (1998) suggestion that religious sites have undergone a process of *musealization*, in which reli-gious sites are viewed by both tourists – and to a lesser extent pilgrims – as museums. Many tourists view religious sites as opportunities for educational or cultural experiences (Winter and Gasson 1996). Therefore, pilgrims, tourists, and residents want religious sites to appear unaffected by the modern world, which is impossible because the sites will have been changed since their construction. While pilgrimage centers have long func-tioned as museums and repositories for art, and bear tangible witness to the history of a religious culture or group (Shackley 2002: 346), tourists see religious sites as symbols of a past age (Daskalakis 1984), and therefore

> space[s] to be preserved rather than used, to be gazed upon but not
> changed ... Thus, when attempts are made to radicalize the use of

that space, whether by the physical modification of the site or by the introduction of charging, a dissonance arises.

(Shackley 2002: 350)

In addition to dealing with various outside stakeholders, site managers also must deal with other external factors. For example, religious sites do not exist in a socio-political and spatial vacuum, as they are affected by the politics and social trends of the area in which they are located. Shackley (2001b: 7) suggests that national, regional, and local instability, ranging from terrorism to national government policies, to vested interests in a site by religious, social, or political groups can heighten management problems. Shackley (2001b: 8) acknowledges that "the easiest sites to manage are those where stability is high even if visitor numbers are high (such as the Vatican or Canterbury Cathedral) since this stability and control permits the development and implementation of effective visitor management systems." For example, sites seen as sacred or being an ancestral homeland have long been the focus of strong emotional attachments by various socio-religious groups, leading to conflict not just between different devotional groups, but also between divisions within a single group, as well as between religious and secular authorities (Olsen 2000, 2003; Kong 2001). In extreme cases, violence has erupted between religious factions, the most obvious examples being Jerusalem, Ireland, and Ayodhya, India. Indeed, religious sites, like most other places endowed with cultural or historic meaning, are repositories of history and memory and "contain multiple levels of sedimented history" and "layers of meaning" (Yeoh and Kong 1996: 55). Competing interests between religious groups over the ownership and maintenance of sacred sites can cause additional difficulties not only over how to interpret them when there are competing discourses over the history and value of a site, but also when religious violence, such as the Al Aska Intifada in September 2000 between Israel and Palestine, can lead to drastic decreases in tourist arrivals. In some instances the threat of terrorism or violence has forced pilgrims and other tourists to seek alternative spiritual destinations (Osborne 2001).

Conclusion

To be competitive in a global tourism market, many national and regional governments use religious heritage sites to attract tourists. These sites, then, are treated as a key component of the cultural landscape and in some cases, flagship tourist attractions. Owing in part to increased marketing efforts, religious heritage sites are being visited more and more by tourists who come for educational and leisure reasons rather than strictly religious motives. Religious site caretakers must focus not only on the internal management issues that arise when tourism grows at their sites, but also on various external factors and stakeholders that play a growing role in the complexity of management that is necessary to conserve the spiritual or religious sense of place and the physical fabric. In some cases increased visitation can easily be managed by hardening a site or by providing better interpretation and instruction. However, this is not possible in all cases. As Graham *et al.* (2000: 24) argue, "landscapes of tourism consumption [that] are simultaneously people's sacred places is one of the principal causes of heritage contestation on a global scale."

Ideally cooperative tourism planning between all shareholders involved should be the goal (Timothy 1998). This is particularly true in urban areas, where the distribution of attractions and facilities is unequal and where the functions and demands on urban forms change rapidly over time. However, cooperation between the stakeholders can be difficult because "tourism is frequently influenced by local power relationships which favour the political or economic elite" (Sharpley 2000: 10). Indeed, the goals of large-scale tourism operations are often incompatible with the goals of other stakeholders (du Cros 2001). In fact, some religious authorities may refuse to cooperate with outside agents when it comes to the use of their religious sites or parts thereof when issues arise that may compromise their particular worldviews.

References

Askew, R. (1997) *From Strangers to Pilgrims: Evangelism and the Church Tourist*, London: Church Army and the Grove Evangelism Series.

Bremer, T.S. (2001a) *Religion on Display: Tourists, Sacred Place, and Identity at the San Antonio Missions*, Princeton, NJ: Princeton University Press.

Bremer, T.S. (2001b) "Tourists and religion at Temple Square and Mission San Juan Capistrano," *Journal of American Folklore* 113: 422–435.

Bremer, T.S. (2004) *Blessed with Tourists: The Borderlands of Religion and Tourism in San Antonio*, Chapel Hill: University of North Carolina Press.

Cohen, E. (1998) "Tourism and religion: a comparative perspective," *Pacific Tourism Review* 2: 1–10.

Coleman, S. and Elsner, J. (1994) "The pilgrim's progress: art, architecture and ritual movement at Sinai," *World Archaeology* 26(1): 73–89.

Daskalakis, G. (1984) "Tourism and the architectural heritage – cultural aspects," in AIEST (ed.) *Tourism and the Architectural Heritage – Cultural, Legal, Economic and Marketing Aspects*, St Gall: AIEST.

de Sousa, D. (1993) "Tourism and pilgrimage: tourists as pilgrims?," *Contours* 6(2): 4–8.

Dietvorst, A.G.J. and Ashworth, G.J. (1995) "Tourism transformations: an introduction," in G.J. Ashworth and A.G.J. Dietvorst (eds) *Tourism and Spatial Transformations: Implications for Policy and Planning*, Wallingford: CABI.

Digance, J. (2003) "Pilgrimage at contested sites," *Annals of Tourism Research* 30: 143–159.

du Cros, H. (2001) "A new model to assist in planning for sustainable cultural heritage tourism," *International Journal of Tourism Research* 3: 165–170.

116   *Daniel H. Olsen*

Eliade, M. (1961) *The Sacred and the Profane: The Nature of Religion*, New York: Harper & Row.

Estabrook, C.B. (2002) "Ritual, space, and authority in seventeenth-century English cathedral cities," *Journal of Interdisciplinary History* 32(4): 593–620.

Fish, J.M. and Fish, M. (1993) "International tourism and pilgrimage: a discussion," *Journal of East and West Studies* 22(2): 83–90.

Graham, B. and Murray, M. (1997) "The spiritual and the profane: the pilgrimage to Santiago de Compostela," *Ecumene* 4(4): 389–409.

Graham, B., Ashworth, G.J., and Tunbridge, J. (2000) *A Geography of Heritage: Power, Culture, and Economy*, London: Arnold.

Grimwade, G. and Carter, B. (2000) "Managing small heritage sites with interpretation and community involvement," *International Journal of Heritage Studies* 6(1): 33–48.

Hall, C.M. and McArthur, S. (1996) *Heritage Management in Australia and New Zealand: The Human Dimension*, Melbourne: Oxford University Press.

Heintzman, P. and Mannell, R.C. (2003) "Spiritual functions of leisure and spiritual well-being: coping with time pressure," *Leisure Sciences* 25: 207–230.

Hobbs, J.J. (1992) "Sacred space and touristic development at Jebel Musa (Mt. Sinai), Egypt," *Journal of Cultural Geography* 12(2): 99–112.

Hughes, G. (1998) "Tourism and the semiological realization of space," in G. Ringer (ed.) *Destinations: Cultural Landscapes of Tourism*, London: Routledge.

Jackson, R.H. and Henrie, R. (1983) "Perception of sacred space," *Journal of Cultural Geography* 3(2): 94–107.

Jackson, R.H. and Hudman, L. (1994) "Pilgrimage tourism and English cathedrals: the role of religion in travel," *The Tourist Review* 4: 40–48.

Kong, L. (2001) "Mapping 'new' geographies of religion: politics and poetics in modernity," *Progress in Human Geography* 25(2): 211–233.

Koskansky, O. (2002) "Tourism, charity, and profit: the movement of money in Moroccan Jewish pilgrimage," *Cultural Anthropology* 17(3): 359–400.

Langley, C.L. (1999) "Churches and chapels on open-air museum sites," in C. Paine (ed.) *Godly Things: Museums, Objects and Religion*, London: Leicester University Press.

Lloyd, D.W. (1998) *Battlefield Tourism*, New York: Berg.

McGettigan, F. and Burns, K. (2001) "Clonmacnoise: a monastic site, burial ground and tourist attraction," in G. Richards (ed.) *Cultural Attractions and European Tourism*, Wallingford: CABI.

Marshall, J. (1994) "The mosque on Erb Street: a study in sacred and profane space," *Environments* 22(2): 55–66.

Millar, S. (1999) "An overview of the sector," in A. Leask and I. Yeoman (eds) *Heritage Visitor Attractions: An Operations Management Perspective*, London: Cassell.

Mitchell, L.S. (1994) "Research in the geography of tourism," in J.R.B. Ritchie and C.R. Goeldner (eds) *Travel, Tourism and Hospitality Research: A Handbook for Managers and Researchers, 2nd Edn*, New York: Wiley.

Nolan, M.L. and Nolan, S. (1992) "Religious sites as tourism attractions in Europe," *Annals of Tourism Research* 19: 68–78.

O'Guinn, T.C. and Belk, R.W. (1989) "Heaven on Earth: consumption at heritage village, USA," *Journal of Consumer Research* 16: 227–238.

Olsen, D.H. (2000) "Contested heritage, religion, and tourism," unpublished Masters thesis, Bowling Green State University, Bowling Green, Ohio.

Olsen, D.H. (2003) "Heritage, tourism, and the commodification of religion," *Tourism Recreation Research* 28(3): 99–104.

Olsen, D.H. (2006) "Religion, tourism, and managing the sacred," unpublished PhD dissertation, Department of Geography, University of Waterloo, Waterloo, Ontario.

Olsen, D.H. and Timothy, D.J. (2002) "Contested religious heritage: differing views of Mormon heritage," *Tourism Recreation Research* 27(2): 7–15.

Orland, B. and Bellafiore, V.J. (1990) "Development directions for a sacred site in India," *Landscape and Urban Planning* 19(2): 181–196.

Osborne, C. (2001) "A facelift for Egypt's sacred sites," *The Middle East* 312: 42–45.

Philip, J. and Mercer, D. (1999) "Commodification of Buddhism in contemporary Burma," *Annals of Tourism Research* 26: 31–54.

Post, P., Pieper, J., and van Uden, M. (1998) *The Modern Pilgrim: Multidisciplinary Explorations of Christian Pilgrimage*, Leuven: Uitgeverij Peeters.

Prabhu, P.P. (1993) "Religion's responsibilities in cultural tourism," in W. Nuryanti (ed.) *Universal Tourism: Enriching or Degrading Culture?*, Yogakarta: Gadjah Mada University Press.

Rimmawi, H.S. and Abdelmoneim, A.I. (1992) "Culture and tourism in Saudi Arabia," *Journal of Cultural Geography* 12(2): 93–98.

Santos, M.G.M.P. (2003) "Religious tourism: contributions towards a clarification of concepts," in C. Fernandes, F. McGettigan, and J. Edwards (eds) *Religious Tourism and Pilgrimage*, ATLAS Special Interest Group, 1st Expert Meeting, Fatima, Portugal: Tourism Board of Leiria/Fatima.

Schouten, F. (1995) "Improving visitor care in heritage attractions," *Tourism Management* 16(4): 259–261.

Shackley, M. (1996) "Visitor management and the monastic festivals of the Himalayas," in M. Robinson, N. Evans, and P. Callaghan (eds) *Culture As the Tourist Product*, Sunderland: Centre for Travel and Tourism.

Shackley, M. (1998) "A golden calf in sacred space? The future of St. Katherine's Monastery, Mount Sinai (Egypt)," *International Journal of Heritage Studies* 4(3/4): 123–134.

Shackley, M. (1999a) "Managing the cultural impacts of religious tourism in the Himalayas, Tibet and Nepal," in M. Robinson and P. Boniface (eds) *Tourism and Cultural Conflicts*, Wallingford: CABI.

Shackley, M. (1999b) "Managing the visitor experience: the case of cultural World Heritage Sites," in W. Nuryanti (ed.) *Heritage, Tourism and Local Communities*, Yogyakarta: Gadjah Mada University Press.

Shackley, M. (2001a) *Managing Sacred Sites: Service Provision and Visitor Experience*, London: Continuum.

Shackley, M. (2001b) "Sacred World Heritage Sites: balancing meaning with management," *Tourism Recreation Research* 26(1): 5–10.

Shackley, M. (2002) "Space, sanctity and service: the English cathedral as Heterotopia," *International Journal of Tourism Research* 4: 345–352.

Shackley, M. (2003) "Management challenges for religion-based attractions," in A. Fyall, B. Garrod, and A. Leask (eds) *Managing Visitor Attractions: New Directions*, Oxford: Butterworth Heinemann.

Sharpley, R. (2000) "Tourism and sustainable development: exploring the theoretical divide," *Journal of Sustainable Tourism* 8(1): 1–19.

Shaw, B.J. and Jones, R. (1997) *Contesting Urban Heritage: Voices from the Periphery*, Aldershot: Ashgate.

Smith, V.L. (1992) "Introduction: the quest in guest," *Annals of Tourism Research* 19: 1–17.

Stevens, T. (1988) "The Ministry of Welcome: tourism and religious sites," *Leisure Management* 8(3): 41–44.

Timothy, D.J. (1998) "Cooperative tourism planning in a developing destination," *Journal of Sustainable Tourism* 6(1): 52–68.

Timothy, D.J. and Boyd, S.W. (2003) *Heritage Tourism*, Harlow: Prentice Hall.

Urry, J. (1990) *The Tourist Gaze: Leisure and Travel in Contemporary Societies*, London: Sage.

Vukonić, B. (1996) *Tourism and Religion*, Oxford: Elsevier.

Vukonić, B. (2002) "Religion, tourism and economics: a convenient symbiosis," *Tourism Recreation Research* 27(2): 59–64.

Willis, K.G. (1994) "Paying for heritage: what price for Durham Cathedral?," *Journal of Environmental Planning and Management* 37(3): 267–278.

Winter, M. and Gasson, R. (1996) "Pilgrimage and tourism: cathedral visiting in contemporary England," *International Journal of Heritage Studies* 2(3): 172–182.

Yeoh, B. and Kong, L. (1996) "The notion of place in the construction of history, nostalgia and heritage in Singapore," *Singapore Journal of Tropical Geography* 17(1): 52–65.

Young, M. (1999) "The social construction of tourist places," *Australian Geographer* 30(3): 373–389.

Part II
Religious traditions and tourism

9   Tourism and the spiritual philosophies of the "Orient"

Chao Guo

In recent years, the importance of East Asia to the global tourism industry as both a destination and generating market has grown considerably. Being part of the fastest growing destination region over the past 30 years, East Asia continued its vigorous performance with an impressive 8 percent increase of international tourist arrivals in 2002 despite the negative consequences of the region's 1997 financial and economic crisis and the global effects of the 2001 terror attacks (World Tourism Organization 2003). The mid- to long-term prospects for Asian outbound and inbound flows also remain strongly positive. In the year 2020, international tourist arrivals to East Asia and the Pacific are expected to reach 397 million; outbound tourist trips from the region are expected to reach 405 million. Both statistics represent an annual growth rate of 6.5 percent over the period 1995–2020, over two percentage points above the global average (World Tourism Organization 2001).

As East Asia is increasingly taking over a higher share of global tourism, it is imperative that aspiring tourism professionals broaden their understanding of the cultures of this region. East Asian cultures bear the imprint of the spiritual philosophies of the "Orient" such as Confucianism, Daoism, and Shintoism, as well as various popular religious practices, such as *Feng Shui*. Confucianism represents a core set of values for Chinese people and other Asians, accounting for a quarter of the world's population; it is also seen as the cultural foundation of Asian societies and the source of social and political stability that facilitated the explosive economic growth in Hong Kong, Japan, Korea, Taiwan, and Singapore in the 1960s and 1970s (Elman *et al.* 2002; Tu 1996). Daoism stands alongside Confucianism as one of the great religious/philosophical systems of China (Hansen 2003). Primarily an indigenous Chinese religion and intricately tied in with Chinese culture, Daoism has pervasively and continuously shaped East Asian cultures in important ways (Kohn 2001). Although consensus among scholars has yet to be reached with regard to the question of whether these spiritual philosophies should be treated as coherent religious systems in a strict (or Western) sense, they nonetheless present unique religious worldviews that have largely influenced the cultures and lifestyles of China, Japan, Korea, and many other Asian countries.

While the relationships between tourism and religion in general, and pilgrimage tourism in particular, have been a persistent theme in the tourism literature (e.g. Vukonić 1996; Cohen 1992; Collins-Kreiner and Kliot 2000; Rinschede 1992; Smith 1992; Timothy 2002), there is surprisingly very little research on the effects of Oriental spiritual philosophies (i.e. Confucianism, Daoism, and Shintoism) on tourism and people's travel behavior. One exception, however, is a study by Arcodia (2003), which outlined some of the key tenets of Confucianism that underpin human relations and interactions, and discussed the implications of Confucian values to tourism management and marketing in three specific areas: individualism, protocol, and business and social ethics. This study reflects a valuable first step in this promising research direction, yet it falls short of covering all the key principles of Confucianism that are of relevance to travel and tourism. Moreover, it does not include Daoism and other popular religions in its coverage.

The goal of this chapter, therefore, is to introduce the basic principles of Confucianism, Daoism, and Chinese popular religion in general, followed by discussions of the implications of these religious worldviews for people's lifestyles and travel behaviors, tourism management and marketing, resort and hotel design, and environmental ethics. Because of space constraints, Shintoism will not be included.

Oriental spiritual philosophies: an overview

Confucianism

Confucianism is a moral and religious belief system founded by Confucius in the sixth century BC. Confucianism manifests a religious worldview in its cosmological orientation, which has been described as "encompassing a continuity of being between all life forms without a radical break between the divine and human worlds" (Tucker 2003: 4). In the Confucian view, humanity is not inherently separate from Heaven – "Heaven and humans are one" (*tianrenheyi*) (Yao 2000: 44). Unlike in so-called otherworldly religions, such as Buddhism and Christianity, which focus on the "other shore" or the Kingdom of God, the primary spiritual concern of Confucianism rests on ordinary human existence. This conscious refusal to find spiritual shelter beyond the human world reflects a strong faith in human beings' capacity to transform the human condition through their own personal efforts (Tu 1999).

The attainment of this cosmological orientation requires the connection of the microcosm of the self to the macrocosm of the universe through spiritual practices of communitarian ethics, moral self-transformation and ritual relatedness (Tucker 2003). First, humans are connected to one another and to the larger cosmological order through an elaborate system of communitarian ethics embodied by the matrix of five human relations: ruler–subject, father–son, elder–younger brother, husband–wife, and friend–friend. Enlightened self-interest or reciprocity – a moral context that combines one's own ambition with concern for others – is a key to communitarian ethics. It is the means by which Confucian societies develop a communitarian basis so that they can become bonded "fiduciary" communities, as distinguished from Western, individualistic, "adversarial" communities. This Confucian view has shaped the traditional Chinese emphasis on the collective over the individual. As Louis Henkin (1986: 27) pointed out, in traditional China "[t]here was no distinction, no separation, no confrontation between the individual and society, but an essential unity and harmony, permeating all individual behavior." He further argued that "China begins with the society, the collectivity, and concentrates on general (not individual) welfare."

Second, the cultivation of virtue in individuals and the moral transformation of humans are the basis for the interconnection of self, society, and the cosmos. Confucianism holds that "humanity can achieve perfection and live up to heavenly principles" and that "humans have their mission in the world" (Yao 2000: 46). Yet the fulfillment of this mission requires continual self-examination, rigorous discipline, and the cultivation of virtue. Consistent with the communitarian ethics, this process of spiritual self-transformation is not an individual spiritual path aimed at personal salvation but rather an ongoing process of rectification needed to cultivate one's "luminous virtue" through a communal act (Little and Reed 1989; Tucker 2003; Tu 1985).

The third practice, acting in accordance with rituals that establish patterns of relatedness, provides mechanisms through which humans are attentive to one another, responsive to the needs of society, and attuned to the natural world. Rituals are not only a means of affirming the emotional dimensions of human life, but also link humans to one another and to the other key dimensions of reality – the political order, nature's seasonal cycles, and the cosmos itself (Tucker 2003: 4–5).

Needless to say, many of the fixed patterns of family, social, and political life that Confucius advocated some 2,500 years ago are no longer relevant to modern society, but many of the moral and spiritual values that the Confucian religious worldview carries still apply. While it is beyond the scope of this chapter to discuss all Confucian values, the following two doctrines on which the substance of Confucianism is centered are particularly relevant to tourism professionals and researchers.

The first principle is benevolence (*ren*). Also translated as "goodwill" or "human-heartedness," this concept is the cardinal virtue of Confucianism. It identifies the capacity of the individual to extend generosity and compassion to all of humanity (Arcodia 2003). Confucius considered benevolence as something people cultivate within themselves before it can affect their relations with others. The best way to approach benevolence is putting oneself in the position of others and then treating others accordingly. This idea can be best expressed with Confucius' own words: "Not to do to others as you would not wish done to yourself" (The Analects 12:2; see Legge 1991: 167).

The second is propriety (*li*), or, more specifically, the rituals or rules of proper conduct. These refer to the norms and unwritten laws that guide thought and action in society and regulate human behavior. Benevolence can be cultivated by observing and acting in accordance with these rituals or rules. One important aspect of propriety is respect for family honor and ancestor worship (Chi-Ping 1989; Hwang 1999; Li 2000). Followers of Confucian teachings are taught to love their families first and then to extend this love and respect to the rest of society. Furthermore, they are also taught to perform a filial duty to pay respect and remembrance to their ancestors regularly, both in life and after death. The ultimate goal in establishing and maintaining these family relationships is the maintenance of social harmony, as a harmonious family was seen by Confucius as the foundation of a harmonious society (Clarke 2000).

Finally, it is also worth mentioning that many Confucian adherents conduct learning and self-cultivation through the practice of quiet-sitting (*jing-zuo*). Quiet-sitting, or silent meditation, is a practice that supports the disciplined, regulated, and exhaustive search for principles, and the proper attitude of mind and reverence (Taylor 1990: 93).

Daoism

Daoism is the indigenous religion of traditional China. Founded in the sixth century BC, it is best known in the West as "Taoism." While Confucianism is more ethico-politically oriented, Daoism focuses more on the promotion of an individual's inner peace and harmony with his or her surroundings.

The concept of "way" (*Dao*) is central to the understanding of Daoism. According to Daoism, *Dao* is the force or law that governs the universe. It is the origin of two basic elements that form the universe, namely *Yin* and *Yang*. *Yin* represents the negative elements (e.g. night, moon, cool, and, traditionally, female), while *Yang* represents the positive elements (e.g. day, sun, hot, and, traditionally, male). Although these two elements conflict, they also complement one another. *Dao* is always seen as lying at the root of creation yet manifests in all that exists in the mundane landscape. Mediated through cosmic, vital energy (*qi*), *Dao* is essential and accessible to human beings in their everyday lives. The self-transformation of humans is inherent in Daoist teachings in the sense that aligning oneself with *Dao* creates harmony and equilibrium, brings out the best in people and attains a state of overall wellbeing in cosmos, nature, society, and the human body (Kohn 2001).

The Daoist approach to self-transformation, however, is quite different from that of Confucianism. For Confucianism, self-transformation is realized within the context of human society and its network of human-to-human relationships; yet for Daoism, self-transformation needs to be understood in light of the embeddedness of humanity and its social organization in a natural, ecological context (Callicott 1994: 78). Accordingly, while Confucians believe that it is through cultivation and education that humans realize self-transformation, Daoists hold that the only key to this transformation is the notion of non-deeming action (*Wu-Wei*). Owing to its seemingly passive or naturalistic nature, *Wu-Wei* later became the target of attack at the beginning of the twentieth century among Chinese intellectuals who regarded Daoist "non-striving" or "purposelessness" as the source of Chinese passivity (Hansen 2003). Yet a *Wu-Wei* approach does not entail simple quietness or passivity but rather a doing that is neither coercive nor assertive in order to achieve a harmony between humans and nature (Callicott 1994: 75).

Politically, a Daoist approach to *Wu-Wei* motivates a reaction against the moralistic and elitist inclinations of Confucianism. Confucianism establishes a rigid, detailed, traditional pattern of hierarchical social behavior that assigns duties to all of an individual's social roles (Hansen 2003). By contrast, Zhuangzi, one of the founders of Daoism, challenged the legitimacy of such dominant social morality and hierarchical structure. He asked what a turtle would choose if offered the option of being nailed in a place of veneration and honor in some place of worship or staying at the lake and "dragging his tail in the mud" (Hansen 2003: n.p.). This nature orientation and the preference of anarchy and free-spiritedness over social hierarchy and ritual roles suggest a kind of Daoist ethos that is antithetical to Confucianism in China.

Similar to Confucians, Daoists also practice meditation, yet for different reasons. To Confucians, quiet-sitting provides a means for the advancement of learning through such self-examination and inward exploration in order to achieve self-improvement in the practice of virtues and elimination of vices. To Daoists, the purpose of meditation is to exclude all distracting thoughts to attain a blending of self with nature and to cultivate health and longevity (Ching 2003). Other popular Daoist practices include *Taiji quan* and *Qigong*, which involve techniques of deep breathing, slow motion, and gentle stretches (Kohn 2001). In recent years, traditional Daoist rituals and local cults have revived and increased activities in China, especially in rural villages.

Popular religion

Chinese popular religion refers to forms of spirituality or religion practiced in the daily lives of Chinese people across all social boundaries, but especially by people of the lower socio-economic classes and illiterates. In quantitative terms, popular religious beliefs and practices have long formed the mainstream of Chinese religion since the earliest periods of Chinese history, constituting the basic support for traditional society, culture, and moral values (Fan 2003; Harrell 1979).

While Chinese popular religious practices vary in their forms and generally lack a systematic structure, there are certain beliefs and practices that are widespread throughout popular religion. These include a belief in heaven, worship of ancestors and deities, funeral and burial rituals, temple festivals of community renewal, various forms of divination, geomancy, spirit-possession, and the exorcism of harmful forces, among others (Fan 2003: 450).

Heaven worship is likely the oldest form of Chinese religion, as evidenced by inscriptions on oracle bones and inside bronze sacrificial vessels that are thousands of years old (Chiang 2003). It originally involved a set of aristocratic rituals performed by the emperors and their highest ministers on behalf of the state. For instance, in ancient times, the Chinese emperors had to offer sacrifices to the Heavenly God (*Shang Di*) on the Altars of Heaven and Earth at the time of the summer and winter solstices (Shaw and Curnow 2004). In addition, cults emerged on a local level to worship the local gods. Many gods worshipped in village temples are deified local historical figures, although others are nature or astral gods (Dean 2003). Deities which have played a particularly important role in Chinese popular religion are the Jade Emperor (from Daoism), Guan-Yin (from Buddhism), and Confucius (from Confucianism).

Ancestor worship is arguably the most important form of Chinese and East Asian popular religion. Influenced heavily by Confucian principles, it is widely practiced across the entire region (Kim 1997; Schrecker 1997). The traditional rituals concerning ancestor worship were very complicated and had to be followed precisely. Although the rituals have been simplified in modern times, they never fade in importance in the Chinese community (Chiang 2003).

Although there are people who believe in popular religions to harmonize themselves with cosmic forces, the goals of most worshippers are probably to seek immediate practical assistance, such as healing, good marriages, and safe childbirths. In Taiwan today, even politicians seek the support of gods and temples for their campaigns (Overmyer 2003). Thus, it is not surprising to note that many believers in popular religion seem to adhere to the principle of reciprocity. The sacrifices offered are often accompanied by petitions addressed to a local god specifying what is desired in return. If a local god is proved to be effective in fulfilling the worshippers' prayers, sacrifices will continue to be offered and more pilgrims will be attracted from more distant areas; without such success, a new god will be chosen (Chiang 2003).

Traditional Chinese festivals are one important form of popular religious practices. The beginning 15 days of a lunar year witness the Chinese New Year and the Lantern Festival, both symbolic of discarding the old and ushering in the new. On these festival days, children gather to set off firecrackers, which serve the religious function of frightening away evil spirits, while adults bring offerings to set before the images of the local gods and their ancestors (Holley 1989). Occurring on the fifth day of the fifth lunar month, the Dragon Boat Festival represents a time for protection from evil and disease for the rest of the year, and is celebrated by dragon boat races and eating rice dumplings (*zongzi*). The Mid-Autumn Festival, celebrated on the fifteenth day of the eighth lunar month, is considered a holiday of family reunion as well as celebration of harvest.

Another important popular religious practice is geomancy (*Feng Shui*). Literally meaning "wind and water," *Feng Shui* is "the skill and art of design and placement of cities, buildings, and interior spaces used to achieve balance and harmony with nature" (Henderson *et al.* 2003: 300; see also Schmitt and Pan 1994). Reflecting a Daoist cosmology and based on *Yin* and *Yang* principles, *Feng Shui* asserts that, to produce vibrant *Qi* (or cosmic health), it is vitally important that sites are favorably oriented and protected from evil influences (*sha*) by other buildings, walls, hills, water, or even insects. Once regarded as superstitions and repressed by the government, *Feng Shui* and other forms of popular religion have regained their popularity in mainland China, as well as in Hong Kong and Taiwan in recent years. In Hong Kong, for instance, every public housing development reportedly must seek a *Feng Shui* consultant for help on the site location and design aspects of the development (Hobson 1994).

A synthesis of the Oriental spiritual philosophies

The Oriental spiritual philosophies discussed previously are not isolated traditions. Throughout history, they have been in constant exchange with one another. For instance, during the Han Dynasty, Confucians incorporated some of the ideas of Daoism (e.g. the *Yin-Yang* and the *Five Elements*) to create a new version of Confucianism that eventually became the state orthodoxy that dominated social life, while Daoism became a subordinate philosophy (Yao 2000). Moreover, Confucianism and Daoism, along with Buddhism, are traditionally known as the 'three doctrines' or 'three religions' (*sanjiao*) of China. The three-in-one perspective is not only a purely theoretical consideration. For many centuries it was the foundation of social and political life in China. Attempts to harmonize Confucianism, Daoism, and Buddhism were made by scholars and ordinary people alike. In many temples or monasteries, whether Confucian, Daoist, or Buddhist, there were often tablets or statues of Confucius, Laozi, and the Buddha, and they were worshipped together (Yao 2000).

Referred to by some scholars as 'religions of harmony', Confucianism and Daoism both base their doctrines on the unity of heaven and humanity (Yao 2000; Kolodner 1994). Both traditions view humanity as inherently inseparable from and equal to the natural world. Rather, humanity is simply a part of the natural system (Campbell 1962: 3–9). In many cases, however,

Confucianism and Daoism are not viewed by some observers as religions per se (Ching 2001). This failure to appreciate the religious dimensions of Confucian and Daoist traditions is largely a result of the restricted definition of religion that requires an element of transcendent otherness and organizational structure (Tu 1999).

Throughout history, Chinese popular religious beliefs have been modified by the development of Confucianism and Daoism, as well as by the diffusion of exogenous religions (e.g. Buddhism) into China (Schipper 1994; Shaw and Curnow 2004). In fact, modern scholars of Chinese religions often find it difficult to define clearly most popular beliefs and sects without resorting to a combination of the three religious views already noted (Yao 2000).

Implications of Oriental spiritual philosophies for tourism

Leisure and travel behavior

Oriental spiritual philosophies not only offer unique religious worldviews but also represent distinguishable lifestyles in many ways. Therefore, a range of leisure and travel behaviors and attitudes may be influenced by these spiritual philosophies, as well as by the cultures and norms that bear the imprint of these spiritual beliefs and religious practices. The following paragraphs discuss the implications of the spiritual philosophies of East Asia for the leisure and travel behaviors of Asian people.

First, both Confucianism and Daoism emphasize the ultimate value of harmony between humankind and nature and the important role of human self-transformation in achieving this goal. To many Confucians, self-transformation involves self-cultivation and experiential learning (Wang and Stringer 2000). As pointed out by Yao (2000), the Confucian educational philosophy places emphasis not on the recitation of the classics but rather on a personal understanding of the proverbs and personal experiences in the wisdom embodied in the texts. Hence Confucians participate in various leisure activities such as ritual practices, music performances, poem reading, and meditation not merely for enjoyment but for the pursuit of understanding the value and meaning of life. This attachment of educational purposes to daily leisure can be vividly expressed in Confucius' own words: "The wise find pleasure in water; the virtuous find pleasure in hills" (*Ren zhe le shan, zhi zhe le shui*) (The Analects 6:23; see Legge 1991: 113). Here Confucius suggested a didactic role of leisure experiences by attaching human characters and merits (e.g. wisdom and virtue) to natural surroundings (e.g. water and hills). Although Daoists differ from Confucians in their approach to self-transformation, they nevertheless take a similar stance toward the meaning of nature for human development. For instance, Laozi, the founder of Daoism, once said: "The highest efficacy is like water" (*Shang shan si shui*) (Dao De Jing 8; see Ames and Hall 2003: 87).

Second, influenced by the communitarian ethics manifested in the Confucian religious worldview, people in East and Southeast Asia have developed a collective culture that emphasizes group orientations and close social relationships. Perceived as part of a network of social relations, individuals find their own identities with reference to others around them and adopt group goals and opinions in exchange for reciprocal care and protection. The American culture of individualism is often regarded as inappropriate, for it represents a selfish and unnatural condition that isolates the self from the group and places personal interests over those of the group (Arcodia 2003; Gudykunst and Nishida 1994). Walker *et al.*'s (2001) research supports the above arguments, finding that autonomy/ independence was a significant motivation for Europeans and North Americans visiting a national park, while group membership and humility/ modesty were significantly more important motivations for Chinese Canadians visiting the same park.

Third, and related to the collective culture, Asian people in general and Chinese people in particular maintain strong family ties and intense social networks. Strongly influenced by Confucian values, this respect for family traditions and the worship of ancestors influence a range of travel behaviors and attitudes, including motivations for travel, time of travel, choice of destinations, decisions about who should travel first, and who should make travel decisions (Nguyen and King 2004). The strength of family ties and ancestor worship is clearly indicated in an axiom by Confucius that states: "Man should avoid traveling far while his parents are still alive; if he has to travel he should be clear about where he is going" (The Analects 4:19; see Legge 1991: 93). Although it may no longer be realistic to apply this ancient principle to all Asians today, family reunions are still a very important concern for those who are away from home. Taking China as an example, tens of millions of people who leave the countryside to work in the large urban centers travel back home for Chinese New Year via different modes of transportation. This causes a huge wave of short-term traffic that imposes pressures on the national transportation system every year.

Globally, family and community help provide a sense of home and belonging for those who live in the Chinese diaspora, which is maintained and strengthened through the medium of travel and tourism. Family ties provide security and order to the diasporic population, which often faces cultural alienation and insecurity in their host countries. The norms and obligations associated with family ties often precipitate trips and influence destination choices (Nguyen and King 2004). In a study of the travel behavior of Vietnamese migrants (*Viet kieu*) residing in Australia, for example, Nguyen *et al.* (2003) found that the *Viet kieu* maintain certain traditional Vietnamese cultural values and Confucian ideals, and these values and ideals influence their travel behaviors and attitudes. In a similar vein, Lew and Wong (2004) reported that for those overseas Chinese who practice Confucian values, existential travel becomes an important means of meeting one's social relationship (*guanxi*) obligations, and that filial piety in particular serves as a motivational element to travel back to China. Besides existential travel, people living in the diaspora also stay in touch with their homelands and maintain ethnic identities through nostalgic festivities and patriotic commemorations carried out in their new host countries (Prevot 1993, cited in Nguyen and King 2004).

In addition to these behaviors and characteristics influencing Asian leisure and tourism, Confucian and Daoist (and Shinto in Japan) shrines and temples are undeniably one of the most significant tourist attractions across East Asia. Nearly all tourists who visit Japan, Hong Kong, or China take advantage of opportunities to visit shrines and temples, and places such as Hong Kong and Japan typically emphasize these sacred places as important attractions for international visitors (*China Tourism* 1999; Nelson 1996; Yu 2000; Zhang 1995).

Tourism management and marketing

From the previous discussion of the effects of eastern spiritual philosophies on people's leisure and travel behaviors, it is not surprising to sense the existence of a variety of challenges and opportunities for tourism management and marketing. The first cohort of issues comes from the tensions between the individualistic and collective cultures. These challenges and opportunities center on how the tourism product or service is consumed and ways in which it may be experienced. Arcodia (2003) suggested that experiences that are group oriented and do not single out individuals may be more comfortable to travelers with a Confucianist mindset. Furthermore, the development of tourism products and services within Confucian cultures, or for Confucian outbound travelers, may have a better chance of commercial success if the negotiating process reflects the collective orientation of Confucian tourism consumers. In a study of the relationship between Chinese cultures and casino customer services, Galletti (2002) similarly suggested that casino managers should know that Chinese cherish group identity and group relations; they see themselves as a group, not as individuals. He recommended that, for example, it is more appropriate to offer a gift to the group as a whole rather than to one group member, or to give gifts to individuals in private and not in front of group members to avoid embarrassment.

The second cohort of challenges and opportunities lies in the recognition and understanding of the Confucian concept of protocol and its requirements for hierarchical respect in contemporary tourism management and marketing situations that attract Confucian-influenced business people. This is particularly critical when negotiating macro and micro international ventures with tourism professionals who may live in any of the countries with a Confucian heritage, but also for understanding some of the motivations behind political decisions that may impact favorably or unfavorably the global environment (Dirlik 1995, cited in Arcodia 2003). Practical examples of responding to Confucian protocol requirements would include the way Confucianist visitors are greeted and welcomed, decisions about seating arrangement at meetings, appropriate gift giving practices, and negotiation etiquette (Irwin 1996, cited in Arcodia 2003). Again, Galletti (2002) similarly suggested that casinos should develop training programs to increase staff awareness of Chinese cultural values and particularities, as well as communication methods so that appropriate protocol (e.g., seniority, rank, and title) can be followed to facilitate the interaction between staff and Asian customers.

The third cohort of challenges and opportunities is associated with diasporas and existential tourism. Existential tourism is the predominant model among the vast majority of ethnic Chinese living outside of China today (Lew and Wong 2004) as well as other ethnic Asian groups living in the diaspora (Nguyen and King 2004). Existential tourism provides a mechanism for the homeland to overcome the vast geographic spaces of diaspora and through face-to-face interactions convert shared ethnicity into social actions that lead to an enhanced quality of life for all. In view of the persistent importance of existential tourism and face-to-face interactions in Chinese society, Lew and Wong (2004) suggest that tourist programs might be properly designed and structured to allow interpersonal relationships to be created upon which social capital can evolve to benefit the homeland. Moreover, there is anecdotal evidence to suggest that these programs might benefit the host land as well. In the past few years, an increasing number of major cities worldwide, from New York to San Francisco and from Paris to Melbourne, have begun to host festival-type special events to celebrate Chinese New Year. In 2004, for instance, a huge Chinese New Year parade was held in Paris as part of France's "Year of China" promotion. Hundreds of Chinese citizens and thousands of Chinese émigrés, along with tourists from all over the world, joined the parade, watched the dragon dances, and participated in a variety of other social and cultural activities (Smith 2004).

The next cohort of challenges and opportunities is associated with the Confucian emphasis on balancing rights and responsibilities. Practical examples of responding to Confucian business and social ethics could include the way in which business relationships are nurtured, a focus on workplace rights and responsibilities in hotels that employ workers with a Confucian heritage, and methods by which tourism can be used to alleviate poverty (Arcodia 2003).

Finally, the revival of Oriental spiritual philosophies in China and other Asian countries has implications for the growth of tourism through the renovation of old pilgrimage centers and the building of new ones.

Lai (2003), for example, noted that since its restoration and reopening in the late 1980s, the Maoshan temple, a famous Daoist pilgrimage center in China, has attracted large numbers of pilgrims and other tourists and has become an important income source for the local community.

Resort and hotel design

The harmony of humankind and nature and the attachment of educational elements to leisure activities have implications for the design of leisure and tourism establishments. In their study, Gotõ and Ching (1998) discussed Confucian influences on two gardens designed a hundred years ago and separated by a considerable distance: Korakuen in Tokyo and Worlitzer Park near Berlin. Gotõ and Ching demonstrated how Confucian ideas affected the designs of these two parks, both of which stand out from other gardens of their time in their concern for the Confucian ideal of self-cultivation through the integration of history, literature, and art with nature. In the case of Japan, a new relationship between humans and nature was developed, which lent nature a practical and political dimension. In the case of the German park, the inseparability of humans and nature was also established. Rather than viewing nature as a force to resist or fight and brought under the control of human will, nature in the German garden was considered to be in harmony with the economy and with human social systems. These examples suggested the possible didactic roles gardens can play. The philosophical foundations underlying the design of these gardens have led to a heightened sensitivity to the importance of gardens and green space in society. These foundations still actively inform contemporary notions about self-cultivation through nature and the role of nature in urban environments in both East Asia and the West.

The design of leisure and tourism establishments is also likely to be influenced by various forms of popular religion. It may be argued, for example, that understanding *Feng Shui*, a significant component of Chinese popular religion, is important for the design, planning, and operation of hotels in the Asia-Pacific area (Hobson 1994; Zetlin 1995). Hobson (1994: 23) outlined five categories under which *Feng Shui* can affect a hotel and its management: location of the hotel; exterior physical design (e.g. the placing of windows and entrances, etc.); interior physical design (e.g. the placing of doors, rooms, and stairways; the layout and positioning of furniture; etc.); marketing of the hotel to those that believe in *Feng Shui*; and employees of the hotel who believe in the principles of *Feng Shui*. In a similar vein, *Feng Shui* might also need to be taken into consideration in resort development and casino floor plans and interior/exterior design. In fact, some casino companies have already started to hire *Feng Shui* consultants to ensure that the environment will be arranged accordingly. Galletti (2002) noted that many Chinese people do not like the Luxor Casino in Las Vegas because of its odd shape and the predominant light on top. From a *Feng Shui* point of view, the light could have negative consequences on people's health, while square rooms are preferred because unusual edges and odd shapes can disturb the flow of energy. While popular religion (and *Feng Shui* in particular) might need to be carefully reviewed in tourism-related businesses, it should be noted that all Chinese do not believe in *Feng Shui* and that following *Feng Shui* principles and guidelines can present developers and managers with considerable challenges in terms of design, operations, and budget (Hobson 1994).

Environmental ethics

The cosmology of the Oriental spiritual philosophies also offers insights to tourism professionals in terms of environmental ethics and sustainable tourism development. Both Confucianism and Daoism share the notion that the universe is an organic system of interdependent components that exist in balance with one another. Hence, all lives are interdependent, because one's life is sustained by relationships with other people, animals, plants, and a more general life-sustaining environment (Capra 1991). In contrast to the Western externalistic point of view on the environment, which is based on separation and confrontation between humankind and the world, the Oriental spiritual philosophies have developed an internalistic view of the environment that rests on interdependence and harmony. The internalistic view focuses on humans as the consummator of nature rather than the conqueror of nature, as a participant in nature rather than nature's predator (Chen 1995; Cheng 1986).

Many environmental scholars and activists have examined the Oriental spiritual philosophies, particularly Confucianism and Daoism, in search of insights to help with the current environmental crisis (e.g. Callicott 1987; 1994; Tucker and Berthrong 1998; Tucker and Grim 1994). Callicott (1994) discussed the great potential for the contribution of Confucianism and Daoism to deep ecology, ecofeminism and, more generally, to a global ecological consciousness and conscience. He suggested, for instance, the Daoist ideal of *Wu-Wei* may "serve as an infinitely adaptable pattern and venerable tradition of wise living for environmentally sound contemporary applications" (Callicott 1987: 75). Adler (1998) maintained that the Neo-Confucian concept of "moral responsiveness" (*ying*) addresses questions of human nature and destiny and questions of moral responsibility – two major themes of environmental ethics – in an integrated, systematic manner, and eases the tension between anthropocentric and ecocentric perspectives on environmental ethics.

There are several examples from the literature on the Oriental spiritual philosophies and their connections to ecological concerns. For example, Zhou Dun-Yi (1017–1073), a famous Confucian believer in Chinese history, refused to cut the grass in front of his window, arguing that he and the grass share a common nature. He suggested that the feelings of the grass and his own feelings were one and the same, which restrained him from cutting the grass. By extension, the feelings of the trees and the emotions of animals are also the same, as are the feelings of nature. From a religious perception of the nature of the world as a unified body comes the explicit prescription for action that suggests care and respect, not just for other people, but also for all living things (Taylor 1998: 53).

These emphases on the goodness of nature and humankind's stewardship over it have clear implications for tourism, particularly in relation to sustainable tourism development (Kalland 2002). Harmony, balance, and equity are three principles of sustainability that fit closely with the Eastern philosophical views of nature. One of the most prominent concepts involved in sustainable tourism development is the preservation of eco-logical integrity by utilizing scales and types of development that are less consumptive and by adopting management techniques that minimize the negative ecological impacts of tourism.

Conclusions

The spiritual philosophies of the Orient have clear and direct implications for tourism, and several relationships have been highlighted here. While it was not a major focus of this chapter, perhaps the most obvious relationship is the shrines and temples scattered throughout China, Hong Kong, Korea, Taiwan, and Japan that function as an important part of the tourism product of East Asia. These structures not only draw adherents of Confucianism and Daoism from across the region, but they also create an important part of the cultural landscape that appeals to non-Asian visitors as well (Peterson 1995).

Daoism, Confucianism, and other forms of Chinese popular religion clearly mediate the manifestations of tourism in East Asia, particularly in terms of human nature and social relations. These Eastern values are more about nature and ancestor worship, as well as social, environmental, and humanitarian ethics than they are about religious services and formal pilgrimages. The Oriental focus on humanity, nature, ordinary human exist-ence, social goodness, and the individual's place in society manifests unique travel patterns among Asians. For instance, travel to natural areas for communion with nature through exercise and meditation is common in the domestic context. Likewise, ancestor worship leads many people to visit shrines built to honor forebears as a way of beseeching the deceased for inspiration and help in matters related to marriage, business, social relations, and education. On an international level, such practices bring Asians of the diaspora from all around the world to visit the lands of their ancestors.

In the Western world, and in most of the other major religions, such as Christianity, Judaism, and Islam, the focus of spiritual growth and devotion is God and His kingdom. In most cases life's journey and the search for God's kingdom is an individual course based on one's faith and good deeds. In addition to religion, the growing materialism of the Western world feeds the socio-cultural roots of individualism even for people who claim no membership in any official religious organization. The Eastern philosophies, however, are an antithesis to these practices. Instead, life's journey is one of highly group-oriented structures, where individuals understand their place in the larger social order and how their individual actions will reflect on the group to which they belong. In this sense, then, they follow a set of communitarian ethics wherein there is no separation between individuals and society. This is perhaps most apparent in the preference for group tours among Chinese, Japanese, and Korean people instead of exclusive individual travel, which is more common among Western culture groups.

Finally, this chapter has argued that the social/human focus of Confucianism and the ecological core of Daoism heavily influence the management of tourism throughout East Asia, particularly in terms of social, environmental, and business ethics. Resort design and staff relations are two examples where this influence is most obvious. There is a great deal of potential for the spiritual philosophies of the East to inform tourism decision makers, destination planners, business managers, and marketing specialists in developing tourism in a socially and ecologically sustainable manner.

References

Adler, J.A. (1998) "Response and responsibility: Chou Tun-I and Confucian resources for environmental ethics," in M.E. Tucker and J. Berthrong (eds) *Confucianism and Ecology: The Interrelation of Heaven, Earth, and Humans*, Cambridge, MA: Harvard University, Center for the Study of World Religions.

Ames, R.T. and Hall, D.L. (2003) *Dao de jing: A Philosophical Translation*, New York: Ballantine Books.

Arcodia, C. (2003) "Confucian values and their implications for tourism industry," paper presented at the annual CAUTHE Conference, Coffs Harbour, Australia.

Callicott, J.B. (1987) "Conceptual resources for environmental ethics in Asian traditions of thought: a propaedeutic," *Philosophy East and West* 37: 115–130.

Callicott, J.B. (1994) *Earth's Insights: A Survey of Ecological Ethics from the Mediterranean to the Australian Outback*, Berkeley: University of California Press.

Campbell, J. (1962) *The Masks of God: Oriental Mythology*, New York: Viking Press.

Capra, F. (1991) *The Tao of Physics: An Exploration of the Parallels between Modern Physics and Eastern Mysticism* (3rd edn), Boston: Shambhala Publications.

Chen, E.M. (1995) "Taoism and ecology," *Dialogue and Alliance* 9(2): 5–15.

Cheng, C. (1986) "On environmental ethics of the two: Tao and the Ch'I," *World and I* 1: 577.

Chiang, A. (2003) *The Elements: A Brief Look at Chinese Folk Religion*, available at www.fccj.edu/library/chi-reli/chi-elem.htm (accessed October 19, 2004).

China Tourism (1999) "Religious culture and art: crystallisation of Buddhism, Taoism and Confucianism," *China Tourism* 233: 66–71.

Ching, J. (2001) "The ambiguous character of Chinese religion(s)," *Studies in Interreligious Dialogue* 11(2): 213–223.

Ching, J. (2003) "What is Confucian spirituality?," in W. Tu and M.E. Tucker (eds) *Confucian Spirituality* (Volume 1), New York: Crossroad Publishing.

Chi-Ping, Y. (1989) "Theology of filial piety: an initial formulation," *Asian Journal of Theology* 3(2): 496–508.

Clarke, I. (2000) "Ancestor worship and identity: ritual, interpretation, and social normalization in the Malaysian Chinese community," *Sojourn* 15(2): 273–295.

Cohen, E. (1992) "Pilgrimage centers: concentric and excentric," *Annals of Tourism Research* 19: 33–50.

Collins-Kreiner, N. and Kliot, N. (2000) "Pilgrimage tourism in the Holy Land: the behavioural characteristics of Christian pilgrims," *GeoJournal* 50(1): 55–67.

Dean, K. (2003) "Local communal religion in contemporary south-east China," *The China Quarterly* 174: 338–358.

Dirlik, A. (1995) "Confucius in the borderlands: global capitalism and the reinvention of Confucianism," *Boundary* 2(2): 229–273.

Elman, B.A., Duncan, J.B., and Ooms, H. (eds) (2002) *Rethinking Confucianism: Past and Present in China, Japan, Korea, and Vietnam*, Los Angeles: University of California, Los Angeles.

Fan, L. (2003) "Population religion in contemporary China," *Social Compass* 50(4): 449–457.

Galletti, S. (2002) "Chinese cultures and casino customers services," *Business Times* 1 November, available at www.jackpots.com.sg/article (accessed October 4, 2004)

Gotõ, S. and Ching, J. (1998) "Confucianism and garden design: a comparison of Koishikawa Kõrakuen and Wörlitzer Park," in M.E. Tucker and J. Berthrong (eds) *Confucianism and Ecology: The Interrelation of Heaven, Earth, and Humans*, Cambridge, MA: Harvard University Center for the Study of World Religions.

Gudykunst, W.B. and Nishida, T. (1994) *Bridging Japanese/North American Differences*, Thousand Oaks, CA: Sage.

Hansen, C. (2003) "Taoism," in E.N. Zalta (ed.) *The Stanford Encyclopedia of Philosophy* (Spring 2003 edn), available at www.plato.stanford.edu/archives/spr2003/tntries/taoism (accessed September 15, 2004).

Harrell, S. (1979) "The concept of soul in Chinese folk religion," *Journal of Asian Studies* 38(3): 519–528.

Henderson, P.W., Cote, J.A., Leong, S.M., and Schmitt, B. (2003) "Building strong brands in Asia: selecting the visual components of image to maximize brand strength," *International Journal of Research in Marketing* 20: 297–313.

Henkin, L. (1986) "The human rights idea in contemporary China: a comparative perspective," in R.R. Edwards, L. Henkin, and A.J. Nathan (eds) *Human Rights in Contemporary China*, New York: Columbia University Press.

Hobson, P. (1994) "Feng Shui: its impacts on the Asian hospitality industry," *International Journal of Contemporary Hospitality Management* 6(6): 21–26.

Holley, D. (1989) "Revival in China: resurgence of traditional folk religion sweeping countryside," *Los Angeles Times* 19 August: 6.

Hwang, K.K. (1999) "Filial piety and loyalty: two types of social identification in Confucianism," *Asian Journal of Social Psychology* 2(1): 163–183.

Irwin, H. (1996) *Communicating with Asia*, St Leonards: Allen & Unwin.

Kalland, A. (2002) "Holism and sustainability: lessons from Japan," *Worldviews* 6(2): 145–158.

Kim, D. (1997) "Suwon's city wall: monument to filial piety," *Koreana* 11(1): 17–27.

Kohn, L. (2001) *Daoism and Chinese Culture*, Cambridge, MA: Three Pines Press.

Kolodner, E. (1994) "Religious rights in China: a comparison of international human rights law and Chinese domestic legislation," *Human Rights Quarterly* 16(3): 455–490.

Lai, C. (2003) "Daoism in China today, 1980–2002," *The China Quarterly* 174: 413–427.

Legge, J. (trans) (1991) *The Chinese English Four Books*, Changsha, China: Hunan Publishing House.

Lew, A.A. and Wong, A. (2004) "Sojourners, *guanxi* and clan associations: social capital and overseas Chinese tourism to China," in T. Coles and D.J. Timothy (eds) *Tourism, Diasporas and Space*, London: Routledge.

Li, L. (2000) "Ancestor worship: an archaeological investigation of ritual activities in Neolithic north China," *Journal of East Asian Archaeology* 2(1/2): 129–164.

Little, R. and Reed, W. (1989) *The Confucian Renaissance*, Sydney: The Federation Press.

Nelson, J.K. (1996) "Freedom of expression: the very modern practice of visiting a Shinto shrine," *Japanese Journal of Religious Studies* 23(1/2): 117–153.

Nguyen, T.H. and King, B. (2004) "The culture of tourism in the diaspora: the case of the Vietnamese community in Australia," in T. Coles and D.J. Timothy (eds) *Tourism, Diasporas and Space*, London: Routledge.

Nguyen, T.H., King, B., and Turner, L. (2003) "Travel behavior and migrant cultures: the Vietnamese in Australia," *Tourism Culture & Communication* 4(2): 95–107.

Overmyer, D.L. (2003) "Religion in China today: introduction," *The China Quarterly* 174: 307–316.

Peterson, Y.Y. (1995) "The Chinese landscape as a tourist attraction: image and reality," in A.A. Lew and L. Yu (eds) *Tourism in China: Geographic, Political, and Economic Perspectives*, Boulder, CO: Westview.

Prevot, H. (1993) *Social Policies for the Integration of Immigrants: The Changing Course of International Migration*, Paris: OECD.

Rinschede, G. (1992) "Forms of religious tourism," *Annals of Tourism Research* 19: 51–67.

Schipper, K.M. (1994) "Sources of modern popular worship in the Taoist canon: a critical appraisal," in *Proceedings of International Conference on Popular Beliefs and Chinese Culture*, Taipei: Center for Chinese Studies Research Series

Schmitt, B.H. and Pan, Y. (1994) "Managing corporate and brand identities in the Asia-Pacific region," *California Management Review* 36: 32–48.

Schrecker, J. (1997) "Filial piety as a basis for human rights in Confucius and Mencius," *Journal of Chinese Philosophy* 24: 401–412.

Shaw, E. and Curnow, T. (eds) (2004) *Electronic Encyclopedia of Religion*, available at www.philtar.ucsm.ac.uk/encyclopedia (accessed September 18, 2004).

Smith, C.S. (2004) "With fanfare and a grand parade, Paris celebrates France's ties to China," *New York Times* 25 January: 10.

Smith, V.L. (1992) "Introduction: the quest in guest," *Annals of Tourism Research* 19: 1–17.

Taylor, R.L. (1990) *The Religious Dimensions of Confucianism*, Albany: State University of New York Press.

Taylor, R.L. (1998) "Companionship with the world: roots and branches of a Confucian ecology," in M.E. Tucker and J. Berthrong (eds) *Confucianism and Ecology: The Interrelation of Heaven, Earth, and Humans*, Cambridge, MA: Harvard University, Center for the Study of World Religions.

Timothy, D.J. (2002) "Sacred journeys: religious heritage and tourism," *Tourism Recreation Research* 27(2): 3–6.

Tu, W. (1985) *Confucian Thought: Selfhood as Creative Transformation*, Albany: State University of New York Press.

Tu, W. (ed.) (1996) *Confucian Transitions in East Asian Modernity: Moral Education and Economic Culture in Japan and the Four Mini-Dragons*, Cambridge, MA: Harvard University Press.

Tu, W. (1999) "Confucius: the embodiment of faith in humanity," *World and I* 14(11): 292–306.

Tucker, M.E. (2003) "Introduction," in W. Tu and M.E. Tucker (eds) *Confucian Spirituality* (Volume 1), New York: Crossroad Publishing.

Tucker, M.E. and Berthrong, J. (eds) (1998) *Confucianism and Ecology: The Interrelation of Heaven, Earth, and Humans*, Cambridge, MA: Harvard University Center for the Study of World Religions.

Tucker, M.E. and Grim, J.A. (1994) *Worldviews and Ecology*, Maryknoll, NY: Orbis Books.

Vukonić, B. (1996) *Tourism and Religion*, Oxford: Pergamon Press.

Walker, G., Deng, J., and Dieser, R. (2001) "Ethnicity, acculturation, self-construal, and motivations for outdoor recreation," *Leisure Sciences* 23: 263–283.

Wang, J. and Stringer, L.A. (2000) "The impact of Taoism on Chinese leisure," *World Leisure* 42(3): 33–41.

World Tourism Organization (2001) *Tourism 2020 Vision: Executive Summary*, Madrid: WTO.

World Tourism Organization (2003) *Tourism Highlights*, Madrid: WTO.

Yao, X. (2000) *An Introduction to Confucianism*, Cambridge: Cambridge University Press.

Yu, B. (2000) "The mystery of Qaidam Basin," *China Tourism* 234: 44–63.

Zetlin, M. (1995) "Feng Shui: smart business or superstition?," *Management Review* 84(8): 26–27.

Zhang, Y. (1995) "An assessment of China's tourism resources," in A.A. Lew and L. Yu (eds) *Tourism in China: Geographic, Political, and Economic Perspectives*, Boulder, CO: Westview.

10 Nature religion, self-spirituality and New Age tourism

Dallen J. Timothy and Paul J. Conover

Since the mid-twentieth century rapid modernization and technological growth in the western world have brought with them fast-paced consumer societies, where people get caught in a time crunch feeling stressed and burned out. Few people in the developed world have time to relax and appreciate nature, develop personal interests, and improve their mental and spiritual health (Lengfelder and Timothy 2000; Schor 1993). At the same time, for various reasons there has been a wave of dissatisfaction with some aspects of traditional organized religion, resulting in breakaway sects, changes in theological viewpoints, transformations of politico-religious views, and varying levels of adherence (Allitt 2003; Houtman and Mascini 2002). These two factors, the frenetic pace of contemporary life, and varying levels of commitment to traditional religion, have caused people to seek alternative lifestyles and spiritual worldviews, particularly in the United States, Canada, the United Kingdom, Japan, and many other developed countries.

One solution to many people's uneasiness has been a turn toward the New Age movement, which stresses the sanctity of nature, harmony of the cosmos, resurrection of ancient spiritual traditions, and self improvement in the realms of spirit, mind, and body. The New Age faction has grown significantly since its emergence in the 1950s and 1960s. Owing to New Agers' beliefs in earth powers, nature spirits, extra-terrestrial visitations, and self-spiritual enhancement, they are passionate travelers and devote much of their effort and income to visiting places associated with spiritual growth. The aim of this chapter is to examine the tenets of the New Age spiritual movement in terms of nature and self religion and how these play out in tourism terms. First the chapter describes the New Age movement and its association with other "alternative religions," followed by an examination of tourist-related activities that believers undertake. New Age spirituality is extremely place-specific, and therefore manifests in tourism terms as being very enthusiastic about pilgrimages of various sorts. Several of the most important sacred locations in the world and the reasons for their spiritual significance are highlighted. Finally, this chapter discusses the main social and ecological concerns associated with the New Age movement as a result of its newness and reliance on pristine environments.

Nature religion, self-spirituality and the New Age

Several spiritual philosophies, or religious orders, focus not on an individual god or multiple gods, but rather channel devotion to the earth and the realm of nature. These so-called 'nature religions' or 'earth religions' advocate respect for the universe and harmonious human-ecology relationships (Hooper 1994; Ibrahim and Cordes 2002). Animism, the belief that spirits inhabit everything in nature (plants, minerals, air, mountains, water, earth, fire, and animals), Native American and other indigenous belief systems, neo-paganism, and East Asian Daoism are prominent examples.

Nature-based belief systems focus on humans as part of nature, not separate from it. From this viewpoint, humans do not have dominion over the earth, nor should they control it. Instead, they are simply one part of a larger system wherein all things on the earth, including the earth itself, interact, have spirits, live, and are capable of feeling (Ball 2000; Charlesworth 1998; Dobbs 1997; Forbes 2001; Gottlieb 1996; Marty 1997; Tapia 2002; Turner 1991; Tyler 1993). Many people in traditional societies and among the pagan sects, worship an Earth Goddess (Gaia, or Earth Mother), who keeps harmony among all living things and whose healing powers are found in sacred locations throughout the world. Such belief systems are common in indigenous societies throughout the world and are believed to be one of the oldest forms of worship.

In most Native American traditional worldviews, there are three basic principles. First, all nature is sacred, including everything dwelling on the earth. Second, even pests and dangerous creatures have their role to play in the circle of life and can teach many important lessons. Finally, everything has a right to live. If a life must be taken (e.g. for food), it is usually taken only after asking the spirits for forgiveness (Albanese 1990; Redmond 1996). Essentially, Native Americans and other indigenous groups have spiritual-based explanations for nearly everything they encounter in nature. They often select unique geomorphological features imbued with their own spirits as places of spiritual energy (e.g. canyons and mountains) (Ball 2000; Ivakhiv 2003).

Neo-paganism is generically a set of related modern beliefs informed and inspired by ancient (pre-Christian) religious practices. Paganism respects nature and weaves in ancient British traditions and various ideas of individualism. Pagan worshippers believe that some archeological sites are sacred places, where the spirit of place inspires meditation, celebration, and communication with deceased ancestors and other spirits (Powell 2003: 36). Pagan groups include Druids, Wiccans (modern witches), and other sects of witchery, and they all have in common various uses of ancient mythology, Goddess (Earth Mother) veneration, magic, reincarnation, and reverence for nature and active ecology. Paganism has seen considerable growth in recent years, and some estimates place adherents at around one million, a quarter of whom live in the United Kingdom (Powell 2003).

A branch of paganism, Druids in ancient times were comprised of a pre-Christian intellectual and religious caste in the British Isles, thought to have originated from Noah's family following the flood. They worshipped Celtic gods and various elements of nature, including the sun, and viewed the earth as divine (Almond 2000). However, with the Roman Christianization of Great Britain, Druidism was supplanted by Christian doctrines and beliefs. Modern Druidism represents the eighteenth-century revival of the ancient order, which centers on pre-Christian gods and goddesses, and wherein the sun is seen to represent divine light. Over the years, many splinter groups have formed several factions of Druidism, and today there are approximately 10,000 Druids in Great Britain (English 2002: 8).

The New Age movement, which falls under the rubric of "alternative spirituality" or "secular religion," and draws heavily from neo-paganism and other forms of nature worship, began to flourish in the United States during the 1970s in response to the counter-cultural mysticism of the 1960s (Hanegraaff 1998, 1999; Ivakhiv 1997; Kyle 1995; Tucker 2002). It rapidly spread to Europe and other parts of the world, including Australia, New Zealand, and Japan (Brodin 2003; Shimazono 1999), fueled by people's dissatisfaction with life, growing stress levels, and struggles with the fast pace and uncertainties of contemporary society (Aldred 2000; Lengfelder and Timothy 2000; Reisinger 2006). Some recent and very broad estimates suggest that there are between ten and twenty million New Agers throughout the world, although precise numbers are difficult to determine (Aldred 2000).

The so-called 'New Age' is typically seen as both a timeframe and a personal journey toward spiritual transformation. As a timeframe, it is seen as the years from the 1960s onward – a period of enormous change when people have, and will continue to live harmoniously with each other and with nature (Aldred 2000; Spangler 1996).

New Age spirituality is not a religion in the formal and organizational sense. Instead, it represents a personal spiritual quest that typically eschews traditional monotheistic religions. New Age philosophies tend to concentrate on what is not associated too closely with traditional theologies and churches (Hanegraaff 1999). O'Neil (2001: 456) describes New Age spirituality as a movement rather than a religion or sect, because, in common with other nature religions, there is no "structural . . . ecclesiastical institution, but instead a proliferation of classes, worships, and seminars focusing on some aspect of New Age teaching." It constitutes an eclectic set of beliefs and practices and is not led by any authority that declares or interprets rituals, canons, or the parameters of orthodoxy. Instead, each person must work out for him/herself what applies individually from a wide range of inspirational writings (Aldred 2000; Hanegraaff 1999; O'Neil 2001; Tucker 2002). Others have referred to it as a spiritual "subculture" (e.g. Attix 2002).

The New Age movement may be seen as both a form of nature religion and self-spirituality. It embraces the oneness of humanity, nature, and the cosmos and is essentially animist in perceiving that the earth and the cosmos are alive and conscious (Albanese 1990; O'Neil 2001). Goddess worship, in the form of Earth Mother (Gaia) has become a popular medium for supplicating the divine spirit of the earth (Ivakhiv 2003; Rountree 2002). For New Age adherents, nature promotes spiritual growth and provides mystical oneness, mysteries beyond the ordinary, and transcendence beyond the limitations, structures, and laws of the physical world (Davis 1996; Drovdahl 1991; Knopf 1988).

The New Age movement merges nature religion with self/individual religion (as opposed to god religion) in that each person must find his or her own subjective truth, for what is true for one person might not be true for everyone (O'Neil 2001; York 2001). In this sense, in "individual religions ... the individual institutes for himself [or herself] and celebrates for himself [or herself] alone" (Durkheim 1995: 44). Self, rather than an objective deity, is thus seen as the symbolic center of New Age religion (Heelas 1996). Each person's duty is to find some form of god and truth within, which is entirely possible, since each person is considered divine. Despite individual divine nature, however, everyone is incomplete and in need of progressive healing from cosmic imbalances and past life wrongdoings. As a result, believers often speak of the need to develop a connected self, true self, whole self, or integrated self (Tucker 2002: 47). The self-god perceptions are enhanced by heavy emphasis on self-esteem development, positive thinking, self-improvement, mental and physical health, and spiritual transformation. Adherents are decisively non-judgmental and generally believe there is no sin. Subsequently there is no supreme being who exercises authority over humans, but rather every person has his/her own gods and spirits who function as friends and helpers (Tucker 2002: 49).

New Ageism is a way of gaining personal meaning in life and effecting self-transformation by living a simpler, more ecologically oriented lifestyle (Ivakhiv 1997). Adherents find value in borrowing freely from ancient cultures of the Americas and East Asia, including elements and symbols of Daoism, Buddhism, and Native American spiritual practices because these are seen as being more harmonious with nature and self-spirituality (Cooper 1999; Hanegraaff 1999; York 2001).

New Agers endorse the use of chants, meditation, yoga, astrology, tarot cards, Ouija boards, fortune telling, palm reading, channeling spirit beings, shamanism, past-life regression, extraterrestrial communication, séances, Tai Chi, out-of-body experiences, acupuncture, and crystals to achieve their goals of self-transformation and holistic living. The body is viewed in energy terms, and a healthy body will allow a freer flow of life energy. It involves holistic, vegetarian, organic, and naturopathic approaches to eating and health care and criticizes conventional medicine (Albanese 1990; Aldred 2000; O'Neil 2001; Reisinger 2006).

This self-centered approach to spiritual living does not, as traditional religion does, bind people to larger social groups or require submission to a higher authority. In fact, according to Tucker (2002: 50), "it does just the opposite ... New Agers mostly reject the social world and any kind of authority beyond the self." Tucker goes on to note that this worldview creates a fairly distinct demographic associated with the movement. In his study, Tucker found that most New Agers are middle-class, white women between 30 and 50 years old. Most have been married at least once, but they tend to have few long-term relationships of any kind. Many have even severed all ties with parents and siblings, and those who are married tend to view marriage as a more open relationship than traditional marriages. In common with their romantic relationships and friends, they tend to change jobs and residences with regularity (Tucker 2002: 50).

New Age spirituality and tourism

New Agers and other nature and self-religionists comprise a major world market for tourism. Millions of New Age trips are taken every year and typically involve activities and destinations that teach people to become more aware of themselves, spiritually tuned, and less materialistic (Brown 1998; Reisinger 2006). With the growth in New Age spirituality and the noteworthy levels of travel that follow, a growing number of tour operators have discovered this valuable market during the past twenty years (Cogswell 1996; Lange 2001). Likewise, several US states have begun to realize the potential economic impact of this lucrative niche and are trying to promote themselves as centers of spiritual energy and important destinations for New Agers. A recent promotional campaign in New Mexico, for instance, attempted to lure travelers to the state's native and energy sacred sites to discover themselves and receive enlightenment (Associated Press 2002).

The growing popularity of New Age tourism has resulted in the proliferation of many guidebooks and websites in recent years to lead people to sacred sites where earth's energy grids abound (Attix 2002). Many books focus on how to do Native American rituals, where to go for optimum energy flows, and ways to explore oneself in places where earth's energy flows abundantly (Andres 2000; Attix 2002; Barlow 1997; Lamb and Barclay 1987). Some even demonstrate, as in the case of Dannelley's (2000) guide to Sedona's sacred energy, the best locations for viewing UFOs. Barlow's (1997) handbook is probably the most comprehensive and lists dozens of sites throughout the western United States that have some kind of spiritual significance for Native Americans and New Agers.

As noted earlier, New Age travel is very place-oriented. During the counter-culture of the 1960s and early 1970s, many people began to adopt elements of mysticism and spiritual philosophies, including Druidism, into their "alternative lifestyles." A strong interest grew in "earth mysteries," including theories of sacred geography and geometry, which focused on powerful energies and unseen forces at ancient places. According to one perspective, powerful archeological sites are aligned with one another, representing linear, prehistoric power lines, which radiate earth's energies. These ancient places "became important to some groups who viewed prehistoric monuments as living places imbued with sacred energy and not as relics from a completed past" (English 2002: 8). For New Age spiritualists, sites of ancient ritual are among the most important destinations because they are believed to have been built in accordance to the energies of nature.

> Essential to spiritual travel is the "sacred site" – sometimes known as a "power place" – a spot endowed, for a variety of reasons, with a special dollop of genius loci. There may be mysterious geological features, such as artesian wells or anomalous currents, enchanted groves or cliffs with strange carvings. Perhaps Druid sorcerers are buried there, or Incan kings and queens, or Pythian sibyls, or a great saint or bodhisattva. Maybe there are pyramids, monasteries, dolmens, sacred serpent mounds. Whatever the reason, these sites – including Delphi, the Egyptian Pyramids, Machu Picchu, parts of the American Southwest and the entire island of Bali, to name a few – are seen as cracks in the universe where the eternal is revealed.
>
> (Hooper 1994: 72)

As a result of ancient location veneration, among the most common New Age destinations are pre-Christian era sites, such as Stonehenge (especially on the summer solstice), the Pyramids of Egypt, Machu Picchu, the Neolithic ruins of southwestern England, and Easter Island. Pagans and New Agers view prehistoric sites as central to their belief systems, and access to places, including Native American ruins, is considered an essential part of their worship (Cogswell 1996; Coles 2003; Duffaut 1998; English 2002).

New Age tourism is also uniquely defined by the activities and types of trips undertaken, combining many elements of cultural, religious, nature-based, and health tourism (Gee and Fayos-Solá 1997). Most New Agers are deep tourists in the sense that they do not simply go to see a place, lay out on the beach or take pictures – they participate and become part of the destination through meditation, prayer, and other rituals (Attix 2002; Ivakhiv 2003; Reisinger 2006; Timothy 2002). This is evident in tour packages that cater specifically to the needs of New Agers. According to one spiritual tour operator (cited in Hooper 1994: 72), his tours are

> not for people who just want to have a good time ... when we go to these sacred places we don't just go there to wave at the Dalai Lama

... We work with some of his head monks, who train us for three days in the Tibetan Buddhist tradition. We stay in the monastery and eat with the monks, sleep with them, meditate with them at three in the morning.

At least four types of New Age tours or activities can be identified: education, health, spiritual growth/personal development, and volunteer. Educational tourism, or "edutourism," as Strutt (1999) calls it, entails traveling for the purpose of learning and getting hands on experience. Many of these focus on photography, art, kayaking, gardening, weaving, Gaelic music and dance, and studying ancient languages. These "alternative vacations" educate and allow people to develop specific skills and "are aimed at immersion rather than observation" (Strutt 1999: 17).

Health holidays, or "holistic tourism," focus on activities such as yoga and spa treatments, shaman visits, metaphysics, tarot, nature hiking, reflexology, crystal healings, meditation, and aromatherapy. Unlike most forms of mass tourism that focus on escapism, holistic holidays emphasize getting travelers to engage with their inner selves and reconcile internal discord through deeper personal and spiritual experiences (Lange 2001; Smith 2003; Timothy 2002). These fitness and healing tours operate on the notion that good physical health leads to strong spiritual health (Cogswell 1996).

While all forms of New Age travel ultimately aim to increase spiritual growth, there are various types of journeys that concentrate more overtly on the spiritual elements. Specific rituals and activities, such as attending seminars and workshops, praying with Tibetan monks at a monastery in the Himalayas, meditating in Glastonbury, England, or undergoing a ceremonial death and rebirth inside one of the pyramids in Egypt are common examples. Package tours of a more spiritual nature are often guided by experienced New Age authors, shamans, or other revered persons and teach people to slip into deeper dialogue with nature, to receive earth's powers and show the best way to live their lives (Hooper 1994).

The fourth form of New Age organized travel involves volunteering in service-oriented endeavors, primarily in the developing world (Smith 2003; Strutt 1999). Teaching people to read, practice personal hygiene, build houses, and grow vegetables are typical volunteer activities.

New Age destinations

Ivakhiv (2003: 99) suggests that New Age travelers can be divided into "mere tourists" and "genuine pilgrims," the difference being that the mere tourists plan to return home rather quickly, while the genuine pilgrims drift from place to place, seeking longer-term spiritual connections. The latter form of New Age travel, Ivakhiv argues, gives rise to the development of networks of healing centers, spiritual communities, retreats, and places of New Age commerce. Perhaps the best two examples of this are Sedona, Arizona (US), and Glastonbury (UK).

Sedona was formally established in 1902 as a farming community. Its population grew steadily following the Second World War, but in the 1970s and early 1980s, when psychics identified the power spots, or vortexes, and particularly the Harmonic Convergence in 1987 there, the population began to boom with people moving in to be close to their "elective centers," or earth power sources (Ivakhiv 2003). Today the community has approximately 17,000 residents, many of whom (*c*.15 percent) gave up well-paying urban employment to move to Sedona, where they live simpler lives, working primarily in manual labor jobs that allow them to work closely with nature (O'Neil 2001), although many of the town's New Agers are also involved in alternative health, psychic services, spiritual guiding, tour guiding, real estate sales, and other service industries (Ivakhiv 1997). These "pilgrim-migrants" are an important part of the economic and social life of the community and outside of work spend much of their time hiking, meditating, seeing psychics and other spiritual counselors, chanting, channeling spirits, and conducting rituals inside stone circles (medicine wheels) (Ivakhiv 1997, 2003).

While not all tourists to Sedona are New Agers, many of the four million visitors each year are, and they seek out the same activities as the pilgrim-migrants who live there now. Allen (1999) refers to Sedona as the "New Age Lourdes," and in popular lexicon it is commonly referred to as the "capital of New Age tourism."

Sedona and its surrounding region have long been seen as sacred territory by the Yavapai and Apache Indian tribes owing to its dramatic topography and colorful sandstone, which have many ancient legends associated with them (Ruland-Thorne 1993). In the eyes of New Agers, the red rock landscape of Sedona is sacred and powerful, a place of high energy and source of spiritual strength and guidance (Ivakhiv 1997).

In the early 1980s, a famous psychic, Page Bryant, proclaimed the existence of several intense energy "vortexes" in Sedona, making it one of the highest concentrations in the world. Following this proclamation, New Age tourism began to boom (Allen 1999; Attix 2002; Lansky 2001; Lengel 1999; Shorey 2002). It became even more well known when another psychic/writer, Arguelles, predicted a "harmonic convergence," which would occur in August 1987. This convergence of earth energy would be felt strongly in Sedona and many other sacred locations around the world, including Stonehenge, Machu Picchu, and the Pyramids of Egypt. In Sedona alone, more than 10,000 pilgrims converged from around the world to experience what Arguelles had envisioned (Allen 1999).

According to Page Bryant, the famous Sedona psychic, vortexes are the "points at which [the earth's] energy currents meet or become coagulated into funnels of energy" (quoted in Ivakhiv 1997: 373). Likewise, McGivney and Archibald (1997: 46) describe vortexes as "natural power spots where psychic energy gushes from the earth like a geyser" and are usually manifested in areas of "tremendous natural beauty created by the elements of land, light, air and water" (Andres 2000: 12). The recognition of the vortexes in Sedona is based on Gaia principles. If the earth is alive, then one can assume that life-energy flows through its body and that there are bound to be "energy centers" where the flow is more intense and concentrated than elsewhere (Ivakhiv 2003: 111). Several ancient monuments, such as Stonehenge and the Eygptian Pyramids, are said to be built on energy vortexes.

The outpourings of energy associated with vortexes act as amplifiers; that is, they magnify, or amplify, what each individual brings to them in terms of physical, mental, or spiritual desires (Andres 2000; Sanders 1992). A vortex, therefore, ultimately influences each person differently based on his/her needs and desires. Each of Sedona's vortexes is said to be a little different from the others – some help old memories resurface, while others enhance artistic abilities and help with decision making (Shorey 2002).

An indirect result of the vortexes is the existence of UFOs, which are also an important part of the New Age product of Sedona. Many people report seeing spirits of ancient Indians, non-human races from outer spaces, UFOs, and mysterious "rock people" (Allen 1999). UFO sightings are a common occurrence, and many of the faithful believe there may be an underground space terminal at Sedona (McGivney and Archibald 1997).

Tours in Sedona focus overwhelmingly on the vortexes and the potential for extraordinary/extraterrestrial experiences (Ivakhiv 1997, 2003). The town is sated with tour companies that offer four-wheel drive, off-road tours ($30–$100) to sacred shrines, areas of intense energy flows, Native American ruins, and mystic canyons. Likewise, dozens of businesses offer palm and tarotcard readings, past-life interpretation, spirit channeling, spas and naturopathic healing. Native American-style encampments, or workshops, are also held periodically and draw people from around the world. Encampment activities include drumming circles, medicine wheel ceremonies, sweat lodges, and lectures by well-known New Age promoters and self-proclaimed shamans.

Visitor information centers in the area pass out information on tours, psychic services, and other New Age-related activities. There is also a Center for the New Age, which provides a focal point for the metaphysical community and acts as a quasi-chamber of commerce. Many guidebooks have also been published in recent years to direct New Age visitors and residents around Sedona and advise where and how best to gain inspiration from the vortexes (Andres 2000). The cover of one booklet assures that visitors will learn how to "tune in, find your personal power place, and take the magic home!" (Lamb and Barclay 1987).

Glastonbury, England, is Europe's center of New Age spirituality. In common with Sedona, self and nature religious adherents come to this medieval market town to mingle with neo-pagans, occultists, theosophists, goddess worshipers, and extraterrestrial contactees (Digance and Cusack 2001; Hutton 1992; Ivakhiv 2003; Jones 1991; Rountree 2002). They visit to experience the "convergence of earth energies brought about by Glastonbury's key position on several major ley lines, interact with a sacred site of remote Celtic antiquity and participate in the worship of the Goddess" (Digance and Cusack 2002: 264).

The New Age movement in Japan is known as "World of the Spiritual" and essentially follows the same practices and belief systems as those of North American and European New Age, although with slightly more influence from China and India. In tourism terms, followers of the World of the Spiritual also tend to travel in search of truth and spiritual enlightenment. While they do sometimes travel to other parts of the world (e.g. Sedona and Stonehenge), most of their worship takes place in Japan and other parts of East Asia. One example is Tenkawa Benzaitensha, a famous pilgrimage site of the World of the Spiritual and an old shrine located in the mountains of Yamato in central Japan. Since its emergence as an important shrine in the 1980s, it has become the center of the Japanese New Age movement, endowed with mystical powers. Like Sedona, it attracts thousands of spiritual mediums and psychics, eccentric artists, and young people on spiritual quests (Shimazono 1999).

Machu Picchu is a popular destination for New Agers from the world over who value its ancient role as a ceremonial site (as some scientists believe) and an abode of Inca high priests and holy Virgins of the Sun (Barnard 1993). New Age pilgrims believe that the white granite of Machu Picchu vibrates with earth energy. Meditation, shaman-led chants, ancient medicine rituals, fire ceremonies, water purification rites, healings, and pipe ceremonies with indigenous leaders are among the most popular activities here (Hooper 1994).

Two additional sacred sites are Stonehenge and Avebury Circle in southwestern England. As part of their sun veneration, Druids, Wiccans, and other neo-pagans began to worship the solstice at Stonehenge in the late nineteenth century. Druids in particular see Stonehenge and Avebury as the center of the ancient Druidic tradition and an important part of their identity, and many believe the mysterious henges scattered throughout Britain were original Druidic temples and played an important astronomical role in the lives of their builders (Almond 2000; Bender 1999; Darvil 1997). While this is only one view of their origins, pagans, and by extension New Agers, see these sites as their spiritual heritage and expect to have free access to them. Nonetheless, English Heritage, the public body that controls Stonehenge and several other neolithic henges, began curtailing access to Stonehenge in the late 1970s as a result of wear and tear on the physical structure. This led to major confrontations between pagan worshippers and government officials/police in 1984. As a result, the UK government restricted access to the site, not allowing believers to celebrate the solstice as they had done since the 1800s. Following years of dispute and negotiations, the monument was re-opened in 1998 for limited use on summer solstice, but in 2000 open worship by Druids, Wiccans and New Agers was allowed, albeit with highly restricted hours and strict behavioral guidelines (Almond 2000; English 2002; Roberts and Everton 2002; Whitelaw 2000). In 2000, thousands of solstice pilgrims flocked to Stonehenge. Since then the number has nearly doubled each year, with 30,148 pagans and New Agers reported worshipping the sun at the monument in 2003 (Powell 2003).

Controversy in New Age tourism

Perhaps more than any of the other spiritual and religious worldviews discussed in this book, the New Age faction has seen tremendous conflict and experienced the most controversy in western societies. The majority of this controversy has focused on the ecological and social impacts of New Age behavior and tourist activities. Such concerns have grown in spite of the movement's emphasis on environmental stewardship (e.g. deep ecology, ecofeminism, nature worship, and nature conservation) and cross-cultural tolerance. The root of the problem lies in the fact that the movement is so young, beginning in earnest only in the 1950s, and that it has no formal organizational structure at its roots. The group's relative youth, therefore, means that there are no sites New Agers can truly claim to be their own – no ancient monuments, temples, or shrines built early on by adherents and passed down through the centuries. Neither are there areas of natural significance that are not already under the control of public land agencies, indigenous groups, or private individuals.

It is a common practice for New Age spiritualists to leave crystals, rock arrangements, fire remnants, candles, and other offerings at sites they consider powerful or sacred. In Sedona, medicine wheels are made by the faithful as a way to channel earth's energy, much to the dismay of the National Forest Service, which controls much of the land surrounding the town and which New Agers consider holy. This "religious graffiti" (Ivakhiv 2003) or "ritual litter" may contribute to the deterioration of cultural and natural sites. At some locations, the soil is rather thin above archeological relics, yet worshippers build fires on them. Likewise, candles are often placed too closely to heat-sensitive limestone (Powell 2003).

At most New Age-venerated spots of archeological importance there have been problems associated with worshippers chipping away at the stones to take bits and pieces with them as sacred souvenirs. This is an especially severe problem at Stonehenge, Machu Picchu, and the Pyramids, although some of the more influential believers are beginning to discourage this type of behavior, since it could result in the group losing the privilege of adulating at these shrines (Powell 2003).

In northern Arizona, Native Americans argue that New Age believers are a physical threat to their sacred sites. The National Forest Service agrees and has attempted to devise ways of getting the hundreds of thousands of Sedona area hikers and disciples from damaging natural and cultural wilderness resources (McGivney and Archibald 1997). In response to these problems, some observers have suggested offering prayers or gestures instead of material offerings (Ivakhiv 2003).

On the socio-cultural side, there is a great deal of conflict between New Age believers and Native Americans. Leaving crystals, candles, or other offerings at sacred Indian shrines upsets most Natives. Many Hopis and Navajos in Arizona, for example, are offended by this lackadaisical pursuit (from their perspective), even though New Agers consider it a sincere form of worship, because the Natives see it as desecration of sacred space and a form of mockery toward indigenous traditions (Jenkins *et al.* 1996).

The biggest complaint, however, by the Indians is the commercialization or commodification of their spiritual heritage for non-indigenous use and profiteering. Aldred (2000: 330–336) provides a great deal of insight into the controversies surrounding the New Age utilization of Native American spiritual heritage. For example, New Age paraphernalia are now being marketed in mass quantities and promoted as "Native American sacred objects." According to some commentators such as Aldred, these practices of borrowing and bastardizing traditions that many people hold sacred perpetuates the history of social oppression in the United States and trivializes an important spiritual heritage.

In addition to issues of material culture, many Native Americans are offended by the use of their sacred ceremonies for commercial gain among New Age retailers. Many New Age resorts have borrowed ideas related to sweat lodges, pipe ceremonies, sage smudgings, and nature dances. Perhaps the most offensive, however, are the "Sun dances held on Astroturf, sweats held on cruise ships with wine and cheese served, and sex orgies advertised as part of 'traditional Cherokee ceremonies'" (Aldred 2000: 333). Likewise, people can now order "sweat tents" by phone or online rather than bothering to build their own, and smudge sticks and herb tea can be ordered with desert CDs, which are promised to provide "the opportunity to experience Native American ritual and wisdom" (Aldred 2000: 334). Merchants have even begun blending Indian themes with other New Age objects, including "Native American Tarot Cards."

Many New Age shamans claim to be able to contact Indian spirits and perform Native rituals. They write best-selling books and lead expensive workshops and claim they can instruct people how to perform Native American spirituality, earning large sums of money, while many Native peoples still live below the poverty line (Aldred 2000). The commercialized practices of these self-proclaimed New Age shamans, or "Shake and Bake Shamans" or "Plastic Shamans," as they are derogatorily known to many Native Americans, are offensive, particularly when they claim to be "authentic."

The New Age response to Native American outcries has been to argue that freedom of religion in the United States assures them the right to borrow indigenous traditions and symbols. Their claim is founded on the idea that everyone has a right to such traditions, because spirituality and truth cannot be owned. According to one activist, "spirituality is not something which can be 'owned' like a car or a house. Spiritual knowledge belongs to all humans equally" (quoted in Aldred 2000: 336).

The final controversial element of tourism in this context relates to the commercialization of spirituality. Observance of New Age is underscored by a lifelong inner journey for spiritual health and enlightenment, best realized through travel and consumption of various products and services (e.g. crystals and psychic readings). As such, "seeking," which takes on a very physical and spatial form, is the primary mode of participation. Whereas in other religions, a church or other organization forms the structural element of spiritual life, in New Age practice, commercialism itself is seen as the organizational element that guides the New Age (Redden 2005). This, according to Redden, is realized in material terms through the selection and consumption of commoditized goods and services through New Age businesses, such as tour operators and merchandisers. Zaidman (2003: 247) agrees and notes that "a close look at a New Age shop reveals shelves loaded with stones, sculptures, oils, body lotions and soaps, candles, books, incense, bells, mirrors, etc." Thus, the foundation on which New Ageism is based, involves the commercial promotion of ideas and practices. In Bruce's (2002: 90 cited in Redden 2005) words, spiritual products, "be they objects (such as crystals) or services (such as a weekend's training or a tarot reading), are sold for a fee."

Conclusion

The New Age movement has grown to considerable proportions in recent years as people have become dissatisfied with traditional religion and as life in western societies has become too frenzied and materialistic. People have begun turning to alternative forms of spirituality and health that rely heavily on nature and self-transformation. New Age religion is defined by a complex and distinctive assortment of activities, rituals, behaviors, and material cultures, borrowed from the ancient traditions of pagans, animists, indigenous societies, and other nature- and self-oriented creeds. This unique approach to spiritual belief systems manifests itself in various ways in the context of tourism.

Adherence requires inner transformations that are based on harvesting earth energies and mystical encounters with spirits and extraterrestrials. These inward journeys are manifest in physical pilgrimages to ancient cultural sites, natural areas, and spas or other healing destinations, believed to be endowed with sacred powers that can cure the mind, body, and soul. These spatial-religious expressions are reflected in the tours purchased, guidebooks acquired, objects consumed, and activities undertaken.

Owing to their status as centers of earth power or cosmic energy, several destinations have developed as major spiritual centers. The most visited among New Agers are Sedona, Glastonbury, and Stonehenge, although there are many other places that are equally important New Age destinations (e.g. Bali, Machu Picchu, and the Pyramids of Giza). During the past decade, other non-traditional locations have begun to realize the lucrative potential of the nature and self-spiritualist market and have started major promotional campaigns.

Several controversies have emerged in recent years with the growth of this post-modern worldview. As a result of New Agers' reliance on indigenous cultures, archeological relics, and natural sites for their own spiritual pilgrimages, the most looming concerns are environmental degradation, commodification of culture, and commercialization of religion.

The example of New Ageism provided here is very different from the other religions examined in this tome, with the exception of the spiritual philosophies of East Asia. This movement lacks a central authority or organization that administers records, conversions, and donations, or that interprets theological precepts. Instead, individuals are the highest authority and the central executive of their own faith.

Despite the abstract and ethereal nature of this system of beliefs and the controversies surrounding it, many people have found peace in following its teachings. As New Ageism continues to grow, its adherents will continue to undertake nature worship and self-deification pilgrimages far into the future.

References

Albanese, C.L. (1990) *Nature Religion in America: From the Algonkian Indians to the New Age*, Chicago: University of Chicago Press.

Aldred, L. (2000) "Plastic Shamans and Astroturf sun dances: New Age commercialization of Native American spirituality," *American Indian Quarterly* 24(3): 329–352.

Allen, F. (1999) "History happened here: the new old west," *American Heritage* 50(5): 30–33.

Allitt, P. (2003) *Religion in America since 1945: A History*, New York: Columbia University Press.

Almond, P.C. (2000) "Druids, patriarchs, and the primordial religion," *Journal of Contemporary Religion* 15(3): 379–394.

Andres, D. (2000) *What is a Vortex? A Practical Guide to Sedona's Vortex Sites*, Sedona, AZ: Meta Adventures.

Associated Press (2002) "New Mexico to unveil spiritual 'essence' tourism campaign," *Arizona Republic* 14 April: T3.

Attix, S.A. (2002) "New Age-oriented special interest travel: An exploratory study," *Tourism Recreation Research* 27(2): 51–58.

Ball, M. (2000) "Sacred mountains, religious paradigms, and identity among the Mescalero Apache," *Worldviews* 4(3): 264–282.

Barlow, B. (1997) *Sacred Sites of the West*, St Paul, MN: Llewellyn Publications.

Barnard, C.N. (1993) "Machu Picchu: city in the sky," *National Geographic Traveler* 10(1): 106–113.

Bender, B. (1999) *Stonehenge: Making Space*, Oxford: Berg.

Brodin, J. (2003) "A matter of choice: a micro-level study on how Swedish new agers choose their religious beliefs and practices," *Rationality and Society* 15(3): 381–405.

Brown, M. (1998) *The Spiritual Tourist*, London: Bloomsbury.

Bruce, S. (2002) *God is Dead: Secularization in the West*, Oxford: Blackwell.

Charlesworth, M. (ed.) (1998) *Religious Business: Essays on Australian Aboriginal Spirituality*, Cambridge: Cambridge University Press.

Cogswell, D. (1996) "Niche for the New Age," *Travel Agent* 21 October: 80–82.

Coles, T.E. (2003) "A local reading of a global disaster: some lessons on tourism management from an annus horribilis in south west England," *Journal of Travel and Tourism Marketing* 15(3/4): 173–197.

Cooper, J. (1999) "Comprehending the circle: Wicca as a contemporary religion," *The New Art Examiner* 26(6): 28–33.

Dannelley, R. (2000) *Sedona Vortex*, Sedona, AZ: Light Technology.

Darvil, T. (1997) "Ever increasing circles: the sacred geographies of Stonehenge and its landscape," *Proceedings of the British Academy* 92: 167–202.

Davis, J. (1996) "An integrated approach to the scientific study of the human spirit," in B.L. Driver, D. Dustin, T. Baltic, G. Elsner, and G. Peterson (eds) *Nature and the Human Spirit: Toward an Expanded Land Management Ethic*, State College, PA: Venture.

Digance, J. and Cusack, C. (2001) "Secular pilgrimage events: Druid Gorsedd and Stargate alignments," in C.M. Cusack and P. Oldmeadow (eds) *The End of Religions? Religion in an Age of Globalisation*, Sydney: University of Sydney, Department of Studies in Religion.

Digance, J. and Cusack, C. (2002) "Glastonbury: a tourist town for all seasons," in G.M.S. Dann (ed.) *The Tourist as a Metaphor of the Social World*, Wallingford: CAB International.

Dobbs, G.R. (1997) "Interpreting the Navajo sacred geography as a landscape of healing," *The Pennsylvania Geographer* 35(2): 136–150.

Drovdahl, R. (1991) "Touching the spirit: the spiritual benefits of camp," *Camping Magazine* 63(7): 24–27.

Duffaut, P. (1998) "More tourists at Machu Picchu," *Tunnels and Tunnelling International* 39(4): 18–19.

Durkheim, E. (1995) *The Elementary Forms of Religious Life*, New York: Free Press.

English, P. (2002) "Disputing Stonehenge: law and access to a national symbol," *Entertainment Law* 1(2): 1–22.

Forbes, J.D. (2001) "Indigenous Americans: spirituality and ecos," *Daedalus* 130(4): 283–300.

Gee, C.Y. and Fayos-Solá, E. (1997) *International Tourism: A Global Perspective*, Madrid: World Tourism Organization.

Gottlieb, R.S. (1996) *This Sacred Earth: Religion, Nature, Environment*, New York: Routledge.

Hanegraaff, W.J. (1998) "Reflections on New Age and the secularization of nature," in J. Pearson, R. Roberts, and G. Samuel (eds) *Nature Religion Today: The Pagan Alternative in the Modern World*, Edinburgh: Edinburgh University Press.

Hanegraaff, W.J. (1999) "New Age spiritualities as secular religion: a historian's perspective," *Social Compass* 46(2): 145–160.

Heelas, P. (1996) *The New Age Movement: The Celebration of the Self and the Sacralization of Modernity*, Oxford: Blackwell.

Hooper, J. (1994) "The transcendental tourist," *Mirabella* 5(8): 71–73.

Houtman, D. and Mascini, P. (2002) "Why do churches become empty, while New Age grows? Secularization and religious change in the Netherlands," *Journal for the Scientific Study of Religion* 41(3): 455–473.

Hutton, R. (1992) *The Pagan Religions of the Ancient British Isles: Their Nature and Legacy*, Oxford: Blackwell.

Ibrahim, H. and Cordes, K.A. (2002) *Outdoor Recreation: Enrichment for a Lifetime*, Champaign, IL: Sagamore.

Ivakhiv, A. (1997) "Red rocks, 'vortexes' and the selling of Sedona: environmental politics in the new age," *Social Compass* 44(3): 367–384.

Ivakhiv, A. (2003) "Nature and self in New Age pilgrimage," *Culture and Religion* 4(1): 93–118.

Jenkins, L., Dongoske, K.E., and Ferguson, T.J. (1996) "Managing Hopi sacred sites to protect religious freedom," *Cultural Survival Quarterly* 21(1): 36–38.

Jones, K. (1991) *The Ancient British Goddess – Her Myths, Legends and Sacred Sites*, Glastonbury: Ariadne Publications.

Knopf, R.C. (1988) "Human experience of wildlands: a review of needs and policy," *Western Wildlands* 14(3): 2–7.

Kyle, R. (1995) *The New Age Movement in American Culture*, Lanham, MD: University Press of America.

Lamb, G.J. and Barclay, S.N. (1987) *The Sedona Vortex Experience: How to Tune In, Find Your Personal Power Place and Take the Magic Home!*, Sedona, AZ: New Leaf and Light Technology.

Lange, D.P. (2001) "Yoga-plus vacations," *New Age* 18(1): 38–41.

Lansky, D. (2001) "Sedona visitor gives 'power vortex' a whirl," *Arizona Republic* November 25: 4.

Lengel, K. (1999) "Ring in the new millennium, New Age style," *Arizona Republic* November 25: 13.

Lengfelder, J. and Timothy, D.J. (2000) "Leisure time in the 1990s and beyond: cherished friend or incessant foe?," *Visions in Leisure and Business* 19(1): 13–26.

McGivney, A. and Archibald, T. (1997) "Wizards of odd," *Backpacker* 25(8): 44–51.

Marty, M.E. (1997) "American nature religion," *Whole Earth* 91(1): 6–24.

O'Neil, D.J. (2001) "The New Age movement and its societal implications," *International Journal of Social Economics* 28(5): 456–475.

Powell, E.A. (2003) "Solstice at the stones," *Archaeology* 56(5): 36–41.

Redden, G. (2005) "The New Age: towards a market model," *Journal of Contemporary Religion* 20(2): 231–246.

Redmond, L. (1996) "Diverse Native American perspectives on the use of sacred areas on public lands," in B.L. Driver, D. Dustin, T. Baltic, G. Elsner, and G. Peterson (eds) *Nature and the Human Spirit: Toward an Expanded Land Management Ethic*, State College, PA: Venture.

Reisinger, Y. (2006) "Travel/tourism: spiritual experiences," in D. Buhalis and C. Costa (eds) *Tourism Business Frontiers: Consumers, Products and Industry*, Oxford: Butterworth Heinemann.

Roberts, D. and Everton, M. (2002) "Romancing the stones," *Smithsonian* 33(4): 86–96.

Rountree, K. (2002) "Goddess pilgrims as tourists: inscribing the body through sacred travel," *Sociology of Religion* 63(4): 475–496.

Ruland-Thorne, K. (1993) *Yavapai: The People of the Red Rocks, The People of the Sun*, Sedona, AZ: Thorne Enterprises.

Sanders, P.A. (1992) *Scientific Vortex Information*, Sedona, AZ: Free Soul Publishing.

Schor, J. (1993) *The Overworked American: The Unexpected Decline of Leisure*, New York: Basic Books.

Shimazono, S. (1999) "'New Age movement' or 'new spirituality movements and culture'?," *Social Compass* 46(2): 121–133.

Shorey, A. (2002) "Sedona: where crystal crunchers and vortex seekers meet," *TimesHerald* September 1: C6.

Smith, M. (2003) "Holistic holidays: tourism and the reconciliation of body, mind and spirit," *Tourism Recreation Research* 28(1): 103–108.

Spangler, D. (1996) *A Pilgrim in Aquarius*, Forres: Findhorn.

Strutt, R. (1999) "Pack your bags and learn," *New Age* 16(7): 17–20.

Tapia, E.S. (2002) "Earth spirituality and the people's struggle for life: reflection from the perspectives of indigenous peoples," *Ecumenical Review* 54(3): 219–227.

Timothy, D.J. (2002) "Sacred journeys: religious heritage and tourism," *Tourism Recreation Research* 27(2): 3–6.

Tucker, J. (2002) "New Age religion and the cult of the self," *Society* 39(2): 46–51.

Turner, D.H. (1991) "Australian Aboriginal religion as 'world religion'," *Studies in Religion* 20(2): 165–180.

Tyler, M.E. (1993) "Spiritual stewardship in aboriginal resource management systems," *Environments* 22(1): 1–8.

Whitelaw, K. (2000) "The sorcery of the stones," *U.S. News and World Report* July 24: 35–38.

York, M. (2001) "New Age commodification and appropriation of spirituality," *Journal of Contemporary Religion* 16(3): 361–372.

Zaidman, N. (2003) "Commercialization of religious objects: a comparison between traditional and New Age religions," *Social Compass* 50(3): 345–360.

11 Global Jewish tourism
Pilgrimages and remembrance

Mara W. Cohen Ioannides and
Dimitri Ioannides

In *The Past is a Foreign Country*, geographer David Lowenthal argues that nostalgia reflects a powerful desire to relive the past, albeit a sanitized past to be viewed from a safe distance. The need "to know how and why things happened is a compelling motive for witnessing past events" (Lowenthal 1985: 22). As the authors have discussed elsewhere (Ioannides and Cohen Ioannides 2002, 2004), it is this very nostalgic need that compels Jews, wherever they may be, to visit the sites where their ancestors or famous Jewish people lived, or where important events in the history of Judaism took place.

This chapter examines the idea of nostalgic pilgrimage as it relates to why Jews choose to travel where (and when) they do. Central to this discussion is an examination of the long history of Jewish diaspora. Because over the centuries Jews have settled and resettled in numerous places throughout the world, the localities where their ancestors lived and where important events took place have emerged as "pilgrimage" sites regardless of whether or not these have direct ties to religious events or persons. Additionally, the chapter discusses how religion itself influences the act of travel for many Jews since religious dictates in Judaism constitute a key interpretation as to when and where they choose to travel and even provide proscriptions as to when one cannot or should not travel.

Creating the diaspora

Jewish history regularly refers to traveling. Potok (1978) even entitled his history of the Jews *Wanderings: Chaim Potok's History of the Jews*. The Hebrew Bible begins with the book of Genesis, telling the first nomadic story, that of the expulsion of Adam and Eve from the Garden of Eden. The book of Exodus concerns the Jews' escape from slavery in Egypt and their subsequent 40-year travels in the Sinai desert. Thus, the Jewish nomadic tradition had its seeds thousands of years ago. Indeed, the people from whom modern-day Jews are descended are believed to have been nomads, perhaps the Amorites or the Hapiru, a group of wandering outcasts (of undetermined ethnic background) in the Near East (Potok 1978).

After their long-term desert wanderings, the Jews entered Canaan and created a nation. However, 700 years later (in 586 BCE), the Babylonians conquered the nation. Jerusalem was destroyed and most of the population was sent into exile and slavery. Nearly 50 years later many exiles returned to Jerusalem, although some remained in Babylonia forming the very first diasporic group of Jews.

Jerusalem was eventually rebuilt and by the time of Julius Caesar, Roman Jews constituted nearly a tenth of the total Roman population (Potok 1978). The second Jewish empire finally collapsed in 70 CE when Titus took Jewish captives to Rome. This marked the beginnings of the great Jewish diaspora. By 600 CE, Jewish communities existed from the Kingdom of the Franks through Mesopotamia. During the Middle Ages, Jewish communities relocated frequently. They moved throughout the Roman, Byzantine, and Ottoman empires, settling as far away as India, and by 1000 CE, a Persian Jewish community had resettled in Kaifeng, China. As the various great empires where Jews lived floundered, anti-Semitism often increased and Jews were forced to settle in new places (Chaliand and Rageau 1995).

As Table 11.1 shows, for more than 3,000 years Jews wandered across Europe and Asia en masse and individually created cohesive communities. Because in Western Europe, they were forbidden to own land or join a guild, they often became moneylenders. This created a stronger diasporic community, especially because families intermarried and sent members to other communities to open businesses, thus strengthening their international ties (Chaliand and Rageau 1995). Families became far-flung and visited each other whenever possible: for holidays, life events, or business excursions (Glückel of Hameln 1987). Additionally, the emphasis placed by Jews on education (see Deut. 7:6, 11:19, and 31:12) meant that sons were often sent far away to study at *Yeshivot* (singular *yeshiva* – schools for the study of Jewish law) with an important scholar, a phenomenon that began before the Middle Ages and continues today. Many young scholars stayed on, while some went to other communities to found their own *yeshiva*, work as rabbis, or teach. Thus the communities, regardless of whether they were formed through forced or voluntary migration, developed tight connections between themselves.

Jewish travel

The law and scriptural dictates on travel

Previously the authors argued that beyond Jerusalem, Judaism attaches no specific significance to religious sites in the manner that Catholics, Hindus, and Muslims view places like Lourdes, Varanasi, and Mecca respectively (Ioannides and Cohen Ioannides 2002). The theologian Abraham Herschel (1955: 200) remarked that "Judaism is *a religion of history, a religion of time*"; thus, importance is not normally placed on a particular locality, but on a time when an event took place. Moreover, public prayer is sanctified by the number of people present (ten adults, a *minyan*), not a specific location or building (Glustrum 1988). Contrary to churches, for instance, synagogues are not considered holy spaces. Nevertheless, for many Jews, Jewish law plays a vital role in governing travel regardless of whether the prime motive for the trip is religious, business-oriented, or personal.

The decrees in the *Tanakh* (Hebrew Bible) concerning when one is required to travel are few. Exodus 34:23 states: "Three times a year all your menfolk must present themselves before the Lord Yahweh, the G–d of Israel" at the Temple in Jerusalem. The festival of Passover marks one of these times. This festival celebrates the spring harvest along with the exodus of the ancient Hebrews from slavery in Egypt (retold in the book of Exodus in the *Tanakh*). The Biblical requirements for this pilgrimage include a ban on eating anything with a leavening agent and the presentation of a sacrificial animal to the Temple (Ex. 23:14–15; Deut. 16: 2–6).

The second pilgrimage must take place during *Shavuot* (Feast of Weeks), a festival combining historical and agricultural meanings. Though no specific scriptural reference is made to the former interpretation, from a historical standpoint the festival celebrates receiving the *Torah* (first five books of the *Tanakh*) at Sinai. The *Tanakh* clearly stresses the festival's agricultural meaning. The book of Exodus (23: 16) explains: "The feast of Harvest, too, you must celebrate, the feast of the first fruits of the produce of your sown field." Leviticus (23:10–22) provides more details for the celebration of this pilgrimage festival, including the admonition to send one's first produce from the harvest and (50 days later) an animal sacrifice to the Temple.

Table 11.1 Jewish diasporic history

Date	Event
1286 BCE	Hebrews enter Canaan and create Jewish nation
586 BCE	Jewish nation destroyed by Babylonians, beginning of diaspora
536 BCE	Some Jews return and begin rebuilding Jerusalem
59 CE	Jews make up one tenth of Roman population
70 CE	Second Temple destroyed, beginning of great Jewish diaspora
600 CE	Jewish communities from Kingdom of Franks through Mesopotamia
1000 CE	Jewish community in Kaifeng, China, already established
1066 CE	First Jews enter England
1290 CE	Jews expelled from England
350 CE–1450 CE	Jews move from Western to Eastern Europe
1492 CE	Jews expelled from Spain, fled to Netherlands, England, and Ottoman Empire
1637 CE	First synagogue in New World in Recife, Brazil

The third pilgrimage to Jerusalem must be performed during *Sukkoth* (Tabernacles, Festival of Booths). This festival continues the Passover Exodus story, focusing on the desert wandering of the Hebrews. It also is an agricultural festival and, thus, referred to as the Festival of Ingathering. According to Exodus 23:16: "Three times a year you are to celebrate a feast in my honor" and "the feast of Ingathering also, at the end of the year when you gather in the fruit of your labors from the fields" (Ex. 23:14).

The role these three pilgrimage festivals have played in Judaism following the destruction of the Second Temple in 70 CE has changed dramatically. While many Jews still visit Jerusalem, they go to mourn its destruction, not to celebrate. Additionally, as Jews are spread throughout the world, for many, the annual pilgrimage to Jerusalem has become practically impossible. Today, on Passover, the Temple sacrifice is represented on the ceremonial plate by a roasted lamb shank bone (boiled chicken neck, or roasted beet) and a roasted egg. For *Sukkoth*, Jews build a hut, used as an extension of the house where all meals are eaten. Over time, *Shavuot* has lost its agricultural element, while its historical meaning marking the time of the receiving of the *Torah* has become the celebrated part of the festival. These examples reinforce the idea that in Judaism, pilgrimages to a holy site, namely Jerusalem, have been largely substituted with ceremonies commemorating past events, which can take place virtually anywhere.

Along with the aforementioned pilgrimage festivals, Judaism includes scriptural dictates that influence travel directly and indirectly. In Exodus 31:12–17, G–d reminds Moses that the Sabbath must be observed "because the Sabbath is a sign between myself and you from generation to generation to show that it is I, Yahweh, who sanctified you." Both Exodus 34:21 and Deuteronomy 5:12–15 repeat this admonition. The observance of the Sabbath makes travel challenging because every seventh day labor must not be performed. The *Talmud* (codified interpretations of the laws in the *Tanakh*) addresses the issue of what constitutes rest versus work on the Sabbath and what work must take place, such as feeding one's animals or life-saving surgery. In fact, the *Talmud* explains exactly how far one may walk on the Sabbath before this act becomes work and, thus, forbidden. This is where Talmudic interpretations relating to the Sabbath are important.

> All acts of servile labor and the preparation of food must be done beforehand. Food-preparation is generally assumed to involve heating the food. . . . One may not kindle a flame on the Sabbath, but one may make use of the light of a flame kindled in advance.
>
> (Neusner 1983: 56)

In modern times, this has relevance to starting a car or turning on an electric light.

The *Talmud* explains that before the Sabbath begins one must establish a "Sabbath residence." The idea of residence is crucial because it shows that the traveler intends to stay in the locale for a set period of time – the length of the Sabbath, from the arrival of the first star on Friday evening until the arrival of the first star on Saturday evening. When one has established this domain, wherever it may be, then one can consider it a residence and is allowed to carry on the everyday activities permitted on the Sabbath. Therefore, travelers must arrive at their destination before the Sabbath begins and should they wish to participate in a public worship service they must find one within the designated boundaries. Their accommodation should provide meals without the need to sign for them, as writing is also prohibited.

The question of *kashrut* poses a particularly serious problem for travelers. *Kashrut*, or keeping kosher, are food laws governing what one may and may not eat, when and with what. Prohibited foods include shellfish, carnivores, herbivores without cloven hooves and two stomachs, animal blood, and the mixing of dairy and meat in one meal (Deut. 14:3–20; Lev. 17:12; Deut. 14:21). Because dietary restrictions also apply to dishes and kitchenware they have major impacts on commercial kosher kitchens. These establishments are divided into two categories: those supervised by a rabbi and those that are not. How strictly a traveler keeps kosher dictates which restaurant he or she will choose to dine in for a meal. The strictest Jewish travelers will only eat at rabbinically supervised establishments, while one who is religious but less strict will eat only fish or vegetarian meals at any restaurant. As Elaine Goldman (1968: 20) explained in her early review of kosher places to stay and eat along the British coast: "There are three alternatives for the Jewish holidaymaker who limits his diet. Either he goes vegetarian, or manages on fish and eggs for the duration of his stay, or confines himself to choosing one of the . . . kosher establishments."

The specific religious-based requirements of many Jewish travelers have influenced the business practices of various providers, such as airlines, hotels, and restaurants that depend to varying degrees on this travel segment. For instance, major hotels and restaurants in various countries seek to provide the level of service required by sophisticated Jewish travelers, while simultaneously adhering to their specific dietary and other restrictions imposed by religious law. Additionally, many international airlines offer the option of pre-ordering kosher meals.

Although they strictly follow the laws of *Shabbat*, most large Israeli hotels work around some of the requirements about not performing work on Saturdays by operating special automatic, or *Shabbat*, elevators (Hyatt Regency Jerusalem 2002; de la Roca 1997). These adapted elevators adhere to a number of restrictions like not allowing the weight of the passengers to influence travel, a typical feature of most elevators, and ensuring the floor indication light does not illuminate (Ask the Rabbi 2003).

Scholars of *halacha* (law) warn travelers that using electronic door keys is forbidden on the Sabbath; guests must find willing gentiles to open their doors for them. Additionally, while using standard keys is permitted, these cannot be taken from the hotel because of rules prohibiting carrying objects, but they can be dropped off at the front desk before leaving the hotel (Weinberger 2000).

Following these restrictions is never easy, but adhering to them outside of Israel is especially difficult, a situation that often troubles religious Jews. To help in this endeavor, Jewish-oriented travel guidebooks usually describe more than a locality's attractions. Bloomfield and Moskowitz's (1991) *Traveling Jewish in America* and Israelowitz's (1999a) *Guide to Jewish Europe*, recommend kosher restaurants, stores, hotels, synagogues, and *mikvah* (ritual baths). *Jewish Action*, a publication of the Orthodox Union, included a list of kosher hotels in its Passover 2000 issue. Bloomfield and Moskowitz (1991) provided names of private individuals whom religious travelers may contact for assistance if no other suitable institution is available, and as noted earlier, Goldman (1968) advised Jews about places to stay and eat in Great Britain, including hotels and restaurants in Bournemouth, Brighton, Margate, Westcliff, and Blackpool.

Tour organizers are well aware of the problem of ensuring the laws of *kashrut* are maintained by facilities catering to Jewish travelers and, thus, offer kosher tours in unlikely places. Africa Kosher Safaris (2001: n.p.) provides "the kosher traveler with a truly exciting and comfortable African experience without the hassle of looking for kosher food in the bush." Based in both the United States and South Africa, this company caters to a wide segment of the market including student travelers and luxury-oriented individuals. The company promises its tour participants the option of gender-specific swimming facilities and communal prayer arrangements.

Entrepreneurs have also invested in this specialized market. Blumenhotels has opened a kosher village outside of Zell am See, Austria. This includes a hotel and holiday apartments; two kosher restaurants, a café, and a pizzeria; a *mikvah* for men; separate pools for male and female guests; childcare; and Sabbath services (Blumenhotels 2004). Shamash (2004: n.p.) touts this as the first kosher hotel in Austria, promising that: "art treasures and sights are no longer difficult to visit for religious Jews."

Nostalgic pilgrims

Nostalgic pilgrimage, the need to visit a place influenced by a strong yearning to connect with one's history, is a relatively new phenomenon albeit one with considerable significance for Jews whether they consider themselves religious or secular (Ioannides and Cohen Ioannides 2002). Eisen (1998: 158) attributes the earliest forms of nostalgic journeys to the middle of the 1800s when various publications invited "Jews back to the sites where their ancestors had practiced Judaism." This phenomenon then grew into the "mitzvah of nostalgia." As Eisen (1998: 185) explains: "Pilgrimages are conducted to the sites of piety ... the journeys made more compelling still – and the resulting 'merit' to the descendants all the greater – by [for example] the ancestors' martyrdom at the hands of the pogromists or Nazis." Nostalgic pilgrimages then, are about visiting one's ancestral homeland but also can be much more, such as going to places holding special significance for Jews in general (e.g. the home of a famous Jew, a historical synagogue, or a Holocaust site) (Collins-Kreiner 1999; Ioannides and Cohen Ioannides 2004; Kosansky 2002; Krakover 2005; Mitsuharu 2003).

Nostalgic pilgrimage encompasses two other pilgrimage types: religious and social, or group, pilgrimage. Religious pilgrimage, Rinschede (1992: 52) explains, "includes the visit of religious ceremonies and conferences, above all the visit of local, regional, national, and international religious centers." This reflects part of what religious, *but also* secular, Jews do on a nostalgic pilgrimage: they visit synagogues, school sites, cemeteries, and mass-grave locations. In a sense, however, such a pilgrimage is also a social pilgrimage because families or groups, as part of a larger social identity, commonly plan excursions to other specific locales associated with Judaism (Rinschede 1992).

It is worth noting briefly the idea that a pilgrim is also a tourist, a theme that was extensively investigated in a special issue of *Annals of Tourism Research* in 1992 (Smith 1992). Cohen (1992) used the term "pilgrim-traveler" to cover both views of religious travelers. In many respects, nostalgic tourism contradicts stereotypical perceptions of the religious pilgrim. One tends to assume that pilgrims must follow some pious path. However, as Smith (1992: 2) argues: "Tourist encounters can be just as compelling and almost spiritual in personal meaning." Nostalgic pilgrimages, then, match Smith's (1992: 4) concept of pilgrimage "in lieu of piety, the visitors seek to experience the sense of identity with sites of historical and cultural meaning." Thus, the motive of nostalgic tourists and pilgrims to travel is not purely religion-based but can also be socially oriented.

Throughout time, Jews have had, as Webber (1992: 246) explains, "an underlying belief in unity," and it is this unity that draws them to "travel Jewishly," which is to: "1) journey to a quintessentially Jewish place (like Israel or even Golders Green, Brooklyn, and Miami Beach); 2) make a concerted effort to search out Jewish points of interest in non-Jewish contexts; or 3) travel with a Jewish purpose" (Pogrebin 1995: 24). An earlier study indicated that at least a quarter of Jewish tourists, regardless of the primary reason for their trip, visited a Jewish-themed place (e.g. store, synagogue, museum) once at their destination, even if their chosen destination could not itself be considered Jewish (Ioannides and Cohen

Ioannides 2002). For many Jews, the experience of visiting Jewish heritage sites provides an opportunity to rediscover their past and their traditions (Gruber 2002). While Jews of the diaspora have traditionally traveled to the religious homeland, namely Israel and specifically Jerusalem, they have also discovered places worldwide with historical and cultural significance for the communities of the diaspora, both existing ones and those that have been lost.

Undoubtedly, Israel itself remains a major pilgrimage site for Jews. Of the 862,000 visitors to Israel in 2002 (Israel Ministry of Tourism 2003), 55 percent were Jews. Jews made up two-thirds of the visitors from France, the United States, the United Kingdom, and Canada. Of the total number of arrivals, 7 percent came on a traditional pilgrimage (regardless of religion), 15 percent for leisure, 7 percent for sightseeing, and 44 percent to visit family and friends. Spending by these tourists contributed some $1.2 billion to the Israeli economy (Israel Ministry of Tourism 2003).

After arriving in Israel, many Jewish travelers visit religiously significant sites, such as the Wailing Wall (the Temple's one remaining wall) (Collins-Kreiner 1999; Shacher and Shoval 1999). However, it should be stressed that these visits are not performed for purely religious purposes, but for cultural and historical ones as well (Epstein and Kheimets 2001; Vukonić 1996). Thus, secular Jews also visit these sites because of their cultural and historic value. This is also true for visits outside of Israel. For example, Jews often visit the Synagogue in Malmø, Sweden, famous in a historical sense as the place where the last Danish Jewish refugees arrived in Sweden in 1943 (Israelowitz 1999a). The Rambam Synagogue in Cordoba, Spain, is visited as a house of worship but more importantly as an historic site where Maimonides, a great Jewish philosopher and physician, studied and prayed. Also, the Touro Synagogue in Rhode Island, the oldest continuously used synagogue in the United States, is a National Historic Site that today attracts many visitors, including secular and religious Jews (National Park Service 2004).

Other sites have also emerged as obvious destinations for nostalgic pilgrimages as Jews increasingly seek to discover their past and/or remember their ancestors. Ancestral gravesite visits constitute a common form of Jewish nostalgic pilgrimage. Although during the first year following a death mourners usually do not visit the grave, since no law specifically prohibits such visits, it has become traditional to visit graves the month before the High Holy days of *Rosh Hashanah* (New Year) and *Yom Kippur* (Day of Atonement). This practice grew from the custom in Talmudic times of visiting graves (*Kolatch*).

Visiting graves of rabbis and other important religious figures has also become common practice for Jews. While no laws specifically require visits to graves, scriptures prescribe that Jews must honor the dead (Epstein 1961).

Some Jews visit these tombs believing they can be healed of various afflictions, and it has become common when visiting a grave to place upon it a pebble or a written petition (Sered 2002). There are several explanations for the custom of placing pebbles on graves. The holy writings comment several times on the links between stones and gravesites. Genesis 29:35, for example, remarks about the tombstone on Rachel's grave. The tradition also alludes to the notion that visitors may build a cairn of stones with each pebble representing an individual's contribution to the cause. Since the graves of many Biblical figures (e.g. King David, Rachel, and Elijah) and Talmudic scholars (e.g. Akiva and Meir) are in Israel, during the Middle Ages, visiting these burial places became part of the Holy Land pilgrimage (Jacobs 1964). Today, however, visits to the tombs of holy persons have been extended to include graveside pilgrimages in destinations outside of Israel (see Table 11.2).

With the annihilation of many Jewish centers in Europe following the Holocaust, many Jews, especially Americans, now visit the sites of these lost communities (Ashworth 2003). Most often they return to localities their families abandoned, such as the *shtetls* (small Jewish villages), or at least the regions where the *shtetl* once existed. For most of these pilgrims the experience is "religious" in the sense that finding one's roots provides insight into one's cultural and spiritual past (Plaut 2003). This personal nostalgic pilgrimage differs from the communal one. When visiting an extermination site, for instance, one is part of a community remembering a community. However, when visiting an ancestor's home one is remembering a personal history and to some degree reliving a personal past (Timothy 1997).

American Jews are not the only ones engaged in nostalgic tourism. According to Cohen (1983), pilgrimages to their own ancestral homelands in places like Eastern Europe and North Africa have become popular for Israeli Jews. In the 1980s, Moroccan-born Israelis became interested in visiting their own native land. The itinerary of these tours is typically highly structured, including visits to tombs of rabbis (e.g. Rabbi Raphael Ankawa). Participants on these tours also visit the graves of their parents and other family members (Kosansky 2002; Levy 1997).

A relatively new type of pilgrimage is one performed to celebrate a b'nai mitzvah (literally: children of the commandment, singular bar/bat mitzvah) in Israel. The most common location of this coming of age event is the Wailing Wall in Jerusalem, while Masada has also become quite popular (Yoken 1994). Tour companies specializing in travel to Israel have recognized b'nai mitzvah as an important market segment (Collins-Kreiner and Olsen 2004; Kustanowitz 1998; Olsen and Timothy 1999) and offer packages for families where the b'nai mitzvoth travel free. By traveling to Israel, the experience is made all the more memorable and replete with meaning and provides an opportunity for the entire family to visit the holy places throughout Israel (Francis 2000).

Table 11.2 Gravesites for pilgrimage

Personage	Location
Rachel (Jacob's wife)	Sin Burak (outside Bethlehem), Israel
Ezekiel (prophet)	Iraq
Nahum (prophet)	Kurdistan
Mordecai and Esther (heroes of the festival of Purim)	Hamadan, Persia
Jeremiah (prophet)	Fostat (Old Cairo), Egypt
Rebbe Nachman (founder of the Breslov Hasidic movement)	Uman, Ukraine
Rabbi Menachem Schneerson (former leader of the Lubavitch Hasidic movement)	New York, New York
Joseph	Shechem (Nablus)
Absalom's Tomb	Kidron Valley (Josaphat Valley), Israel
King David	Mount Zion, Israel
Zachariah	Kidron Valley (Josaphat Valley), Israel
Aaron, the High Priest	Awarta (Samaria), Israel
Rabbi Simon Bar Yohai and Rabbi Eleazar	Meirun, Israel
Maimonides, Rabbi Yohannon Ben Zakai	Tiberias, Israel
Rabbi Meir Baal ha-Ness	Tiberias Hot Springs (Hammath), Israel
Rabbi Yehouda Benatar	Fez, Morocco
Rabbi Chaim Pinto	Mogador, Morocco
Rabbi Amram Ben Diwane	Ouezzan, Morocco
Rabbi Yahia Lakhdar	Beni-Ahmed, Morocco
Moses ben Israel Isserles, aka Remuh	Krakow, Poland
Rabbi Yitzchak Luria (kabbalist)	Safed, Israel
Abraham, Sarah, and their sons	Cave of the Patriarchs, Israel

Promoting nostalgic pilgrimages

While Jews search for suitable destinations, site managers seek to find ways to attract visitors. Numerous travel books describe sites of special interest to potential Jewish travelers but also, as already mentioned, kosher establishments in which to stay and eat. Gruber (1994: vii) explains that her book is

> a guide to these many remaining traces of Jewish culture and civiliza-
> tion. It is designed both as a practical guide for travelers to the various
> countries and as a sourcebook for armchair travelers interested in
> learning about a vital part of European and Jewish history – past and
> present.

The travel guide by Postal and Abramson (1962: vi) also aimed to help people recapture their past. They go so far as to suggest that their work, *The Landmarks of a People*, addresses the needs of business travelers, pleasure-seekers, students, and pilgrims as they come to honor parents and other kinfolk buried abroad, as Second World War veterans who want to see the postwar Continent, or as former refugees from the Nazi terror visiting their old homes.

In the United States, the heritage promoted to attract Jewish travelers is not exclusively European, or non-American. As American Jews, for instance, have spread around the country, the early Jewish settlements in places like New York, Boston, and Natchez, Mississippi, provide opportunities for Jewish Americans to discover their own American roots. Books like *The Jewish Friendship Trail: Guidebook – Jewish Boston History Sites* (Ross 2003) and *The Jewish Heritage Trail of New York* (Israelowitz 1999b) guide travelers around the United States to see American Jewish historic sites, some of which may also have religious links, such as cemeteries and synagogues.

Generally, Jews travel extensively and sites associated with Judaism receive large numbers of visitors. For example, the National Museum of American Jewish History (2003) in Philadelphia attracts approximately 40,000 visitors annually, two-thirds of whom are Jewish. In Amsterdam, the Anne Frank House (2003) had 908,000 visitors in 2002, reflecting a steady growth in visitors numbers since 9,000 first came to the site in 1960. Jewish Venice (1999) proudly posted on its website that "Venice and the Jewish Quarter attracts over 300,000 tourists a year!" The Auschwitz-Birkenau Concentration Camp catered to 427,000 visitors from over 100 countries, with the United States and Israel being in the top six origin countries of visitors (Memorial and Museum Auschwitz-Birkenau 2004).

European countries are especially aware of the growing demand for Jewish nostalgic pilgrimages. They use this awareness to promote their tourism industries. In 1994, the Hungarian Tourism Board, Malev Airlines, and the Kempinski Hotel sponsored a trip for *The Jewish Monthly* journalists, resulting in a half page article "The Renaissance of Tourism." Many of the former Soviet bloc states have begun catering to Jewish tourists by adding memorials to mass graves that specifically remember the murdered Jews, since these countries realize that more Jews will visit these areas if there is a conscious attempt on behalf of the host community to acknowledge that such atrocities took place (Plaut 2003).

Nevertheless, nostalgic tourists can be too narrow in their focus. Poland discovered that the majority of the 100,000 Jewish tourists from Israel and the United States visit the monument in the Warsaw ghetto and the death camps. The concern, as voiced by Foreign Minister Wlodzimierz Cimosziewicz, is: "These visits result in them [the tourists] perceiving Poland as just one big cemetery" (Green 2003: n.p.). In response, the Polish government has funded a Museum of the History of Polish Jews to be built near the former ghetto in Warsaw. The museum's purpose is to highlight the 800 years of Jewish culture in Poland for the edification of both Poles and foreign tourists (Green 2003).

Many localities throughout the United States have also begun catering to the growing market for nostalgic pilgrimages. For example, The Greater Philadelphia Tourism Marketing Corporation includes a page on "Sacred Places," advertising that Philadelphia has more than 800 houses of worship, including the most "first churches" in the United States (Greater Philadelphia Tourism Marketing Corporation 2003). The list of sacred places includes Mikvah Israel founded in 1740. Other Philadelphia Jewish sites include cemeteries, places of worship, and shrines.

Both Israel and Turkey recognize Izmir, Turkey, as a potential major Jewish attraction. The restoration of the old city of Izmir will include a number of synagogues, the *mikveh,* and the cemetery containing the grave of Rabbi Haim Palagi. According to the Turkish Government Tourism Office, over 229,000 Israelis visited Turkey in 2002 (Shapiro 2002).

Tour companies around the world have jumped into the fray. For example, San Francisco Jewish Landmarks Tours promotes tours of Santa Fe and Pioneer Jews of New Mexico. Margaret Morse Tours runs excursions such as The Jewish South and China & Yangtze River. Additionally, Israel Tour Connection offers a tour to visit Kahal Zur Israel, the oldest synagogue (founded 1630) in the New World, in Recife, Brazil, which was only excavated in 2000 (Popson 2002). There are many other global examples of tour operators targeting Jews from the diaspora to travel to Jewish and non-Jewish sites (Collins-Kreiner and Olsen 2004; Shoval 2000).

Even international organizations encourage Jews to be nostalgic. For example, INEX (Association for Voluntary Service), part of the Centre for International Youth Exchange and Tourism, offers volunteer work camp experiences to the world's youth. In 2002, the program had nearly 400 participants (INEX 2003). Its Czech Republic offerings include a "special project [that] pays attention to the particular former Jewish culture in CR" and encourages participants because "only people aware of cultural traditions and values and with their own relationship to the cultural history may be responsible for the surrounding environment and subsequently able to create a healthy society" (INEX 2003: n.p.).

Conclusion

This chapter has demonstrated that Jewish travel and related behavior are dictated by religious law, and importantly, by a powerful need to perform "mitzvahs of nostalgia." Jews, wherever they may be and regardless of their attitudes towards religion, regularly undertake "pilgrimages" that are not strictly religious but, rather ones taking them back in time, linking them with their Judaic past. This travel for nostalgic purposes, then, blurs the boundaries between purely religious pilgrimage and other forms of travel (business- and leisure-oriented) since it appears that many Jews (religious and secular) regularly visit at least one Jewish-oriented site even if their primary reason for travel has nothing to do with Judaism. The Jewish nostalgic pilgrimage then reinforces Nolan and Nolan's (1992: 77) belief that "there is no evidence to suggest that tourism and pilgrimage are intrinsically incompatible."

The travel industry itself, including tour operators, entrepreneurs, hotels, airlines, governments, international organizations, and the publishers of travel guides, recognize the enormous potential for Jewish nostalgic pilgrimages and have played an important role in promoting this type of travel to various parts of the world. Gruber (2002: 134) highlights this by saying that "tourism providers, entrepreneurs, and local boosters now recognize that Jewish sites and personalities can define the identity of atmosphere of a town or neighborhood as much as do churches, kings, and city halls." Moreover, the travel industry, both in Israel and abroad, has made substantial strides toward catering to the strict dietary and other requirements that religious Jews must follow according to Judaic laws.

In the final analysis, perhaps one of the most telling aspects about the role of Jewish travel within international tourism is that it has become part of popular culture, literature, and folklore. Jonathan Safran Foer's (2003: 477) recent short story *The Very Rigid Search* tells of a tour guide in Odessa and an American Jewish tourist searching for his roots. The main character, Jonathan Safran Foer, describes his father's business in this manner:

> My father toils for a travel agency here in Odessa. It is denominated Heritage Touring. It is for Jewish people, like the hero, who have cravings to leave that ennobled country America, and visit humble towns in Poland and Ukraine. My father's business scores a translator, guide, and driver for the Jews, who try to find places where their families once existed.

References

Africa Kosher Safaris (2001) Available at www.africakoshersafaris.com/main/ index.php (accessed February 19, 2004).

Anne Frank House (2003) "Anne Frank FAQ: How many visitors has it received since it opened?" available at www.annefrank.nl/eng/main/faqantwoordtekst7df0. html?id=41 (accessed February 19, 2004).

Ashworth, G.J. (2003) "Heritage, identity and places: for tourists and host communities," in S. Singh, D.J. Timothy, and R.K. Dowling (eds) *Tourism in Destination Communities*, Wallingford: CABI.

Ask the Rabbi (2003) *Shabbat Elevators*, available at www.ohr.edu/ask_dv/ (accessed April 8, 2004).

Bloomfield, B.A. and Moskowitz, J.M. (1991) *Traveling Jewish in America* (3rd edn), Lodi, NJ: Wandering You Press.

Blumenhotels (2004) *Hotel Knappenhof*, available at www.222.kkosher-hotel.at/ koscher-hotel-de/Newer%20Ordner/en_kappenhof.htm (accessed February 19, 2004).

Chaliand, G. and Rageau, J.-P. (1995) *The Penguin Atlas of Diasporas*, New York: Penguin Books.

Cohen, E. (1983) "Ethnicity and legitimation in contemporary Israel," *The Jerusalem Quarterly* 28: 111–124.

Cohen, E. (1992) "Pilgrimage and tourism: convergence and divergence," in A. Morinis (ed.) *Sacred Journeys: The Anthropology of Pilgrimage*, Westport, CT: Greenwood Press.

Collins-Kreiner, N. (1999) "Pilgrimage holy sites: a classification of Jewish holy sites in Israel," *Journal of Cultural Geography* 18(2): 57–78.

Collins-Kreiner, N. and Olsen, D. (2004) "Selling diaspora: producing and segmenting the Jewish diaspora tourism market," in T. Coles and D.J. Timothy (eds) *Tourism, Diasporas and Space*, London: Routledge.

de la Roca, J. (1997) "The new Sharon Commercial Center," *The Jerusalem Post Internet Edition* 11 May, available at www.jpost.com/com/Archive/11.May.1997/RealEstate/Article_2.html.

Eisen, A.M. (1998) *Rethinking Modern Judaism: Ritual, Commandment, Community*, Chicago: University of Chicago Press.

Epstein, A.D. and Kheimets, N.G. (2001) "Looking for Pontius Pilate's footprints near the Western Wall: Russian-Jewish tourists in Jerusalem," *Tourism, Culture and Communication* 3(1): 37–56.

Epstein, I. (1961) *The Babylonian Talmud*, London: Soncino Press.

Foer, J.S. (2003) "The very rigid search," in P. Zakrzewski (ed.) *Lost Tribe: Jewish Fiction from the Edge*, New York: Harper Collins.

Francis, E. (2000) "Summer journeys to adulthood," *Travel Agent* March 4: 74–76.

Glückel of Hameln (1987) *The Memoirs of Glückel of Hameln* (Marvin Lowenthal, Trans.), New York: Schocken.

Glustrum, S. (1988) *The Language of Judaism*, Northvale, NJ: Jason Aronson.

Goldman, E. (1968) "Jews who stay at home: Britain's kosher coastline," *Jewish Observer and Middle East Review* January 26: 20.

Greater Philadelphia Tourism Marketing Corporation (2003) "Welcome to sacred places in Philadelphia and its countryside," Philadelphia Culture Files, Greater Tourism Marketing Corporations, available at www.gophila.com/culturefiles (accessed February 20, 2004).

Green, P.S. (2003) "Jewish museum in Poland: more than a memorial," New York Times January 9, available at www.nytimes.com/2003/01/09/international/europe (accessed February 19, 2004).

Gruber, R.E. (1994) *Jewish Heritage Travel: A Guide to East-Central Europe*, New York: Wiley.

Gruber, R.E. (2002) *Virtually Jewish: Reinventing Jewish Culture in Europe*, Berkeley: University of California Press.

Herschel, A.J. (1955) *God in Search of Man: A Philosophy of Judaism*, New York: The World Publishing Company.

Hyatt Regency Jerusalem (2002) *Gate 54*, available at www.gate54.com/africa_and/middle/east/IL/jerusalem/hotels/hotel_10204483.html (accessed April 20, 2003).

INEX (2003) *Association for Voluntary Activities*, available at www.sci-ivs.org/workcamp03/czech_sda_about.htm (accessed February 19, 2004).

Ioannides, D. and Cohen Ioannides, M.W. (2002) "Pilgrimages of nostalgia: patterns of Jewish travel in the United States," *Tourism Recreation Research* 27(2): 17–26.

Ioannides, D. and Cohen Ioannides, M.W. (2004) "Jewish past as a 'foreign country': the travel experiences of American Jews," in T. Coles and D.J. Timothy (eds) *Tourism, Diasporas and Space*, London: Routledge.

Israel Ministry of Tourism (2003) *Tourism Statistics*, available at www.tourism. gov.il/english/default.asp (accessed February 19, 2004).

Israelowitz, O. (1999a) *Guide to Jewish Europe: Western Europe* (10th edn), Brooklyn: Israelowitz.

Israelowitz, O. (1999b) *The Jewish Heritage Travel of New York*, Brooklyn: Israelowitz.

Jacobs, J. (1964) *Jewish Encyclopedia*, New York: Ktav Publications.

Jewish Venice (1999) *The Jewish Ghetto*, available at www.jewishvenice.com (accessed February 19, 2004).

Kosansky, O. (2002) "Tourism, charity and profit: the movement of money in Moroccan Jewish pilgrimage," *Cultural Anthropology* 17(3): 359–400.

Krakover, S. (2005) "Attitudes of Israeli visitors towards the Holocaust remembrance site of Yad Vashem," in G. Ashworth and R. Hartmann (eds) *Horror and Human Tragedy Revisited: The Management of Sites of Atrocities for Tourism*, New York: Cognizant.

Kustanowitz, S. (1998) "Ceremonial offerings: several companies provide Holy Land packages for bar and bat mitzvah celebrations," *TravelAge* 33(37): 32–33.

Levy, A. (1997) "To Morocco and back: tourism and pilgrimage among Moroccan-born Israelis," in E. Ben-Ari and Y. Bilu (eds) *Grasping Land: Space and Place in Contemporary Israeli Discourse and Experience*, Albany: State University of New York Press.

Lowenthal, D. (1985) *The Past is a Foreign Country*, New York: Cambridge University Press.

Memorial and Museum Auschwitz-Birkenau (2004) *Latest News, 23 February. Państwowe Muzeum Auschwitz-Birkenau*, available at www.auschwitz.org.pl/html/ eng/aktualnosci/news.php (accessed April 9, 2004).

Mitsuharu, A. (2003) "Hasidic pilgrimage to Uman: history of a Jewish sacred place in Ukraine," *Slavic Studies Hokkaido University* 50: 105–106.

National Museum of American Jewish History (2003) *About the Museum*, available at www.nmajh.org/information/about.htm (accessed February 20, 2004).

National Park Service (2004) *Touro Synagogue*, available at www.nps.gov/tosy (accessed February 20, 2004).

Neusner, J. (1983) *A History of Michnaic Law of Appointed Times. Part 5: Studies in Judaism in Late Antiquity*, Leiden: E.J. Brill.

Nolan, M.L. and Nolan, S. (1992) "Religious sites as tourism attractions in Europe," *Annals of Tourism Research* 19: 68–78.

Olsen, D.H. and Timothy, D.J. (1999) "Tourism 2000: selling the millennium," *Tourism Management* 20: 389–392.

Plaut, J. (2003) *My Grandparents' Belarus: Journey through a World Abandoned*, Omaha: Creighton University.

Pogrebin, L.C. (1995) "Travel and tourism," *Moment* 20(5): 24–25.

Popson, C. (2002) "First New World Synagogue Rediscovered," *Archaeology* March/ April, available at www.archeology.org/0203/newsbriefs/synagogue.html (accessed February 19, 2004).

Postal, B. and Abramson, S.H. (1962) *The Landmarks of a People: A Guide to Jewish Sites in Europe*, New York: Hill and Wang.

Potok, C. (1978) *Wanderings: Chaim Potok's History of the Jews*, New York: Alfred A. Knopf.

Rinschede, G. (1992) "Forms of religious tourism," *Annals of Tourism Research* 19: 51–67.

Ross, M.A. (2003) *The Jewish Friendship Trail: Guidebook – Jewish Boston History Sites*, Boston: BostonWalks.

Sered, S.S. (2002) "Healing and religion: a Jewish perspective," *The Yale Journal for Humanities in Medicine*, available at www.info.med.yale.edu/intmed/hummed/ yjhm/spirit/healing/ssered.htm (accessed February 19, 2004).

Shachar, A. and Shoval, N. (1999) "Tourism in Jerusalem: a place to pray," in D.R. Judd and S.S. Fainstein (eds) *The Tourist City*, New Haven: Yale University Press.

Shamash (2004) *The Jewish Network*, available at www.shamash.org/links. Tourism_and_Travel/more2.shtml

Shapiro, H. (2002) "Izmir may become next Jewish tourist attraction in Turkey," *The Jerusalem Post* November 6: 5.

Shoval, N. (2000) "Commodification and theming of the sacred: changing patterns of tourist consumption in the 'Holy Land'," in M. Gottiener (ed.) *New Forms of Consumption: Consumers, Culture, and Commodification*, Oxford: Rowman & Littlefield.

Smith, V.L. (1992) "Introduction: the quest in guest," *Annals of Tourism Research* 19: 1–17.

Timothy, D.J. (1997) "Tourism and the personal heritage experience," *Annals of Tourism Research* 34: 751–754.

Vukonić, B. (1996) *Tourism and Religion*, New York: Elsevier.

Webber, J. (1992) "Modern Jewish identities: the ethnographic complexities," *Journal of Jewish Studies* 43: 246–267.

Weinberger, D. (2000) *Around the World the Halachic Way*, Lawrence, NY: Weinberger.

Yoken, J. (1994) "My bar mitzvah in Israel," *Calliope* March: 32–33.

12 Buddhism, tourism and the middle way

C. Michael Hall

Buddhism is usually regarded as one of the world's great religions. Predating both Christianity and Islam and having its origins in Hindu cosmology, Buddhism's adherents number well over 300 million people with the BBC giving a figure of 350 million (BBC 2004). However, the exact number of Buddhists is hard to ascertain because of the extent to which it merges with other belief systems such as Taoism and Confucianism as well as the multitude of different traditions and hence, institutional structures in Buddhist thought. Although Buddhists are primarily concentrated in south and eastern Asia, Buddhism has increasingly begun to attract devotees in the West. Although the essential core belief sets of Buddhism carry across its many sects and traditions, the unity of Buddhism disappeared shortly after the Buddha's death (*paranirvana*), and there are now many different traditions of Buddhist faith around the world.

However, in a strictly Western sense Buddhism is not a religion at all. There is no personal god, no unchanging and immortal soul, and no necessity for the salvation of the latter by the former. Buddhism has no criteria for orthodoxy (Humphreys 1993). It has no equivalent to the Pope in the Roman Catholic faith, no dogmas that must be believed, and no Bible in the Christian sense of inspired revelation. There is no equivalence of baptism in Buddhism, and one cannot be excommunicated and expelled from the fold as in many Christian faiths. A Buddhist is as a Buddhist does.

Buddhism, also known as The Way – the middle way between devotion to the pleasures of the senses and self-mortification – was the teaching of Guatama Siddhartha of the Sakya clan, commonly known as Buddha or the Buddha. Adherents of this Buddhist way generally refer to it as Saddharma (the True Law); members of the southern schools of Buddhism (Hinayana, Theravada) use the term "Buddhism Dhamma" ("That Which Upholds"). Buddhism is a system of doctrine and practice built up by the followers of Buddha about what they believed to be his teachings and insights. We do not know precisely what the Buddha taught, for nothing was written down. Instead, his words were committed to memory by his chief disciple, Amanada, and then passed down orally for several hundred years before being committed to text. The purpose of Buddhism is to achieve a state of consciousness known as Enlightenment. For the practicing Buddhist two things at least are beyond argument: that Buddha attained supreme enlightenment, and that he taught the path or way to it to all humankind. However, Guatama Siddhartha was a man not a god, and what one human can do so others can also achieve.

Guatama Siddhartha became Buddha when he became *buddha*, that is "enlightened" or "awakened," and had "shattered the power of self, broken the fetters of the thinking mind, and made his consciousness one with universal consciousness" (Humphreys 1993: 17). In this state Buddha had achieved absolute wisdom and absolute compassion. According to Suzuki (1957: 40): "There are two pillars supporting the great edifice of Buddhism: Mahaprajna, great Wisdom, and Mahakaruna, great Compassion. The wisdom flows from the compassion and the compassion from the wisdom, for the two are one." In seeking the state of *buddha* one is therefore traveling the same paths as the Buddha.

> Treading in the footsteps of this extra-ordinary man, the long and difficult path which he described from suffering to the end of suffering, from limited mind to No-mind, from the world of the born and formed to the "Unborn, Unformed", from relative to absolute awareness.
>
> (Humphreys 1993: 18)

The notion of a journey is therefore intrinsic to Buddhism, as it is, of course, to many other religions. As Adair (1978: 9) observed with respect to pilgrimage, "the pilgrim instinct must be deep in the human heart, although it finds different expressions according to time, place and culture." In the case of Buddhism the concept of life being journey or a travail arguably has even more importance for understanding human movement and tourism, because personal experience becomes an integral component of understanding or "testing" Buddhist teachings. As Humphreys (1993: 125) commented: "A Buddhist truth is a phrase without meaning until it has been found to be true by personal experience."

Born around the year 580 BCE in the village of Lumbini in present-day Nepal into a royal family, travel outside of his privileged surrounds allowed Buddha to view the suffering of others. This in part led Siddhartha to follow various religious paths in attempting to escape the inevitability of death, pain and old age, including those of religious study, meditation and extreme asceticism. However, Siddhartha pursued the middle way and one day, seated beneath the Bodhi tree (the tree of awakening) he finally achieved enlightenment and became the Buddha. For the next 45 years Buddha then set in motion the "wheel of teaching," the *dharma*, which has become the core of the Buddhist belief systems and the way to *nirvana* and enlightenment, the state of being that lies beyond suffering.

According to Buddhist belief, existence is conditioned by the consequences of individual actions, including past actions (*karma*), consciousness continues after death and finds expression in future life (*rebirth*), but by following the Buddha's path it is possible to escape the cycle of suffering (*nirvana*). The four noble truths that the Buddha experienced seated beneath the Bodhi tree and which are at the heart of Buddhist teaching are: *Dukkha* (all existence is unsatisfactory and filled with suffering which is the central fact of living), *Trsna* (the root of suffering that can be defined as a craving or clinging to the wrong things; searching to find stability in a shifting world is the wrong way), *Nirvana* (that it is possible to find an end to suffering) and the *Ariyan* (Noble) eightfold path that is the way to finding the solution to suffering and bring it to an end. Such a path entails Right Views, Right Aspirations, Right Speech, Right Conduct, Right Mode of Livelihood, Right Effort, Right Mindfulness and Right Rapture (Suzuki 1957).

Buddhism is an extremely tolerant faith, which has arguably been one of its attractions, particularly for those sick of the worst excesses of religious dogma and the seeming inhumanity of those who preach fundamentalist views. Given the inherent tolerance of the Buddhist path it should not be surprising therefore that there are many different schools of Buddhism to be found throughout Asia and the developed world. Each of these traditions has its own interpretations of Buddhism, as well as its own institutional structures and centers of learning and teaching. Nevertheless, the multitude of interpretations is compatible with a faith that stresses the importance of individuals finding their own middle way to enlightenment through their own experiences and looking in on themselves. Buddha can only teach the path to follow; individuals have to find it and follow it themselves.

Nevertheless, while each individual has to find his or her own path, pilgrimage-related travel therefore has a tradition in the Buddhist faith that reaches back to its very beginnings (Chan Khoon San 2001; Foard 1982; Pruess 1976). In the *Digha Nikaya 16*, the Lord Buddha stated that there were four places which the pious person should visit, including the places where the Buddha was born (Lumini in Nepal), was enlightened (Bodh Gaya (Buddhagaya, Bihar state in India)), began the wheel of teaching (at Sarnath near Varanasi (Benares)[1] in India), and where he attained *paranirvana* or passed away (Kushinara (Kushnigar or Kushinagar)). Four other sacred sites that are regarded as some of the great places of Buddhist pilgrimage are places sanctified by the Buddha through the miracles he performed. These are Savatthi/Shravasti, where the Buddha performed a twin miracle to silence heretics, after which he ascended to Tavatimsa heaven to preach to his mother; Sankisa/Sankasia, where the Buddha descended from Tavatimsa Heaven accompanied by Brahma and Sakka, after preaching to his mother and the devas for three months; Rajgir (Rajagaha), where the Buddha tamed Nalagiri, the drunken elephant; and Vaishali (Vesali), where a band of monkeys dug a pond for the Buddha's use and offered him a bowl of honey.

The core pilgrimage places associated with the Buddha's life are all located in the northern Indian states of Uttar Pradesh and Bihar, except Lumbini, which is in Nepal. In ancient times, this area was called *Puratthima* (Eastern Tract), but today it is referred to as the Buddhist Circuit. In Buddhism, any act, including pilgrimage, derives its value from its ability to awaken positive mental states. Therefore, visiting the places that Buddha noted, as well as others that he visited during his lifetime (Singh 2003), has come to assume much significance for Buddhists. The significance of the Buddha's suggestions for visits by the pious does not have the same centrality in the Buddhist faith as it does, for example, for Muslims to visit Mecca during their lifetime. However, as in Islam and arguably even in some Christian traditions (Sumption 1975), those who die in such a pilgrimage receive a special blessing for such a sacrifice. In the case of Buddhists, this means being reborn in a realm of heavenly happiness after death (*Digha Nikaya 16, Maha-Parinibbana Sutta*).

As with many religions there is a hierarchy of sites available for pilgrimage in Buddhism (Figure 12.1). The core pilgrimage sites of nearly all religions are those places emphasized by the godhead – or equivalent (as in the case of Buddhism) – to be locations that the pious and faithful should visit. The next level of sites are those associated with the god's presence on earth, which may be locations associated with particular messages, stories or experiences and those mentioned as being visited in key scriptures. In addition, locations that hold "authentic" religious relics associated with the godhead may also be regarded as fitting into this category. The third level sites are the central locations associated with the different traditions of religious faiths as they have dispersed across the globe. They are often seats of religious institutional structures and learning. Other locations that assume religious significance for pilgrimage and visitation are those associated with great interpreters and exemplars of faith including great teachers, as well as those who would be described as saints. Also included in this category would be those places in which individuals have publicly described their visions or enlightenment. The fourth category of pilgrimage location is that of "routine" sites of devotion (e.g. community temples and spaces that are used for religious celebration). Table 12.1 provides an overview of some of the current and potential Buddhist sites of significance for pilgrims and tourists in selected Asian countries. The relative significance of the different elements in this hierarchy will differ from religion to religion, and even within religions depending on the emphasis taken in different traditions. For example, probably the most notable Buddhists in the West, the Dalai Lamas, like most Tibetans, historically remained isolated from the rest of the globe. The current Dalai Lama, His Holiness the fourteenth Dalai Lama Tenzin Gyatso, is only the second ever to have traveled as far from Tibet as India and China, and he is the first to engage in round-the-world travel.

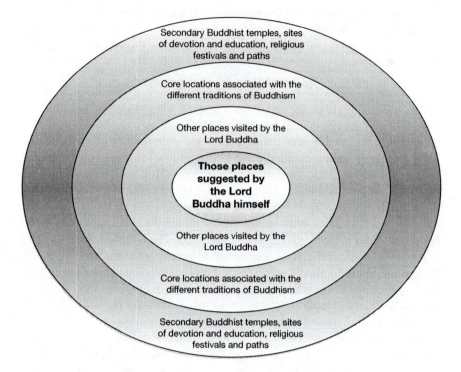

Figure 12.1 A hierarchy of pilgrimage sites in the Buddhist faith

In the case of Buddhism a number of pilgrimage sites became established shortly after the Lord Buddha's death (Singh 2003), including those suggested by the Buddha himself. In the case of Sri Lanka, Dhammika (2004: n.p.) noted that: "By about the 3rd century CE a pilgrim's circuit consisting of sixteen sacred places (solos mahasthana) had developed, most of them associated with the Buddha's legendary visits to the island." As in interpretations of the religious experiences, for the Buddhist the outer journey is regarded as the physical manifestation of an inner spiritual journey, with the path traveled being a framework for the travel within. Enlightenment need not come from such a journey, as the process lies in the mind. However, the path and objects encountered become a focal point for meditation and reflection. Importantly, the journey and the people encountered along the path also provide opportunities to practice the eightfold path.

Clearly not all visits to Buddhist sites of significance can be regarded as pilgrimage in a religious sense. However, an arbitrary divide between "religious" and "secular" pilgrimage or visitation does not provide an adequate representation of the tourist encounter with Buddhism sites either. There is clearly a continuum between secular and religious pilgrimage in

Table 12.1 Current and potential Buddhist sites of significance for pilgrims and tourists in selected Asian countries

Location	Significance
Bangladesh	
Pahapur	Excavated remains of the most important and largest known monastery south of the Himalayas.
Mahasthan	The oldest archeological site in the country with extensive Buddhist ruins.
Mainamati	The area contains over 50 ancient Buddhist settlements dating from the eighth to twelfth century AD. There is also a significant museum.
Kutila Mura	A small hill on which is situated with three stupas side by side representing the three jewels of Buddhism.
Bhutan	
Taktsang Monastery, Paro	The monastery is one of the most sacred sites of Mahayana Buddhism and is included as part of nearly every foreign tourist's itinerary as well as being the most important religious site in Bhutanese domestic travel.
Gangtey Gompa	Situated in the Phubjikha valley this is the most important and prominent Nyingma monastery in Bhutan.
Jigme Memorial Chorten	A modern monastery that combines the function and form of both a stupa and lhakhang (temple).
Cambodia	
Angkor Wat	This ancient temple complex has also been designated a World Heritage Site and is a major tourist destination in Cambodia.
China	
Yung-kang (Shansi) and Lung-men (Honan) caves	The caves are some of the most remarkable Buddhist sites for the massive simplicity of their immense rock-carved Buddhas and their delicate ornamentation.
India	
Bodh Gaya	The place of the Buddha's enlightenment and a spiritual home for all Buddhists. Bodhhi Temple was recently declared a World Heritage Site.
Sarnath	Where Lord Buddha began his wheel of teaching.
Kushinagar	The place that the Buddha chose for his final exit from this earth (*parinirvana*).
Savatthi/Shravasti	Where the Buddha performed a twin miracle to silence heretics, after which he ascended to Tavatimsa heaven to preach to his mother.
Sankisa/Sankasia	Where the Buddha descended from Tavatimsa Heaven accompanied by Brahma and Sakka, after preaching to his mother and the devas for three months.
Rajgir (Rajagaha)	Where the Buddha tamed Nalagiri, the drunken elephant.
Vaishali (Vesali)	Where a band of monkeys dug a pond for the Buddha's use and offered him a bowl of honey.
Rajgir	Site of the first Buddhist council.
Nalanda (Baragaon)	Nalanda, once the most renowned university in ancient India as a center of Buddhist learning, covers an area of 14 hectares and has the ruins of 11 monasteries and 5 temples and has been declared a World Heritage Site.

Table 12.1 (continued)

Location	Significance
	The Buddhist-related sites of Sanchi, and the Caves of Ajanta and Ellora have also been recognized as World Heritage Sites.
Indonesia	
Borobudur	This Buddhist temple complex in Java, which is also a World Heritage Site, still retains spiritual significance for many Buddhists.
Japan	
Nara	The Japanese imperial capital of the eighth century, remains one of the great centers of Buddhist history and contains pagodas, early Buddhist and Shinto shrines, gardens, the Nara National Museum, and the Todai-ji temple, which contains an immense bronze Buddha statue.
Kyoto	Holds numerous Zen temples dating from the Hieian period, as well as significant gardens developed along Zen principles.
Laos	
Vientiane	The Laotian capital contains numerous stupas including That Luang Stupa (which Lao people refer to as Phra That Luang, meaning great stupa), Vat Phra Keo (Ho Phra Keo) and Vat Simuan.
Vat Xieng Thong	Located at the end of the elongated peninsula between the Mekong River and the Nam Khan River in Luang Prabang Province, this former royal temple was called the "Temple of the Golden City."
Vat Mai	Also in Luang Prabang, this was once the residence of Pha Sang Khalat, Lao's most senior monk and Supreme Patriarch of Lao Buddhism.
Myanmar (Burma)	
Yangon (Rangoon)	Shwedagon Pagoda.
Kyaik Htiyo Pagoda	Also referred to as the Golden Rock.
Bagan	The capital of the first Myanmar Empire which contains over 2,000 religious sites.
Mandalay	The Mahamuni Pagoda which enshrines the Mahamuni Buddha image was built by King Bodawpaya in 1784.
Inlay Lake	Famous for the Phaungdaw Oo Pagoda, which contains five Buddha images.
Nepal	
Lumbini	The birthplace of Lord Gautama Buddha and therefore one of the holiest sites in Buddhism. Lumbini is also a World Heritage Site and a notable Hindu pilgrimage site.
Kathmandu Valley	Boudhanath Mahachaitya (Boudhanath Stupa) is the largest stupa in the Kathmandu Valley. The area is also site for Tibetan refugees. Swayambhu Mahachaitya located in the western part of the Kathmandu Valley is also a significant pilgrimage location.
Patan	The area contains numerous Buddhist temples.

Table 12.1 (continued)

Location	Significance
Solu-Khumbu	The district which is part of the Sararmatha zone more usally associated with trekking and mountain climbing contains numerous Gompas (monasteries).
Sri Lanka	
Anuradhapura	Located 205km from the city of Colombo, Anuradhapura was the first capital city of Sri Lanka founded in the fourth century BC. and is a famous Buddhist center of learning which includes a Bodhi tree grown from a sapling taken from the Sri Mahabodhi in Anuradhapura.
Mihintale	Located 11km east of Anuradhapura this was the site of the conversion of King Devanampiyatissa to Buddhism in the year 247 BC.
Kandy	The hill capital of Sri Lanka dating back to the sixteenth century. Dalada Maligawa is the temple where the sacred Tooth Relic is kept.
Sri Pada	Also known as Adam's Peak (2,237m). Buddhists believe that the summit enshrines the footprint of Lord Buddha. Thousands of pilgrims from different faiths climb the mountain to pay homage starting from the full-moon day of December to the full-moon day of April in the following year.
Columbo	Lord Buddha visited Kelaniya Temple, located 12km from present-day Columbo, at the end of the eighth year on attaining enlightenment.
Thailand	
Bangkok	The city contains numerous temples including Wat Phra Kaeo (Temple of the Emerald Buddha), Wat Phra Chetuphon (Wat Pho), Wat Arun, Wat Suthat and Wat Benchamabophit.
Nakhon Pathom Province	Wat Phra Pathom Chedi.
Saraburi	Wat Phra Phutthabat.
Tibet (currently occupied by China)	
Lhasa	The Dalai Lama's Potala Palace is located in this holy city. The palace, like many Tibetan monasteries, is now a state museum.
Jokhang monastery	Located south-east of the Potala, this monastery is the most sacred of all Tibetan pilgrimage sites.
Vietnam	
Huong Pagoda Complex	The Huong Pagoda (Perfume Pagoda) and the Huong Tich Cave are located 70km south-west of Hanoi.
Tay Phuong Pagoda	Also called Sung Phuc Pagoda and located 45km south-west of Hanoi.
Hanoi	Tran Quoc Pagoda.
Hue City	The Thien Mu Pagoda is part of the Hue Ancient Complex of pagodas and temples which are designated as World Heritage Sites.

Primary sources: Chan Khoon San 2001; ESCAP 2003; Singh 2003; Dhammika 2004.

terms of motivations, encounters and the extent to which it fulfills religious beliefs, although arguably from a Buddhist perspective, a secular "path" may still be regarded as an appropriate inner journey to *nirvana*. Buddhism is a way of living and would likely not be troubled even if the Buddha were proven never to have existed (Humphreys 1993).

In Buddhism, life is inherently represented through notions of flow and movement, a concept that has found considerable appeal in the West given insights into physics, ecology and mobility (e.g. Capra 1975, 1997; Zukav 1979). For many Buddhists the act of travel may be regarded as an external indication of the inner spiritual journey. What becomes important in the journey is its reflection of an application of positive human values such as compassion, caring and sharing as part of the improvement of self in this life in preparation for the next as much is it does an avowed desire to follow in the path of the Buddha; in fact the two are inseparable. Indeed, one of the most well-known Buddhists in the West, His Holiness the fourteenth Dalai Lama (2003), urged such a message during his tour of the United States:

> [T]hink about other, or humanity, or world, your mind widen. So, as a result, your own problem then appears not much significant – not like unbearable. But if one think oneself only me, me, me, like that, then your whole sort of attitude, very narrow. So then small problem appear looks very big, very serious, unbearable.

From a Buddhist perspective, therefore, how a person travels in relation to others is more important than the act of traveling itself. Such a position also means that from such a perspective the concept of secular itself needs to be drawn apart to recognize that those who hold secular human values that emphasize compassion, caring and sharing and a recognition of the spiritual need to be distinguished from those who see the world primarily in terms of the material. Arguably, this perspective may also provide some understanding of the relative ease in which Buddhist sites and travelways are opened up to non-Buddhists in that there is great tolerance for other people who are also seeking their own path. Indeed, it is this quality, perhaps as much as the somewhat romantic perception of many Buddhist centers in the West (e.g. Craft 1999; Kleiger 1992; Bentor 1993; Shackley 1994), that has contributed to increasing interest in travel to Buddhist sites of significance (Teague 1997).

One of the great difficulties in assessing the relative importance of Buddhism for tourism and travel is the complexity in gauging an accurate assessment of the proportion of people who undertake travel as an expression of faith. Unfortunately, figures on inbound and outbound tourism are rarely collected on the basis of faith (Russell 1999). Instead, one is forced to examine specific locations imbued with religious significance and attempt to collect relevant data in such places. However, even here one is confronted with difficulties in reconciling the Buddhist worldview of life as journey and the desire to use reductionist empirical approaches to classify and quantify the varieties of tourism experience. Even without such a worldview, accurate data are sparse. For example, a 2003 Economic and Social Commission for Asia and the Pacific (ESCAP) review of Buddhist tourism in Asia was unable to arrive at any comprehensive overview of pilgrimage tourism, although it did argue for the development and promotion of Buddhist tourism circuits noting that:

> Increased prosperity and a better quality of life that is sustainable and beneficial for both present and future generations might remain the aspirations behind the various development goals and strategies. However, a better understanding of the history, culture, religion and aesthetics associated with the Buddhist heritage throughout Asia can do much more than enhance the economic and development contribution of tourism. In terms of heritage, Buddhism can help provide a guiding philosophy, aesthetic and approach to progress that tourists, local communities, planners and policymakers can use in their approach to tourism in Asian countries.
>
> (ESCAP 2003: 4)

The approach of ESCAP to the development of Buddhist tourism routes in East Asia is replicated in the interests of a number of governments in the region in further encouraging international tourism, particularly travel by Buddhists in wealthier countries in the region, such as Japan and Korea, as well as from the Western developed world. However, there are significant differences in approaching Buddhist tourism depending on the political and religious composition of the government. In many ways, some of the issues associated with the development of Buddhist tourism are no different from that of any other form of international tourism development, in terms of the requirements for infrastructure, hospitality services and access. Nevertheless, a number of governments also recognize the need for sensitivity in developing Buddhist tourism sites because of concern over commodification and negatively affecting the spiritual dimension of certain sites. For example, in the case of Bhutan, international tourism is inherently Buddhist tourism, because of the packaging of Bhutanese culture for the visitor. Therefore, a core issue is related to the extent to which Buddhist tourism can be developed, managed and promoted "without diluting the sanctity and sacredness of the festivals and religious establishments" (ESCAP 2003: 18). Yet, arguably, such concerns are not officially articulated to the same extent in countries such as Myanmar and Tibet where Buddhist culture is deliberately commodified by governments not only to attract foreign exchange through tourism and to use international tourism as an indication of political openness, but also as a mechanism of control over local populations (Kleiger 1992; Philp and Mercer 1999).

Nepal is the only country in the region to keep specific statistics on pilgrimage. Holiday and pleasure are the main purpose for visiting Nepal, with trekking and mountaineering being the second most frequent reason. Yet travel for pilgrimages grew substantially in the late 1990s from 4,068 international visitors who stated that pilgrimage was their main purpose in 1997 to over 15,000 in 2000. However, the unstable political situation in Nepal makes it difficult for further development of Buddhist tourism, although there is some irony in this as the Lumbini Declaration issued at the end of the World Buddhist Summit held at Lumbini on December 1–2, 1998 endorsed Nepal's proposal to recognize, declare and develop Lumbini as the fountain of world peace and the holiest Buddhist pilgrimage center (ESCAP 2003). Nevertheless the potential for growth in religious tourism once political stability returns is substantial. Sri Lanka has experienced significant growth in international and domestic pilgrimage since a cease-fire was declared between the Tamil Tigers and the Sinhalese government in 2002. According to Berkwitz (2003: 62):

> For the first time in about two decades of intermittent warfare and the ongoing threat of state-sponsored or terrorist violence, persons of all religious and ethnic background were enjoying the freedom to travel safely throughout the island. As a result, the numbers of pilgrims and other domestic tourists traveling to distant sites of religious import-ance increased dramatically ... Buddhist pilgrimage sites in the East, such as Dighavapi, Buddhangala, Seruvavila and Somavati, are welcoming large numbers of devotees for the first time in many years. Travelers are able to visit with monks living in areas that were subject to the threat of warfare earlier and are encouraged to help fund major rebuilding projects taking place at those sites.

Nevertheless, almost as soon as tourism began to increase following the return of stability, concerns began to be raised with respect to the commercialization of the spiritual experience and environmental damage. Growth in tourism may also potentially contribute to political unease in some situations because of conflicting spiritual and material interests. As Berkwitz (2003: 63) noted, "as more Buddhists flock to shrines located in predominantly Tamil and Muslim areas, there will be more chances for conflicts to erupt at and around these pilgrimage areas."

As well as conflict over spiritual space, conflict may also arise in relation to competing development strategies of sacred space. Within the Buddhist faith debates over the connection between spirituality, development and the environment are becoming increasingly commonplace to the extent that, in some circumstances, it may even lead to action against certain types of tourism development that are seen to be at odds with the sanctity of a Buddhist site. For example, in Thailand, Buddhist monks have become increasingly involved in development issues with the first case in which monks took a conservationist position, being in opposition to the 1985 proposal to build a cable car up Doi Suthep mountain and through Doi Suthep-Pui National Park in Chiang Mai to promote tourism and economic development. The significance of the site for Thai Buddhists was described by Chayant Pholpoke (1998: 265–6):

> Doi Suthep is a mountain lying at the outskirts of the town of Chiang Mai, named for a seventh-century Lawa chieftain who converted to Buddhism, became a monk, and retreated from the world to the mountain which now bears his name. It is the location of an important Buddhist monastery, Wat Pra That, which houses a relic of the Buddha. Local people revere the mountain temple as a destination of spiritual significance for Buddhists. Since the construction of the temple in the fourteenth century, Doi Suthep has been an important pilgrimage site.

Opposition by monks to the cable car project were primarily on heritage and spiritual grounds although wider connections were made to forest conservation issues (Taylor 1996; Darlington 1998). Concerns over development are not isolated to Thai Buddhist monks, and throughout the Buddhist worldview, debate is occurring over how development and environmental concerns can be reconciled (Chatsumarn Kabilsingh 1998; Kaza and Kraft 2000).

Conclusion

The Thai and Sri Lankan cases demonstrate that throughout east Asia, Buddhist tourism-related development is surrounded by the same type of political issues that affect tourism in general as well as contestation of sacred space (see Eade and Sallnow (1991) for a discussion of this in a Christian context). However, as in other faiths, there is usually no clear-cut perspective on what would constitute appropriate tourism development. From a Buddhist perspective such a situation is not necessarily a problem of religious dogma. The Lord Buddha and the Buddhist scriptures teaches that each must find the middle way along the path of experience. The path of tourism development followed in one location will not be the same as that followed in another, and neither will be the journey of the pilgrim tourist. Instead, what is important is that each individual in life undertakes such a journey full of compassion and the other precepts of the eightfold path.

Note

1 Alternate Anglicized spellings are often used for the same locations.

References

Adair, J. (1978) *The Pilgrims' Way: Shrines and Saints in Britain and Ireland*, London: Thames and Hudson.

BBC (2004) *Religion and Ethics, Buddhism, introduction*, available at www.bbc. co.uk/religion/religions/buddhism/intro.shtml (accessed 25 January 2005).

Bentor, Y. (1993) "Tibetan tourist Thangkas in the Kathmandu Valley," *Annals of Tourism Research* 20(1): 107–137.

Berkwitz, S.C. (2003) "Recent trends in Sri Lankan Buddhism," *Religion* 22: 57–71.

Capra, F. (1975) *The Tao of Physics: An Exploration of the Parallels Between Modern Physics and Eastern Mysticism*, London: Fontana/Collins.

Capra, F. (1997) *The Web of Life: A New Synthesis of Mind and Matter*, London: Flamingo Books.

Chan Khoon San (2001) *Buddhist Pilgrimage*, Selangor Darul Ehsan: Subang Jaya Buddhist Association.

Chatsumarn Kabilsingh (1998) *Buddhism and Nature Conservation*, Bangkok: Thammasat University Press.

Chayant Pholpoke (1998) "The Chiang Mai cable-car project: local controversy over cultural and eco-tourism," in P. Hirsch and C. Warren (eds) *The Politics of Environment in Southeast Asia: Resources and Resistance*, London: Routledge, 262–277.

Craft, C. (1999) "Walking around the Buddha," *Cross Currents* 49(2): 197–205.

Darlington, S.M. (1998) "The ordination of a tree: the Buddhist ecology movement in Thailand," *Ethnology* 37(1): 1–15.

Dhammika, S. (2004) *Sacred Island: A Buddhist Pilgrim's Guide to Sri Lanka*, Buddha Dharma Education Association, available at www.buddhanet.net/sacred-island/introduction.html (accessed 25 January 2005).

Eade, J. and Sallnow, M.J. (1991) *Contesting the Sacred: The Anthropology of Christian Pilgrimage*, London: Routledge.

ESCAP (Economic and Social Commission for Asia and the Pacific) (2003) *Promotion of Buddhist Tourism Circuits in Selected Asian Countries*, ESCAP Tourism Review No. 24, New York: United Nations.

Foard, J.H. (1982) "The boundaries of compassion: Buddhism and national tradition in Japanese pilgrimage," *Journal of Asian Studies* 61(2): 231–251.

His Holiness the fourteenth Dalai Lama (2003) "The Dalai Lama in Central Park," transcript of excerpts from the Dalai Lama's speech given in Central Park in New York City, September 21, 2003, available at www.beliefnet.com/story/133/story_13307_1.html (accessed 25 January 2005).

Humphreys, C. (1993) *The Buddhist Way of Life*, New Delhi: Harper Collins Publishers India.

Kaza, S. and Kraft, K. (eds) (2000) *Dharma Rain: Sources of Buddhist Environ-mentalism*, Boston: Shambhala Publications.

Kleiger, P. (1992) "Shangri-La and the politicization of tourism in Tibet," *Annals of Tourism Research* 19(1): 122–125.

Philp, J. and Mercer, D. (1999) "Commodification of Buddhism in contemporary Burma," *Annals of Tourism Research* 26(1): 21–54.

Pruess, J.B. (1976) "Merit-seeking in public: Buddhist pilgrimage in northeastern Thailand," *Journal of the Siam Society* 64(1): 167–206.

Russell, P. (1999) "Religious travel in the new millennuim," *Travel & Tourism Analyst* 5: 39–68.

Shackley, M. (1994) "Tourism in the land of Lo, Nepal/Tibet: The first eight months of tourism," *Tourism Management* 15(1): 17–26.

Singh, R.P.B. (2003) *Where the Buddha Walked: A Companion to the Buddhist Places of India*, Varanasi: Indica Books.

Sumption, J. (1975) *Pilgrimage: An Image of Medieval Religion*, London: Faber and Faber.

Suzuki, D.T. (1957) *The Essence of Buddhism* (2nd rev. edn), London: The Buddhist Society.

Taylor, J. (1996) "'Thamma-chät': activist monks and competing discourses of nature and nation in Northeastern Thailand," in P. Hirsch (ed.) *Seeing Forests for Trees: Environment and Environmentalism in Thailand*, Chiang Mai: Silkworm Books, 37–52.

Teague, K. (1997) "Representations of Nepal," in S. Abram, J. Waldren and D.V.L. Macleod (eds) *Tourists and Tourism. Identifying with People and Places*, New York and Oxford: Berg, 173–195.

Zukav, G. (1979) *The Dancing Wu Li Masters: An Overview of the New Physics*, London: Fontana Paperbacks.

13 Tourism and Islam

Considerations of culture and duty

Dallen J. Timothy and Thomas Iverson

Many of the world's most impressive natural and cultural sites are located in countries where Islam is the dominant religion (e.g. Petra in Jordan, the Great Pyramids of Egypt, the Roman ruins of Syria, Lebanon and Libya, the ancient cities of Yemen, the vast deserts of Saudi Arabia, the rainforests of Malaysia and the volcanoes of Indonesia). Yet with only a few exceptions, Islamic countries generally have not seen significant growth in tourism during the past twenty years, and most have not become major international destinations since the end of the Second World War, when global tourism began to flourish.

On the other hand, however, Muslims are avid travelers. In fact, Islamic doctrine encourages travel. Some Muslims consider themselves closer to God when they travel and believe their prayers are more effective while traveling than when offered at home (Aziz 2001). They can be found touring throughout the Middle East, Asia, Africa, Europe and North America on business, visiting friends and relatives, shopping or simply relaxing. They are also passionate travelers for religious purposes throughout the Middle East, including Mecca and Medina in Saudi Arabia, as well as Jerusalem and other places of religious importance. In their world travels, Muslims wield distinctive behaviors in terms of the foods they eat, the company they keep and the activities they undertake.

This chapter aims to describe these unique matters of tourism development and international travel by examining the tenets of Islam and their influences on travel, tourism and pilgrimage in this religion. The chapter first focuses on patterns of tourism in primarily Islamic nations, followed by an examination of one of the largest travel phenomena in the world, the *Hajj*, and its religious and tourism-related components. Also considered are other forms of travel among Muslims and the negative association that has been created in recent years between Islam and terrorist acts against tourists.

Islam in brief

The prophet Mohammed was born in 570 CE[1] in Mecca. His father died before Mohammed was born, so he was raised by his mother and other relatives. Though he did not have a formal education, Mohammed developed a reputation as a good person and fair trader. His first wife had employed him to manage her caravan trading operation, at which he was successful and known for his honesty. For example, the story is told of a tense moment among the clans of Mecca in the year 605 CE. The "Black Stone," an important religious relic, was removed for a renovation of the holy Kaaba, and when the time came to return it to its place, each of the four major clans desired the honor of replacing it. Mohammed (al-Ameen – he who can be trusted) was given the responsibility of deciding who would place the stone in the sacred edifice. The prophet's clever solution was to place the stone on a rug and have each clan representative hold a corner of the rug to lift the stone, whereupon Mohammed put it into place with his own hands (Kaïdi 1980: 30).

As he matured, Mohammed became very introspective and curious about the idolatry that characterized local religion. He often sought tranquility in nearby caves where he could ponder and meditate. As a result, in 610 CE, he began receiving visits in the caves by the angel Gabriel, who brought revelations from God. Gabriel urged Mohammed to recite and preach God's messages to the public. These revelations were recorded into verses and chapters (*Surahs*) into the compilation known as the Quran. Much of the volume deals with the absolute oneness and omnipotent nature of God (*Allah*), absolute submission to his will (Islam), the spread of God's religious community (*ummah*), social justice (almsgiving to the poor), and the struggle against aggression, injustice and temptation (*jihad*) (Lynn 2004: 44). This fifth concept, *jihad*, is often represented only according to its first two meanings. However, Mohammed argued that the greatest *jihad* is the personal struggle against temptations.

Soon after establishing the movement, a small following began to accept the word of God as revealed to Mohammed. However, their teachings against idolatry and pagan polytheism were threatening to the ways of Mecca's population, who thus began to persecute the prophet and his followers. When the persecution and death threats increased, Mohammed and his followers left Mecca in 622 and settled in Medina, some 420km away. Their trip to Medina (*hijra*) marks the beginning of the Muslim calendar (in Gregorian years it translates to July 16, 622) (Bloom and Blair 2000; Brown 2004; Robinson 1999).

Throughout the following ten years, many converts were won in Medina, and several battles were fought with the aggressive Meccans. However, in 629, Mohammed returned to Mecca as a pilgrim, though with military escorts, and defeated many local persecutors, cleansed the Abrahamic holy sites of pagan gods and rituals, and established the true religion in that city. Mohammed died suddenly in June 632 and was buried under his house in Medina (Bloom and Blair 2000).

During his lifetime, Mohammed revealed many important doctrines and practices that determine the religious (and everyday) lives of Muslims throughout the world. The prophet declared that Islam was built on five pillars, which all Muslims must follow. The first pillar is the profession of monotheistic faith, or testifying that there is no god but God (*Allah*) and that Mohammed was his messenger. Second is the performance of ritual prayers five times each day – at dawn, midday, mid-afternoon, sunset and nightfall. This must normally be done "prostrate" on the ground facing the direction of Mecca. The third duty is giving alms to the poor or providing other forms of charity donations. Fasting from dawn to sunset during the month of Ramadan is the fourth pillar of Islam. This includes abstaining from worldly desires, as exemplified by adults avoiding food and drink between sunrise and sunset each day of the holy month. Finally, for those who have the resources, performing the *Hajj* – the pilgrimage to Mecca – is the fifth requirement in Islam. The *Hajj* is a major effort that requires preparation, study and devotion. A lesser pilgrimage, known as the *Umrah*, is not compulsory but represents a similar mission to holy sites for enlightenment and as devotion to duty. The *Hajj* and *Umrah* are described in more detail later in the chapter.

Islam, with more than one billion adherents, is the dominant religion in forty-six countries, and a major secondary religion in many others (Table 13.1). Most of the world's Muslims are spread throughout North Africa, the Middle East, Central Asia and Southeast Asia, although there are large populations living in Europe, East Africa and North America as well. In most countries where Islam dominates the religious landscape, religion is the center of life. It shapes all social, political and economic behavior (Henderson 2003).

Tourism in Muslim countries

In many nations, particularly in the Middle East and North Africa, Islam is the foundation of society and order of law. Islamic principles thus underpin tourism policy, development objectives, and the management and operation of the industry.

Several Muslim states have become important tourist destinations over the years since the end of colonialism in North Africa, the Middle East and Asia in the mid-twentieth century. Famous Islamic architecture (e.g. Taj Mahal and the Ottoman palaces in Turkey), archeological sites from other empires (e.g. Romans), business meetings, culture and climate are among the most appealing attractions in the Muslim world (Jackson and Davis 1997). Nonetheless, most Islamic nations, particularly in the Middle East, have been reluctant recipients of tourism growth. Many studies have examined the dynamics of tourism in countries where Islam is the majority religion, typically all touching on the implications of religion on tourism (e.g. Abdurrahman 2000; Alavi and Yasin 2000; Alhemoud and Armstrong 1996; Baum and Conlin 1997; Burns and Cooper 1997; Cunha 2004; Domroes 2001; Frembgen 1993; Grötzbach 1983; Henderson 2003; Horsfall 1996; Laporte 2003; Poirier 1995; Seddon and Khoja 2003; Timothy 1999; Wiebe 1980).

Table 13.1 Countries having Muslims as a majority of the population

Country	% Muslim	Country	% Muslim
Afghanistan	99	Niger	95
Albania	70	Nigeria	50
Algeria	99	Oman	>90
Azerbaijan	93.4	Pakistan	97
Bahrain	98	Qatar	95
Bangladesh	88.3	Saudi Arabia	100
Brunei	67	Senegal	95
Burkina Faso	55	Sierra Leone	60
Chad	51	Somalia	99.9
Comoros	98	Sudan	70
Cyprus (North)	>95	Syria	86
Djibouti	94	Tajikistan	85
Egypt	90	Tunisia	98
Gambia, The	95	Turkey	99
Guinea	85	Turkmenistan	89
Indonesia	87	United Arab Emirates	96
Iran	99	Uzbekistan	88
Iraq	97	Yemen	>95
Jordan	95		
Kuwait	85	*Countries with large Muslim*	
Kyrgyzstan	75	*populations*	
Lebanon	59.7	Bosnia & Herzegovina	40
Libya	97	Cote d'Ivoire	40
Malaysia	60.4	Eritrea	48
Maldives	99	Ethiopia	45
Mali	90	Guinea Bissau	45
Mauritania	100	Kazakhstan	47
Morocco	99.9	Tanzania	45

Source: Central Intelligence Agency 2005; US Department of State 2005.

Muslim leadership and citizenry recognize the negative social impacts that tourism typically brings with it – drugs, alcohol consumption, lewd behavior, gambling, immodest dress, open affection between males and females, sexual promiscuity and prostitution (Beckerleg 1995; Henderson 2003; Sindiga 1996), which are all forbidden by Islamic law. Likewise, tourists are often ignorant of local mores and take photographs of local people, which is offensive to many Muslims. They also often behave irreverently in and around sacred places, especially in entering holy places wearing shoes, which should be removed before entering, and scant clothing (Power 2000). Some countries are more tolerant of western callousness and/or ignorance (e.g. Indonesia and Malaysia), but tension is still created when tourists misbehave, such as in Kenya where visitors sometimes offend local Muslims by "kissing or fondling right in front of a mosque when prayers are going on" (Sindiga 1996: 429).

Many Muslims are highly concerned about the immoral influences of tourism. This is why the industry has not been a major development priority in many Islamic nations (Baum and Conlin 1997). Images of the Arab world as portrayed in the western media often negatively depict Islamic societies as oppressive, harsh, violent, and intolerant. These unfortunate stereotypes have curtailed the growth of tourism in many potential destinations which, according to Aziz (2001: 152), "has not disappointed those in the Arab world who fear [tourism's] impact on traditional Islamic culture, for there is a widely held counter stereotype of Western tourists as shameless hedonists corrupting local morals through empty promises of economic benefit."

In most Middle Eastern countries, tourist resorts tend to be more inland focused rather than coastal, for it is typically coastal tourism that encourages bare skin and public sensual contact (Poirier 1995; Ritter 1975). Several Muslim countries have devised two primary methods to cope with the immoral downside of tourism. First is the establishment of segregated tourist areas (Ritter 1982; Sindiga 1996). The Maldives, for example, which has developed a thriving tourist economy, has in a creative way developed the product in the form of segregated resort islands, wherein uninhabited atolls become the domain of tourists. This has resulted in an interesting pattern where each island is a self-contained and separate tourist resort. Foreign tourists are required to spend their time on their resort island of choice and may only visit inhabited islands with restrictions, including that they dress and behave appropriately. Segregation in this example minimizes the contact between the hedonistic tourists and the Muslim population (Domroes 2001).

The second coping mechanism is to minimize cultural conflict by emphasizing domestic tourism and encouraging tourists from other Muslim majority countries to visit (Alavi and Yasin 2000; Henderson 2002; Ritter 1975). For example, Saudi Arabia's international tourism efforts have emphasized visits from nearby Middle Eastern states (Rimmawi and Ibrahim 1992), and only recently has the country started issuing tourist visas for the world's population at large. Even with the new visa policies, Saudi Arabia still shuns hedonistic forms of tourism (e.g. beach resorts). Likewise, while beach resort-based tourism and other forms of international pleasure travel have become popular in Malaysia, there is now a move going on to get more Muslims to visit from other countries (Henderson 2002).

While many Muslim nations desire to develop international tourism, and international arrivals to the Middle East are increasing, Muslims face many significant challenges to their cultural traditions and religious lifestyles (Burns and Cooper 1997). Aziz (2001: 154) acknowledged this and suggested that "no Islamic country has yet managed to accommodate the needs of Western tourists without compromising the religious and cultural expectations of most of its own people." One recent example is the Turkish

Republic of Northern Cyprus, where casinos were recently established in an attempt to grow the tourist economy, and unfortunately, widespread prostitution has followed (Scott 2003). Concerns over such "moral pollution" are not unique to Muslim countries.

Hajj

The most important travel event for Muslims is of course the fifth pillar of Islam, the pilgrimage (*Hajj*) to Mecca, Saudi Arabia, which every Muslim is required to make at least once in a lifetime, inasmuch as he or she is physically and financially able. In many predominantly Muslim countries, national governments and some international organizations (e.g. the UN) have programs to assist citizens in making the pilgrimage by providing loans, grants and other forms of support. In addition, TV and radio shows sometimes offer *Hajj* trips as prizes, while less affluent Muslims from Africa often finance their *Hajj* travels by taking along goods for trade or barter (Aziz 2001).

The *Hajj* begins on the seventh day of the twelfth month (*Zul-Hijja*) of the Islamic calendar. During the past five years, the number of travelers to Mecca has grown considerably. The latest data from 2003 and 2004 show that over two million Muslims traveled from all over the world (including Saudi Arabia) each year to participate in the *Hajj* rituals (Table 13.2). While the number has grown in recent years, the Saudi Arabian government's Ministry of the Hajj restricts the number of pilgrims that can visit the country at any one time by implementing quotas on arrivals from individual countries (Aziz 2001; Din and Hadi 1997; McDonnell 1990). Since the massive influx of pilgrims occurs at a specific time within the Muslim calendar, the logistics of managing the *Hajj* are quite complex.

Table 13.2 Arrivals of *Hajj* pilgrims in Saudi Arabia, 1996–2004

Year	Pilgrims from Mecca	Pilgrims from other areas of Saudi Arabia	Pilgrims from outside Saudi Arabia	Total
1416 AH (1996 CE)	126,739	658,030	1,080,465	1,865,234
1417 AH (1997 CE)	121,516	652,744	1,168,591	1,942,851
1418 AH (1998 CE)	113,928	585,842	1,132,344	1,839,154
1419 AH (1999 CE)	127,146	648,122	1,056,730	1,831,998
1420 AH (2000 CE)	105,369	466,230	1,267,555	1,839,154
1421 AH (2001 CE)	108,463	440,808	1,363,992	1,913,263
1422 AH (2002 CE)	110,592	479,984	1,354,184	1,944,760
1423 AH (2003 CE)	116,887	493,230	1,431,012	2,041,129
1424 AH (2004 CE)	119,364	473,004	1,419,706	2,012,074

Source: Government of Saudi Arabia 2005.

Note: Under the Islamic calendar, it is possible to have two *Zul-Hijja* months (or part months) in the same Gregorian year (e.g. 1973), or the *Hajj* being split between years.

While there is some debate about the historical origins of *Hajj*, most Muslims believe that God revealed its associated rituals to Abraham. Later generations, however, corrupted the practice by adding idolatry and other falsehoods to the *Hajj* until Mohammed purified and re-established the rituals as they were revealed from God in their pure Abrahamic form (Long 1979; Robinson 1999).

The *Hajj* is mentioned several times in the Quran. For instance, in Quran 22: 27–30, God told Mohammed to publicize the *Hajj* to the people:

> Announce to the people the pilgrimage. They will come to you on foot and on every lean camel, coming from every deep and distant highway that they may witness the benefits and recollect the name of God in the well-known days (*ayyam ma'lumat*) over the sacrificial animals He has provided for them. Eat thereof and feed the poor in want. Then let them complete their rituals and perform their vows and circum-ambulate the Ancient House . . . Whoever honors the sacred rites of God, for him it is good in the sight of his Lord.
>
> (see Peters 1994: 7)

Another popular *Surah* (verse) proclaiming the *Hajj* is found in Quran 3: 90–91.

> The first House of Worship founded for mankind was in Bakka (Makkah). Blessed and guidance to mankind. In it are evident signs, even the Standing Place of Abraham . . . and whoever enters it is safe. And the pilgrimage to the temple [*Hajj*] is an obligation due to God from those who are able to journey there.
>
> (see Long 1979: 3)

The region surrounding Mecca is known as the sacred area (*haram*), which includes Arafat, Mina and Muzdalifa, other important locations in the *Hajj* experience. The boundaries of *haram* are visibly marked with signs and checkpoints, and only Muslims are permitted to cross over them (Robinson 1999; Rowley 1997). In the city of Mecca, a large mosque, *al-masjid al-haram*, or the sanctuary, dominates the landscape and is the center of the annual pilgrimage. The mosque has been expanded on several occasions to accommodate the growth of pilgrim numbers and even encloses the two hills Safa and Marwa, which play a critical role in the pilgrimage rituals.

At the center of the mosque complex is the Kaaba, a structure built of granite blocks, covered in a black drape, and standing 15 meters tall, 12 meters long and 10.5 meters wide. It is the holiest site in all of Islam and the core of the pilgrimage experience (Long 1979; Peters 1994). According to Islamic tradition, the Kaaba was the earliest place of worship on the earth, built by Adam himself. According to some Muslim scholars, after Adam and Eve's expulsion from the Garden of Eden: "Adam ... complained to God that he no longer heard the voices of the angels, so God instructed him to build a house on earth and to circumambulate it in the way that he had seen the angels circumambulate his throne" (Robinson 1999: 142). However, it is commonly thought that through intervening years the Kaaba fell into ruin and was finally destroyed in the flood (Long 1979; Robinson 1999). The dominant belief is that the present-day Kaaba was built by Abraham and his son Ishmael on the same site where the original Kaaba had been built by Adam (Peters 1994). However, over the years it fell into pagan use but was later cleansed by Mohammed (Brown 2004). In the east corner of the Kaaba is the Black Stone, which was sheathed in silver in 757 CE (Kaïdi 1980). This stone is believed to have been part of the original building constructed by Adam and plays a very important part in the *Hajj*.

Northwest of the Kaaba is a semi-circular wall known as the *hatim*, which covers the *al hijr* – believed to be the burial place of Ishmael, the son of Abraham, and Hagar, his mother. Near the Kaaba's northeast side is a water well known as Zamzam, and nearby is a stone bearing Abraham's footprints known as the "station of Abraham" (*Maqam Ibrahim*). The station of Abraham is the location where Abraham stood to invite human-kind to participate in the *Hajj*, and as a result, God caused his footprints to remain there (Peters 1994; Robinson 1999).

The small hills Safa and Marwa, which were located near the mosque but are today enclosed within it, are significant because Hagar, the wife of Abraham, was said to have run back and forth between the two hills searching for water so her son Ishmael would not die of thirst. Her prayers were answered when the spring of Zamzam amazingly burst forth under Ishmael's hand (Robinson 1999).

To prepare for the *Hajj*, a person must have put his or her earthly affairs in order and become spiritually prepared. This has traditionally included paying off all debts before departure and making sure that family members remaining at home are cared for while the pilgrims are away. In addition, pilgrims are encouraged to select dependable and virtuous travel companions (Long 1979).

During the *Hajj*, pilgrims must perform certain rites and ceremonies, which represent Abraham and Hagar's experiences. The process is about making covenants to devote oneself to God and pledge loyalty and servitude (Khan 1986). The first step in the *Hajj* is a ritual purification, or cleansing, known as *ihram*, which takes place on the seventh day of *Zul-Hijja*. At this time, street clothes are exchanged for special *ihram* garments, comprised of loose dresses for women and two white wrap-around sheets for men. The pilgrims then enter the *al-masjid al-haram* through the gate of peace and listen to a sermon regarding the rituals they will be per-forming. On the eighth day of the month, the pilgrims carry out the arrival *tawaf* and *sai*. The *tawaf* entails circumambulating the Kaaba seven times and paying homage to the Black Stone. Following the circumambulation procedure a prayer is offered at the station of Abraham, after which participants drink water from Zamzam Well.

Sai is the act of running between Safa and Marwa seven times, which takes place after circling the Kaaba. Upon completing *tawaf* and *sai*, the pilgrims proceed to Mina, where they pray from the middle of the eighth day to sunrise of the ninth day. After dawn on the ninth day, they continue to the plain of Arafat, approximately 15km further east, where they pitch their tents and stand on the plain from midday to sunset praising God and asking for forgiveness. This part of the pilgrimage is considered to be the most important element of the *Hajj*. In the evening the tents are dismantled and participants walk en masse to Muzdalifa, praying as they go. At Muzdalifa, the night is spent praying. The next morning, pilgrims gather small stones and return to Mina, where they throw the stones at three pillars, while uttering the words "In the name of God. God is most great." The casting of stones represents fighting against Satan. Once all stones are gone, pilgrims get their hair shaved off or remove a symbolic portion, and those who can afford it carry on by sacrificing a sheep, goat or camel. Less affluent pilgrims are permitted to sacrifice chickens in place of larger animals (Long 1979). Ordinary clothes then replace the *ihram*, and pilgrims return to Mecca to re-perform the *tawaf* and *sai*. Then they return to Mina to throw more pebbles at the pillars on the eleventh day. The twelfth day is spent in Mina in prayer. On the thirteenth day, the pilgrims again throw more stones and return to Mecca where they go through *tawaf* one last time, followed by a prayer at the station of Abraham. Finally the pilgrims drink more water from Zamzam Well, and the *Hajj* is finished (Bloom and Blair 2000; Long 1979; Robinson 1999).

People who complete the *Hajj* earn the designation "hajji," which is a title of great respect and admiration (Bloom and Blair 2000). To demonstrate that they have completed the pilgrimage at Mecca, people often integrate al-Hajji into their formal name.

As noted previously, strict regulations are in place to assure that non-Muslims do not enter Mecca or participate in the *Hajj*. However, virtual pilgrimages are offered online on many websites for those who are unable to do the *Hajj*, both Muslims and other curious observers.

In common with many other religious organizations and countries that host religious pilgrimages, Islam in general and Saudi officials in particular reject the idea that *Hajj* is any kind of tourism, owing to the association of tourism with stereotypical hedonistic behavior. Nonetheless, from the global tourism perspective, the *Hajj* is seen as one of the world's largest tourist gatherings, which demonstrates many of the traditional components of tourism (Ahmed 1992; Aziz 2001). Residents of Mecca and Medina have earned a living by serving the needs of Muslim pilgrims for centuries.

For instance, Meccan guides are often hired to help adherents perform the rites of *Hajj*, arrange accommodations, and provide food, water and transportation (Ahmed 1992; Long 1979). Many are also hired as interpreters for travelers who do not speak or understand Arabic.

Much of the *Hajj* experience has been modernized to meet the needs of an ever-sophisticated and growing demand. Perhaps the best example of this is the proliferation of air-conditioned motor-coaches that transport some people to and from Mecca and between some of the sacred sites. Even the great *al-masjid al-haram* (mosque) of Mecca is now entirely air-conditioned for the comfort of pilgrims (Ahmed 1992). Likewise, many affluent believers today are replacing tents with three- and four-star, air-conditioned hotels. The speed of modern transportation modes enables hundreds of thousands of people to visit Mecca who might not have otherwise been able. The growth in air access has made the trip much shorter than by traditional means (i.e. caravan, foot and ship) (Ahmed 1992; Metcalf 1990; Rollier 1985). Special charter flights are arranged by various world airlines, often for special pilgrim fares, and Jeddah's airport has its own *Hajj* terminal to process all the pilgrims.

Another modern-day phenomenon is the proliferation of hundreds, if not thousands, of travel agencies and tour operators throughout the world who sell "Hajj tours" (Ahmed 1992). A search of the Internet will reveal that for approximately US$3,500, religious tourists from North America and Europe can have air-conditioned and fully-guided *Hajj* experiences, including visits to Medina. Sightseeing and shopping have also become important components of the pilgrimage to Saudi Arabia (Aziz 2001).

These commercial aspects of the pilgrimage have received considerable criticism in recent years from fundamentalist Muslims and scholars of Islam, suggesting that the *Hajj* has been diluted and lost much of its meaning by extreme commoditization of what should be a laborious demonstration of faith. Some people blame the Saudi ruling family and government for commodifying a sacred experience and secularizing sacred space (Delaney 1990; Siddiqui 1986).

According to Bhardwaj (1998), there are three additional problems associated with the *Hajj*: accommodations, health and transportation, although the transportation and accommodations issues are being dealt with as new flights and motor-coach routes are initiated and new hotels are being built. Health and safety have been persistent concerns among governments and *Hajj* travelers for many years. Every year many people, sometimes numbering in the hundreds, are severely injured or killed as they are crushed and/or suffocated in the rush of huge crowds in and around the great mosque while performing the *tawaf* and *sai*. In common with large gatherings of other religions, such as the Kumbha Mela among Hindus, there are also major concerns about sanitation and the chronic spread of diseases such as smallpox, cholera, malaria and tuberculosis (Ahmed 1992; Balkhy *et al.* 2003; Wilder-Smith 2003).

Other destinations and forms of travel/pilgrimage

In addition to the obligatory *Hajj*, Muslims take many other religiously motivated trips, which are known as *ziarat* and tend to focus on shrines, mosques and tombs associated with Mohammed, saints, martyrs, *imams* and other revered people (Bhardwaj 1998). These voluntary pilgrimages are akin to pilgrimage travel in Christianity, which people do as a way of getting closer to God, gaining blessings in their lives, and celebrating famous figures from history. Bhardwaj (1998: 71) identified two major types of *ziarat*, the first being journeys undertaken purely for sentimental or emotive reasons. These may be a source of spiritual enrichment as people travel to listen to holy discourses by *imams* (religious leaders), participate in festivals, visit famous graves or well-known locations where the prophet was known to have stayed, or to celebrate special days related to marriages and births. By way of example, many Muslims visit St Katherines Monastery in Egypt because Mohammed was known to have stopped there and was offered kindness and hospitality (Shackley 1998).

In Ghazanfar's (2004: 2) travelogue, he describes his visit to Spain in 1998 as "almost like a pilgrimage" and later writes that he has had "the good fortune of having done some travelling here and there, none – except my visits to Mecca and Medina – surpasses the spiritual and emotional experience that I felt upon being immersed for a few days into Spain's Islamic past" (2004: 12). His travelogue, though, raises some issues that are probably germane to many tourism sites that have seen religious strife. Historical interpretation is quite subjective and unfortunately it is likely that areas with minimal political and cultural representation by Muslims are prone to develop culturally biased information and tours.

Ghazanfar's clear joy at discovering the remnants of Islam in the architecture, art, industry and even in the local street names was tempered by his negative experiences with the guided interpretation at major tourism sites. His experience at Cordoba's Grand Mosque was not very encouraging for fellow Muslim tourists. During a tour, Ghazanfar (2004: 7) "felt the urge" to pray a *suna* (a voluntary prayer) and relates this incident:

> I chose a remote corner away from the group and began my prayers. However, as I was ecstatically engaged in my prayers, during the second raka'at, an angry man, trembling with rage, approached me, literally breathing into my face, screaming in Spanish, "No Muslim prayers. No Muslim . . . prayers" (so I understood). At first I resisted the pressure of the security guard; but I had to break my prayers . . . Despite my protests (and the protest of the guide and other members of the group), the guard tightly held my arm and escorted me out of the Mosque.

The second type of *ziarat* is comprised of trips having to do with problems of everyday life. These journeys may be taken as a way of seeking benefits, such as health for a family member, or visiting a grave, which may assist in healing a disease or improving fertility. Some trips of the second type take the form of a contractual arrangement, or vow, with God or a saint and involve supplicating him for blessings. Many of these practices are not Islamic per se (i.e. not supported by the Quran or *hadiths*), but are likely carryovers from traditional indigenous beliefs that have merged with Islamic practice in various countries.

Another type of religious journey and practice occurs in Mecca, namely the *Umrah*. *Umrah*s are not obligatory in Islam, but are encouraged as a demonstration of devotion to God. This ritual is similar to the *Hajj*, but it can be done any time of the year and does not involve as many activities or as much time. Despite the *Umrah*'s deep significance, it does not replace the *Hajj*, but it is seen as being a way of getting closer to God and supplicating him for mercy. *Umrah* entails the same cleansing ritual as in the *Hajj*, donning the *ihram* garments, entering the *Umrah* gate of *al-masjid al-haram* and performing the *tawaf* and *sai* rituals. After performing the *sai*, the rites of *Umrah* are finished. Hair is shaved (men) or cut (women) at Marwa and the *ihram* is abandoned (Robinson 1999: 133).

Finally, festivals are an integral part of Muslim life in terms of both religious devotion and leisure travel (Ibrahim 1982). There are in effect only two official Islamic festivals – *id al-adha* and *id al-fitr*. The sacrificial festival (*id al-adha*) occurs at the end of *Hajj*. At the time when the *hajjis* themselves are offering sacrifices at Mina, Muslim families throughout the world celebrate by sacrificing lambs or other animals, distributing two-thirds of the meat to the needy, and eating the rest. The second festival, *id al-fitr* (festival of breaking the fast) takes place at the end of the Ramadan fast. These festivals typically last three or four days and entail feasting, rejoicing, praying, outdoor meetings, sermons and exchanging gifts (Robinson 1999). Unlike ceremonies in other religions, such as cremation rituals among Hindus in Bali, these events are generally closed to non-Muslims.

In addition to these two events, many Muslims around the world commemorate the prophet's birthday, the deaths of *imams*, the date when Mohammed designated Ali his successor, and the lives of local saints (Martin and Mason 2003, 2004; Robinson 1999). Many Islamic countries have their own adaptations of various festivals and events and celebrate local Muslim heroes (Martin and Mason 2004). Such practices are not supported by the Quran or *hadiths*, but have evolved in various cultures.

In addition to Mecca, there are many locations throughout the Middle East and elsewhere that are considered sacred and historically significant and to which many Muslims undertake *ziarat*. Medina is perhaps the best case in point. While not strictly a part of the *Hajj*, many pilgrims who have not already visited Medina do so during their journey to Saudi Arabia. The city was home to Islam and Mohammed for several years after he departed Mecca in 622 CE, and it is from Medina that the message of Islam quickly penetrated Africa and Asia. Medina is considered the second holiest city after Mecca, because it is the place where Mohammed is buried. The most important site in the city is the Mosque of the Prophet, one of the holiest three shrines in the world (Long 1979; Russell 1999). Many *hajjis* visit Medina for ten days following the *Hajj* so that they can perform fifty prayers, for according to the words of the prophet, praying in the Mosque of the Prophet is a thousand times more effective than praying anywhere else, with the obvious exception being *al-masjid al-haram* in Mecca. As Robinson (1999) notes, the side-trip to Medina is not a requirement, but it is strongly recommended if a person can afford it.

Owing primarily to its role in the life of Mohammed, as well as its status as home to many of the ancient biblical prophets, Jerusalem is the third holiest city for Muslims. In fact, before prayers were directed toward Mecca, they were offered toward Jerusalem (Robinson 1999). The Dome of the Rock in Jerusalem was built in 692 CE over a rocky outcrop that is one of Islam's most sacred places, variously identified as the site where God tested Abraham by demanding he sacrifice Ishmael, the location of the Holy of Holies of the ancient Jewish temple, the possible burial place of Adam, and the location where Mohammed was taken into heaven (night journey) (Bloom and Blair 2000: 68). On the mount, Mohammed led fellow prophets Abraham, Moses and Jesus in prayer and soon after ascended into heaven with the angel Gabriel where he met Adam, Enoch, Joseph, Moses, Aaron, John the Baptist, Jesus and Abraham (Emmett 2001: 121–122).

Other Muslim sacred sites in Jerusalem include a mosque built on the site of the ascension of Jesus, a mosque that marks the burial site of the prophet Samuel, and the tomb of David, as well as various other monuments, mosques, Islamic schools and cemeteries. In addition to its past importance, expected future occurrences in the city also contribute to its sacredness. These events include the resurrection, the day of judgment, the second and final exodus (*hijra*), and the gathering of the righteous (Emmett 2001).

Other implications

Mandatory and voluntary pilgrimages are only two manifestations of Islam in the world of tourism. As noted earlier, Muslims who can afford to travel are enthusiastic travelers (Balasubramanian 1992; Eickelman and Piscatori 1990). The Quran in fact refers often to travel, and Muslims are encouraged to travel the world for cultural encounters, to gain knowledge, to associate with other Muslims, to spread God's word, and to enjoy and appreciate God's creations (Aziz 1995; Din 1989; Hidayat 1992). Despite this encouragement, traditionally Muslims have been reticent about undertaking recreational pursuits, including travel, because these are seen as a waste of precious time that might be better used doing deeds that bring people closer to God (Rimmawi and Ibrahim 1992). However, a change in attitude among most of the faithful has accompanied modernization, as new forms of pleasure travel have become more acceptable in contemporary Islamic societies (Sönmez 2001).

Owing to constraints on improper behavior and immodest dress, few Muslims take holidays to beachfront destinations. Instead, the majority of their travel motives aside from religion are visiting friends and relatives (VFR), business, shopping and sightseeing. According to data from the early 1990s, VFR and leisure sightseeing are the main reasons Muslims choose to travel abroad, followed by business (Balasubramanian 1992). For example, Muslims visiting the United States might travel to Dearborn (Michigan), Toledo (Ohio) or other places that have developed large, and tight-knit Muslim communities through immigration. In conjunction, they might also visit Disneyland and other well-known attractions that maintain wholesome family themes. There is also a considerable cohort of wealthier people from the Middle East who travel to Europe for health treatments.

Researchers have also noted that religious observance manifests in several interesting behavioral patterns associated with tourism. First, Islamic tradition advocates traveling in groups on trips to faraway places. As a result, most Muslims, especially from the Arabian Peninsula, prefer to travel in groups of family members or friends (Rimmawi and Ibrahim 1992). Likewise, religion has a bearing on the types of accommodations the travelers will select and where they will eat. For devout Muslims, hotels that offer gender segregated swimming and recreational facilities (Taylor and Toohey 2001), prayer rooms or that are located near mosques are more desirable, and restaurants must offer *halal* foods prepared in a suitable manner for Muslim consumption. Muslim tourists require patience and perseverance with the lack of knowledge or sensitivity to their religious culture. For example, one of the authors (Iverson) had a learning experience with a young server in a Chinese-American restaurant. She was instructed clearly to assure that there was no pork in the meal being ordered. When the meal arrived there were pieces of ham in the food, and the server's response was "yes, ham, but not pork." When informed that ham *is* pork, she innocently replied: "Oh, I didn't know that, actually I'm a vegetarian." On another occasion, a flight attendant seemed disinterested when asked by Iverson if the meat in a sandwich was pork. She replied simply that she did not know and did not bother to find out. Clearly there is a need for cross-cultural education in the tourism service sectors if there is a desire to treat Muslim tourists with respect.

Because of the importance of travel to Muslims, they are typically given special dispensation while they are away from home as regards fasting and prayer. For instance, while traveling, Muslims are permitted to shorten their prayers and postpone fasting during Ramadan for pragmatic reasons (Aziz 1995; Din 1989).

From the perspective of Muslims as hosts, Aziz (2001) and Henderson (2003) note that kind hospitality is a mark of a good Muslim. The Quran and the words of the prophet both instruct Muslims to treat people with respect and offer hospitality to travelers in the way of food, water, shelter and with the burden of carrying luggage (Din 1989). In fact, being a good host is seen as a religious duty and a way of getting closer to God by following the example of Mohammed. By serving others in this way, Muslims believe in a broad context that they are doing God's errand and will be rewarded for their service (Aziz 2001).

A final relationship between Islam, tourism and terrorism, should also be mentioned briefly. The connection between Islam and terrorism has gained increasing media coverage since the Gulf War of the early 1990s, the US terrorist attacks of 2001, and the Iraqi war of 2003–2005, and various other conflicts in between. The media-fed western image of Muslim societies is one of danger, terror and heavy-handedness (Power 2000). While terrorists have waged a few attacks on tourists themselves, most violence is not aimed directly at tourists, the obvious exceptions being attacks in Egypt in 1995, the Bali bombing in 2001, and attacks on tourists again in Egypt and Bali in 2005. It appears that extremists who wish to do harm to foreign visitors are becoming more brazen, which Aziz (1995) acknowledges as a response to irresponsible tourism development that takes advantage of Muslims and ignores Islamic values. Harming "innocents" is a bold way of getting media attention for a cause.

The media has blown many of these incidents out of proportion, and has created a negative stereotype of Muslims throughout the world. Reporting is often biased and carries implicit assumptions, for example when the 1995 Oklahoma City bombing occurred, early reports falsely attributed the act to Muslim terrorists. Media reports focus on a select few who think they are justified in killing tourists and other innocent bystanders to demonstrate their disapproval of modernization and western influences. While extremists use the Quran and other holy writs as justification to wage war on the west, many more use the same sources to justify peaceful relations. While recent world events accentuate the intolerant views of some Islamic extremist groups, the original writings of Mohammed and the revelations in the Quran encourage tolerance of religions, ethnicities and racial diversity (Brown 2004; Hidayat 1993). The tactics used by terrorists are not Islamic. That is, they violate core principles of Islam, such as harming women, children and non-combatants. The media tends to ignore the majority of the Islamic world that has good relations with the west and encourages cultural contact within the bounds of Islamic law. In the words of Robinson (1999: 2): "Nothing is said about situations in which Muslims live in peace, or about Islam's emphasis on charity and personal integrity. Presumably, such things are not considered newsworthy." In many Muslim countries, Christian and Jewish communities have coexisted peacefully with Muslims from Mohammed's time to today.

The portrayal of Muslims in the media has affected tourism around the world. Following each major event, tourism to the Islamic world has collapsed, even in places as far away from the Middle East as Malaysia and Indonesia. Likewise, these events alter the travel plans of some Muslims. Following September 11, many Muslims were reluctant to travel to the west, fearing some form of official retaliation (i.e. by customs and immigration officers) or unofficial (by destination residents and service providers). To address both of these problems, countries like Malaysia and the Maldives have attempted to counter these images by portraying themselves as tourist-friendly destinations. Malaysia has responded to both of these effects by promoting its product as Muslim-friendly. Tourism Malaysia has intensified its advertising of Malaysia as a safe destination with a familiar culture to other Islamic states, and is cooperating with the Organization of Islamic Countries and Islamic Development Bank to stimulate travel by Muslims within the Islamic region. Tourism Malaysia's webpages are already available in Arabic and religion may be stressed more in future marketing in the Middle East and Asia (Henderson 2002: 77).

Conclusion

The growth of tourism in recent years has, for the most part, bypassed countries that are predominantly Muslim, even though these countries have natural and built heritage that rivals more popular destinations. In part, the slow development of tourism is related to concerns that the culture of tourism will pollute the religious and social values of the destination residents. The crosscurrents of Muslim tourism seem to have quite distinct problems and opportunities. Predominantly Muslim countries must continue to balance the desire for economic growth – the potential of tourism – with the desire to maintain or improve upon the piety of their people. Globalization, the Internet, and "McDonaldization" are already forces seen as potentially dangerous influences. Tourism is usually seen in the same light.

Meanwhile, the tragic events of September 11, 2001, and the war in Iraq have brought Islam into greater focus in non-Muslim countries, and there is greater interest in learning more about the religion. While the initial global response to the terrorist acts was avoidance, of travel generally, but of travel to Muslim nations in particular, the industry has once again begun to flourish and more people than ever before are traveling to the Middle East and other parts of the Islamic world.

There is significant scriptural support for travel in the Quran, where references can be found to travel in the form of the required *Hajj* and several other forms of travel and for various reasons. There is an emphasis on traveling not just for pleasure but also to recognize that, in examining the built architecture of lost civilizations, it is worthwhile to ponder the nature of their demise. Travel for the sake of learning is also supported by a well-known *hadith*, in which Mohammed said "Seek knowledge even in China," with the obvious reference to faraway places.

Travel is recognized as stressful, as there is little control over one's schedule or environment. For this reason there are dispensations for the mandatory prayers, and dispensations for those who are fasting. The *Hajj* is a massive organizational problem and human errors have led to problems with sanitation, crowd control and "stampedes" of worshippers with unfortunate results, and price gouging of pilgrims by unscrupulous officials. It would be unusual not to find such problems in any crowd of two million people, but the treatment of these events by the media often reflects judgmental views towards Saudis or towards Muslims in general.

Finally, for destinations that seek to attract Muslim tourists, increased research on their market characteristics is warranted. By making simple adjustments to menus, providing sanctuaries to facilitate daily prayers, and becoming more aware of the rituals and routines associated with Islam, these tourists may be easily accommodated.

Acknowledgment

The authors would like to thank Dr Farouq Abawi for comments on an early draft of this chapter.

Note

1   We use the Gregorian calendar for consistency with other chapters in this volume. Likewise, while the Quran has remained unchanged in its original Arabic form, there are spelling variations in English translations. There are no "correct" spellings of English equivalents, including Al-Qu'ran, Quran or Koran.

References

Abdurrahman, M. (2000) "On Hajj tourism: in search of piety and identity in the new order Indonesia," unpublished doctoral dissertation, University of Illinois.

Ahmed, Z.U. (1992) "Islamic pilgrimage (Hajj) to Ka'aba in Makkah (Saudi Arabia): an important international tourism activity," *Journal of Tourism Studies* 3(1): 35–43.

Alavi, J. and Yasin, M.M. (2000) "Iran's tourism potential and market realities: an empirical approach to closing the gap," *Journal of Travel and Tourism Marketing* 9(3): 1–20.

Alhemoud, A.M. and Armstrong, E.G. (1996) "Image of tourism attractions in Kuwait," *Journal of Travel Research* 34(4): 76–80.

Aziz, H. (1995) "Understanding attacks on tourists in Egypt," *Tourism Management* 16: 91–95.

Aziz, H. (2001) "The journey: an overview of tourism and travel in the Arab Islamic context," in D. Harrison (ed.) *Tourism and the Less Developed World: Issues and Case Studies*, Wallingford: CABI.

Balasubramanian, S. (1992) "Arabian Gulf outbound," *Travel & Tourism Analyst* 6: 26–46.

Balkhy, H.H., Memish, Z.A. and Osoba, A.O. (2003) "Meningococcal carriage among local inhabitants during the pilgrimage 2000–2001," *International Journal of Antimicrobial Agents* 21: 107–111.

Baum, T. and Conlin, M.V. (1997) "Brunei Darussalam: sustainable tourism development within an Islamic cultural ethos," in F.M. Go and C.L. Jenkins (eds) *Tourism and Economic Development in Asia and Australasia*, London: Cassell.

Beckerleg, S. (1995) "'Brown sugar' or Friday prayers: youth choices and community building in coastal Kenya," *African Affairs* 94(374): 23–38.

Bhardwaj, S.M. (1998) "Non-Hajj pilgrimage in Islam: a neglected dimension of religious circulation," *Journal of Cultural Geography* 17(2): 69–87.

Bloom, J. and Blair, S. (2000) *Islam: A Thousand Years of Faith and Power*, New York: TV Books.

Brown, D. (2004) *A New Introduction to Islam*, Oxford: Blackwell.

Burns, P.M. and Cooper, C. (1997) "Yemen: tourism and a tribal-Marxist dichotomy," *Tourism Management* 18: 555–563.

Central Intelligence Agency (2005) *The World Factbook 2004*, Washington, DC: Central Intelligence Agency.

Cunha, S.F. (2004) "Allah's mountains: establishing a national park in the Central Asian Pamir," in D.G. Janelle, B. Warf and K. Hansen (eds) *WorldMinds: Geographical Perspetives on 100 Problems*, Dortrecht: Kluwer.

Delaney, C. (1990) "The 'hajj': sacred and secular," *American Ethnologist* 17(3): 513–530.

Din, A. and Hadi, A. (1997) "Muslim pilgrimage from Malaysia," in R.H. Stoddard and A. Morinis (eds) *Sacred Places, Sacred Spaces: The Geography of Pilgrimages*, Baton Rouge: Louisiana State University.

Din, K.H. (1989) "Islam and tourism: patterns, issues and options," *Annals of Tourism Research* 12: 542–563.

Domroes, M. (2001) "Tourism in the Maldives: the advantages of the resort island concept," *Tourism* 49(4): 369–382.

Eickelman, D.F. and Piscatori, J. (eds) (1990) *Muslim Travellers: Pilgrimage, Migration, and the Religious Imagination*, Berkeley: University of California Press.

Emmett, C.F. (2001) "Jerusalem's role as a holy city for Muslims," *BYU Studies* 40(4): 119–134.

Frembgen, J.W. (1993) "Die Sehnsucht nach dem irdischen Paradies: Ethnotourismus zu den Kalasha," *Internationales Asienforum* 24(1/2): 45–56.

Ghazanfar, S.M. (2004) *Spain's Islamic Legacy: A Muslim's Travelogue*, Manchester: FSTC Limited.

Government of Saudi Arabia (2005) *Central Department of Statistics – Hajj data*, available at www.planning.gov.sa/statistic/sindexe.htm (accessed January 8, 2005).

Grötzbach, E. (1983) "Der Ausländertourismus in Afghanistan bis 1979: Entwicklung, Struktur und räumliche Problematik," *Erdkunde* 37(2): 146–159.

Henderson, J.C. (2002) "Tourism promotion and identity in Malaysia," *Tourism, Culture and Communication* 4(2): 71–81.

Henderson, J.C. (2003) "Managing tourism and Islam in Peninsular Malaysia," *Tourism Management* 24: 447–456.

Hidayat, K. (1993) "Observations from an Islamic perspective," in W. Nuryanti (ed.) *Universal Tourism: Enriching or Degrading Culture?*, Yogyakarta, Indonesia: Gadjah Mada University Press.

Horsfall, K. (1996) "Islam and tourism in the Middle East: the case of Egypt," unpublished doctoral dissertation, University of Strathclyde.

Ibrahim, H. (1982) "Leisure and Islam," *Leisure Studies* 1(2): 197–210.

Jackson, R.H. and Davis, J.A. (1997) "Religion and tourism in western China," *Tourism Recreation Research* 22(1): 3–10.

Kaïdi, H. (1980) *Mecca and Medinah Today*, Paris: Les editions J.A.

Khan, W. (1986) "Hajj and Islamic da'wah," in Z.I. Khan and Y. Zaki (eds) *Hajj in Focus*, London: Open Press.

Laporte, S. (2003) "Jeunes femmes en voyage: une expérience tunisienne," *Téoros – Revue de Recherche en Tourisme* 22(1): 29–35.

Long, D.E. (1979) *The Hajj Today: A Survey of the Contemporary Makkah Pilgrimage*, Albany: State University of New York Press.

Lynn, G. (2004) "A brief history of Islam," *Metior* 4: 44.

McDonnell, M.B. (1990) "Patterns of Muslim pilgrimage from Malaysia, 1885–1985," in D.F. Eickelman and J. Piscatori (eds) *Muslim Travellers: Pilgrimage, Migration, and the Religious Imagination*, Berkeley: University of California Press.

Martin, W.H. and Mason, S. (2003) "Leisure in three Middle Eastern countries," *World Leisure* 45(1): 35–44.

Martin, W.H. and Mason, S. (2004) "Leisure in an Islamic context," *World Leisure* 46(1): 4–13.

Metcalf, B.D. (1990) "The pilgrimage remembered: South Asian accounts of the hajj," in D.F. Eickelman and J. Piscatori (eds) *Muslim Travellers: Pilgrimage, Migration, and the Religious Imagination*, Berkeley: University of California Press.

Peters, F.E. (1994) *The Hajj: The Muslim Pilgrimage to Mecca and the Holy Places*, Princeton, NJ: Princeton University Press.

Poirier, R.A. (1995) "Tourism and development in Tunisia," *Annals of Tourism Research* 22: 157–171.

Power, C. (2000) "Touring Muslim style," *Newsweek* July 10: 69.

Rimmawi, H.S. and Ibrahim, A.A. (1992) "Culture and tourism in Saudi Arabia," *Journal of Cultural Geography* 12(2): 93–98.

Ritter, W. (1975) "Recreation and tourism in the Islamic countries," *Ekistics* 40: 149–152.

Ritter, W. (1982) "Tourism in Arabia: study in recreational behaviour of guest and host communities," in T.V. Singh, J. Kaur and D.P. Singh (eds) *Studies in Tourism, Wildlife, Parks, Conservation*, New Delhi: Metropolitan.

Robinson, N. (1999) *Islam: A Concise Introduction*, Washington, DC: Georgetown University Press.

Rollier, S. (1985) "The Hajj," *Pakistan Hotel and Travel Review* 7(8): 6–7.

Rowley, G. (1997) "The pilgrimage to Mecca and the centrality of Islam," in R.H. Stoddard and A. Morinis (eds) *Sacred Places, Sacred Spaces: The Geography of Pilgrimages*, Baton Rouge: Louisiana State University.

Russell, P. (1999) "Religious travel in the new millennium," *Travel & Tourism Analyst* 5: 39–68.

Scott, J.E. (2003) "Coffee shop meets casino: cultural responses to casino tourism in Northern Cyprus," *Journal of Sustainable Tourism* 11(2/3): 266–279.

Seddon, P.J. and Khoja, A.R. (2003) "Saudi Arabian tourism patterns and attitudes," *Annals of Tourism Research* 30: 957–959.

Shackley, M. (1998) "A golden calf in sacred space? The future of St Katherine's Monestary, Mount Sinai (Egypt)," *International Journal of Heritage Studies* 4(3/4): 124–134.

Siddiqui, K. (1986) "Foreword," in Z.I. Khan and Y. Zaki (eds) *Hajj in Focus*, London: Open Press.

Sindiga, I. (1996) "International tourism in Kenya and the marginalization of the Waswahili," *Tourism Management* 17: 425–432.

Sönmez, S.S. (2001) "Tourism behind the veil of Islam: women and development in the Middle East," in Y. Apostolopoulos, S. Sönmez, and D.J. Timothy (eds) *Women as Producers and Consumers of Tourism in Developing Regions*, Westport, CT: Praeger.

Taylor, T. and Toohey, K. (2001) "Behind the veil: exploring the recreation needs of Muslim women," *Leisure/Loisir* 26(1/2): 85–105.

Timothy, D.J. (1999) "Participatory planning: a view of tourism in Indonesia," *Annals of Tourism Research* 26: 371–391.

US Department of State (2005) *Background Notes*, available at www.state.gov/r/pa/ei/bgn (accessed January 5, 2005).

Wiebe, D. (1980) "Die heutigen Kultstätten in Afghanistan und ihre Inwertsetzung für den Fremdenverkehr," *Afghanistan Journal* 7(3): 97–108.

Wilder-Smith, A. (2003) "W135 meningococcal carriage in association with the Hajj pilgrimage 2001: the Singapore experience," *International Journal of Antimicrobial Agents* 21: 112–115.

14 Pilgrimage in Sikh tradition

Rajinder S. Jutla

This chapter examines the significance of pilgrimage in the Sikh religion. Pilgrimage refers to a journey motivated by religion and plays a significant role in almost all world religions. Major pilgrimage centers around the globe exert a powerful pull in attracting tens of thousands of believers and tourists all year round. Jews and Christians often journey to Israel. Muslims are mandated to visit Mecca at least once in their lifetime. Sikhs also go on pilgrimage, but to comprehend fully the role of pilgrimage in Sikhism, it is necessary to examine the patterns of pilgrimage on the Indian subcontinent.

India is a country of religious diversity, the birthplace of Hinduism, Buddhism, Jainism and Sikhism. Pilgrimage in India plays an important role in generating mass tourism. More than 20 million Hindus make an annual pilgrimage to sacred rivers and mountains associated with Hindu mythology, scriptures and events (Bhardwaj 1987; Singh 1997, 2005). The Ganges is the most sacred of these rivers, and along its banks, cities such as Rishikesh, Hardwar, Allahabad and Varanasi, in the state of Uttar Pradesh, are important pilgrim centers in Hinduism. The city of Allahabad is perhaps the most revered since it lies at the confluence of the Ganges and the Jamuna, another sacred river. It also hosts the well-known Hindu festival of Kumbha Mela, which involves ritual bathing, singing, feeding the poor and holy men and women, and debating and standardizing doctrine. Sovik (2001) reported that the Kumbha Mela of 2001 at Allahabad drew between 20 and 40 million pilgrims. The state government of Uttar Pradesh spends millions of dollars to attract domestic and international pilgrims and tourists to each festival.

Pilgrimage is equally significant in Buddhism. Buddhists believe that pilgrimage to holy sites will provide them with a shortcut to Nirvana, or enlightenment. The Indian cities of Bodh Gaya and Sarnath attract thousands of Buddhist pilgrims from not only India but also Thailand, Korea, Japan and other countries where Buddhism is prominent. The act of pilgrimage is also deeply rooted in Jainism. Jain pilgrims visit Mount Abu in Rajasthan, and many pay homage to the 58-foot statue of the Jain saint, Bahubali, in Southern India (Farrington 1999: 87).

Important Sikh pilgrimage centers are scattered throughout India and Pakistan and attract millions of Sikhs and other tourists each year. On September 1, 2004, between 3 and 4 million Sikh pilgrims from all over India and from many corners of the globe visited the Golden Temple at Amritsar. This occasion marked the four hundredth anniversary of the installation of *Guru Granth Sahib,* the Sikh scripture, at the Golden Temple (Bhatnagar 2004). The event was also observed in many other parts of the world. In Toronto, Canada, for example, the event was celebrated with a procession in which 50,000 Canadian Sikhs participated (Black 2004).

In the past, researchers, such as Eliade (1969), Turner (1973), Turner and Turner (1978), Vukonić (1996) and Smith (1992) have attempted to explore the relationship between pilgrimage and tourism. Turner and Turner (1978: 20) have shown that there are several similarities between traditional pilgrims and tourists. Although pilgrims and tourists visit places that share the same infrastructure and are largely involved in similar activities, such as sightseeing and shopping, their motivation for undertaking travel is different. Pilgrims, or religious tourists, are on a spiritual quest, whereas the quest of other tourists is recreation and gaining new knowledge of the place in terms of local history, traditions and architecture. Pilgrims are actively involved in various religious rituals, whereas non-adhering tourists act as spectators who observe the activities of pilgrims. Cohen (1992) confirms that pilgrims' behavior becomes very similar to that of tourists when pilgrim destinations are located far from their home or journey origins. He also concludes that when the religious center belongs to another religion, culture or society, the individual is a traveler-tourist.

Relatively little research has been conducted in the area of pilgrimage and tourism. The majority of past studies on pilgrimage and tourism have focused on the issues and traditions of pilgrimage in western religions, such as Judaism and Christianity (Eade 1992; Hudman and Jackson 1992; Jackowski and Smith 1992; Nolan and Nolan 1992; Rinschede 1992; Vukonić 1992), although some studies have highlighted eastern religions such as Hinduism, Jainism, Islam and Buddhism (Bharati 1963, 1991; Bhardwaj 1983, 1991; Dubey 1987; Barber 1991; Hawley and Goswami 1983; Sopher 1987; Cohen 1992). Likewise, a 2002 special issue of *Tourism Recreation Research* focused on sacred journeys. It included a wide range of topics, but the majority of contributions focused on western religious tourism. It included one article on Sikh pilgrimage in which Jutla (2002) examined Sikh attitudes toward travel. Elsewhere, Michaud (1998) dealt with the social anthropological aspect of Sikh pilgrimage. However, the lack of studies on Sikh pilgrimage shows that this area of research needs to be explored further.

This chapter discusses the significance of Sikh pilgrimage in the context of Sikh scripture and practices, whether pilgrimage is prescribed for salvation or spiritual enlightenment. The paper provides a brief background to

Sikhism in terms of its philosophy, beliefs and practices. It also identifies important Sikh pilgrimage centers, and provides tourism planners with an understanding and appreciation of Sikh philosophy and cultural traditions.

Introduction to Sikhism

The world population of Sikhs is approximately 22 million, the majority of whom live in the state of Punjab in India. They form about 2 percent of the entire Indian population. Over 3 million Sikhs live outside India in other countries, from India and Southeast Asia to the Pacific Rim, Africa, Europe and the Americas. Most live in England, Canada and the United States. Many Sikhs of the diaspora travel to the Indian subcontinent and other places for vacations, family reunions, cultural contact and pilgrimage (Hannam 2004). Their travels are an important component of international and local commerce.

Sikhism is a world religion, founded in the fifteenth century by Guru Nanak. India at that time was torn apart by religious fanaticism and the caste system. India's major religions were Hinduism, the religion of the majority, and Islam, the religion of the ruling class. Muslims and Hindus were alienated from each other socially and culturally. Hindus were divided into various castes, while Muslims had no such caste system. Hindus believed in many gods, while Muslims worshipped one God. Hindus looked upon the cow as sacred; Muslims often ate beef. Hindus cremated their dead; Muslims buried them (Duggal 1989: 2). To squelch these differences and transform India into an Islamic state, the Muslim rulers enacted harsh laws and forced conversion on their subjects.

The Hindus were systematically persecuted and treated as second-class citizens in their own country. They were required to pay a *jizia*, a tax, for being Hindu. They also had to pay taxes to visit places of pilgrimage, and could not build new temples nor repair old ones (Duggal 1989: 2). Guru Nanak spoke against this injustice and tried to bring harmony among India's people. He founded a monotheistic religion based on the universal equality of humankind. Guru Nanak stated that there was only one God who belonged to all religions, castes, ethnicities and nationalities. His message was further continued by nine Gurus who shaped the Sikh community by giving it a written language, religious scriptures, and institutions over the centuries. Over the centuries, the Sikh Gurus continued to struggle against oppression and injustice. The last Mughal ruler, Aurangzeb, was a zealous Muslim who attempted to convert the people of his empire forcefully. He persecuted the Sikhs, who represented the only resistance to his authority. Consequently, harsh punishment and torture were dealt to the Sikhs, and bounties were placed on their heads (Jutla 2002).

Guru Gobind Singh, the last living Sikh Guru, preached to his followers that when all peaceful means fail, it is honorable to draw the sword against injustice. In 1699, he founded *Khalsa* or "the pure," free from the taboos of caste, color and social status, always ready to protect the weak and fight for injustice. The Guru encouraged Sikhs to adhere to five practices. First, they are not to cut their hair. *Kesh,* or uncut hair, symbolizes spirituality and saintliness. The four other practices deal with wearing or carrying the following objects: a *khanga* (small wooden comb) symbolizing cleanliness, order and discipline in life; a *kara* (steel bracelet) symbolizing self-restraint and strong will; *kachha* (knee-length shorts), symbolizing purity of moral character; and finally a *kirpan* (a sword), symbolizing truth and justice. In an attempt to remove all vestiges of the caste system, Sikh men were also asked to assume the surname *Singh* (lion) and women were asked to adopt the name *Kaur* (princess). The Guru also prescribed that Sikh men wear a turban, which at that time represented royalty. The Mughal emperor had banned non-Muslims from wearing it. Wearing turbans elevated the status of Sikh men and gave a distinctive identity to the Sikh community. This allowed them to stand out in a crowd and not shirk their duties to defend against injustice. Today, Sikhs wear turbans with a great sense of pride.

In 1708, before Guru Gobind Singh died, he formally passed on his guruship to the *Guru Granth Sahib*, the Sikh holy book. He admonished Sikhs to acknowledge the *Guru Granth Sahib* in his place as their Guru instead of a living person. The *Guru Granth Sahib* is a unique scripture, entirely written and compiled by the Gurus themselves. It is the final authority for Sikhs and no changes to it are permitted. It is an anthology of sacred poems praising God and providing a guide to spiritual enlightenment and true happiness. This sacred work is held in high esteem and treated with great respect by Sikhs.

Guru Granth Sahib is placed in Sikh temples, known as *gurdwaras*. The most important Sikh places of pilgrimage are the *gurdwaras* associated with Sikh history. The word *gurdwara* literally means "gateway to God." *Gurdwaras* typically have entrances on all four sides symbolizing that they were open to people from all four castes of Hinduism and from the four cardinal points (Bhatti 1986). Everyone is welcomed into the *gurdwara* regardless of caste. The four entrances therefore symbolize the universality of Sikhism. Besides meeting spiritual needs, *gurdwaras* also serve as community centers. In many rural areas of Punjab, political meetings are held there. Some *gurdwaras* also run schools and health centers, thereby fulfilling the other needs of their communities.

Pilgrimage and Sikhism

To understand fully the significance of pilgrimage in Sikhism, it is necessary to examine the social and political context of the era in which Sikhism evolved. In the fifteenth century, when Sikhism emerged, both Hinduism and Islam emphasized the need for making pilgrimage for spiritual upliftment. The Sanskrit word used for "pilgrimage place" is *tirath.*

In Hinduism, the journey to a *tirath* is known as *tirath yatra*, which means "journey to a sacred ford" (Fowler 1997: 248). This is in reference to the sacred rivers of India. Water is an integral element of pilgrimage since it represents the life-giving nature of God (Fowler 1997: 248). The meaning of the term *tirath yatra*, however, has changed over time to refer to a journey of believers to any significant religious place, not only a sacred river. The act of *tirath yatra* physically and mentally immerses a person into a sacred space (Coleman and Elsner 1995: 138). Visiting a sacred space is believed to purify the inner self and bring the individual closer to God (Fowler 1997: 248).

For Hindus, the River Ganges is sacred, and individuals try to visit it at least once in a lifetime. Hindus believe that a bath in the Ganges not only cleanses the body outwardly but also is symbolic of an inner spiritual purification. Pilgrims also go there to immerse the ashes of deceased loved ones with the hope that this will release them from the cycle of life, death and rebirth. Islam also encourages Muslims to perform *Hajj*, or to go on pilgrimage to Mecca in Saudi Arabia at least once in their lifetime (Coleman and Elsner 1995). *Hajj*, which is considered to be the fifth pillar of Islam, is symbolic of humankind's journey towards God. Unlike Hinduism and Islam, Sikhism does not encourage its followers to go on pilgrimage for spiritual purification. There are no sacred rivers or mountains in Sikhism. Guru Nanak viewed pilgrimage as an external activity devoid of inner devotion. Sikh scriptures clearly state that pilgrimage is meaningless, that it is more important to have a pure mind and lead a truthful life.

Sikh Gurus have made a number of references to pilgrimage in the *Guru Granth Sahib*. Guru Nanak stated that "to worship an image, to make pilgrimage to a shrine, to remain in a desert, and yet have the heart impure is all in vain" (Singh and Rai 2001: 12). He also noted that ". . . bathing at a place of pilgrimage is futile, it does not please God," but "remembering and meditating on God is equivalent to going on pilgrimage to all sixty-eight holy places" (*Guru Granth Sahib*: 2) – the number of places of pilgrimage visited by Hindus of the time. Guru Nanak reiterated that "any merit from pilgrimage, asceticism, compassion and charity is barely worth a sesame seed" (*Guru Granth Sahib*: 4), but affirmed that "hearing, remembering, and loving the Name of God [is what] immerses us in the true shrine existing within one's self" (*Guru Granth Sahib*: 2). For Sikhs, true pilgrimage is an inner journey. Physically going to a holy place is neither encouraged nor required. It is more important to remember, repeat and reflect on God's name. This is what puts one on the road to union with God. Guru Amar Das, the third Guru also had this to say about pilgrimage: "God is my sacred shrine or place of pilgrimage and pool of purification; I wash my mind in His name" (*Guru Granth Sahib*: 1286). The *Guru Granth Sahib* further states that "wandering through pilgrimage places, does not eliminate one's maladies. There can be no peace without remembering God's Name" (*Guru Granth Sahib*: 906).

The Sikh Gurus discouraged blind ritualism, image worship, and visits to sacred rivers and tombs. In *Guru Granth Sahib*, the ninth Guru, Teg Bahadur, wrote: "Those who make pilgrimage to sacred shrines, observe ritualistic fasts and make donations to charity while still taking pride in their minds are like elephants who take a bath and then roll in the dust. Their actions are in vain" (*Guru Granth Sahib*: 1428). Sikh scripture stresses that having a bath in a sacred river does not cleanse the mind's egoism or break the cycle of rebirth. "Pilgrimages to *Ganges, Gaya and Godavari* are merely wordly affairs" (*Guru Granth Sahib*: 1195). *Ganges* and *Godavari* are sacred rivers for Hindus, whereas *Gaya* is a popular Buddhist pilgrimage center. Guru Nanak further stated: "I know nothing about religious robes, pilgrimages or stubborn fanaticism, I hold tight to the truth" (*Guru Granth Sahib*: 844). The following verse from the *Guru Granth Sahib* questions the significance of pilgrimage: "Shall we go to bathe at the pilgrimage places? No. *Naam* [God's Name] is the only true pilgrimage. Pilgrimage is the contemplation on the Word. That gives inner spiritual light" (*Guru Granth Sahib*: 687). Sikh Gurus have clearly stated that the only pilgrimage center is in one's heart. The inner pilgrimage is not only recommended, but is also in fact the only prescribed path to God.

However, a dichotomy may exist between the authority of the *Guru Granth Sahib* and what is actually practiced by believers. McMullen (1989: 1–2) stated that there are two aspects of religious practice: normative and operative. The former is based on the authority of religious scriptures, whereas the latter is based on the religious practices that are actually held by believers. Although pilgrimage is not prescribed in Sikh scriptures and is in fact commonly discouraged, Sikhs still go on pilgrimage. This can be observed in some of the rituals associated with the *Gurdwara* at Goindwal in Punjab, which was built in 1571, by the third Guru, Amar Das. It has a historic well with 84 steps leading down to the water level. Guru Amar Das, who founded the city of Goindwal, also constructed this well to provide a reliable water supply to the local village as droughts were common in those days.

Nowadays, to promote Goindwal as a pilgrimage center, legends have been added (Singh 2001). According to one legend a person reciting *Japji Sahib*, the morning prayer, on every step, is promised an escape from the cycle of rebirth (Dogra and Mansukhani 1996). This promise is not consistent with Sikh teachings. Guru Amar Das stated that one can experience the infinite (God) only by bathing in the realization of inner knowledge and self realization (*Guru Granth Sahib*: 250). He also stated that "the wandering holy men are tired of making pilgrimages to sacred shrines; they have not been able to conquer their mind. Truth is the place of pilgrimage" (*Guru Granth Sahib*: 753). "Bathing at the sacred shrines of pilgrimage, the filth of the mind is not washed" (*Guru Granth Sahib*: 890). There are no verses in the *Guru Granth Sahib* that encourage Sikhs to bathe in the well at Goindwal for salvation. Nevertheless, many people undertake this activity. The teachings of the *Guru Granth Sahib* very clearly state that a pilgrim should mentally immerse in the pool of remembrance of God to wash away all sins. Pilgrimage is thus the awareness of the infinite within each individual. The journey to the infinite within is the only pilgrimage that can cure all ills (Singh 2001).

The Sikh Gurus themselves visited Hindu and Muslim pilgrimage sites not for spiritual enlightenment but to indicate to people that performing blind ritualism, bathing in sacred rivers, or visiting temples or mosques will not help them attain spiritual salvation. They wanted people to be rational and to understand that, to realize God, one does not have to travel long distances to specific places. Salvation is instead achieved through hard work, honest living and remembering God. In Sikh scriptures, the act of pilgrimage is not emphasized at all, contrary to the writings and doctrines of Hinduism, Islam and various other religions.

Sikh travel behavior and practices

Despite the lack of encouragement to travel, Sikhs are avid travelers. *Ardas* is a prayer that is recited by Sikhs at the conclusion of religious ceremonies and functions. The prayer, however, is not part of the *Guru Granth Sahib*, but it provides a sense of cohesiveness among the members of the Sikh community. The verses of *Ardas* instruct Sikhs to remember the five *Takhts,* the seats of the Sikh religion, and all *gurdwaras. Takhts,* or thrones, are of particular importance to the Sikh community since these are the places where decisions about the religion and issues concerning the community are made. Three *takhts* are located in Punjab, India: the *Akal Takht Sahib* (Golden Temple) in Amritsar, *Takht Keshgarh Sahib* at Anandpur Sahib, and *Takht Damdama Sahib* located at Talvandi Sabo. Two additional *takhts*, the *Takht Patna Sahib* and the *Takht Hazoor Sahib*, are located in Bihar and Nanded, Maharashtra, respectively (Figure 14.1). The first, the *Akal Takht*, at Amritsar is the oldest and most important. Although all five *takhts* are equal in sanctity, the *Akal Takht* at Amritsar is the central nerve center of the Sikh community worldwide. Here national and international social, religious and political issues concerning the Sikhs are discussed and debated.

Ardas also requests that Sikhs contemplate the deeds of martyrs and heroes who sacrificed their lives for the sanctity of the *gurdwaras*. On January 25, 1952, *Akal Takht Sahib* at Amritsar added verses imploring Sikhs to visit, manage and worship at the *gurdwaras*, which had been separated from India since its partition in 1947. These verses very clearly emphasize the importance of not only visiting the five important *takhts* but also *gurdwaras* around the country and those that were left outside India after its partition (e.g. Pakistan and Bangladesh).

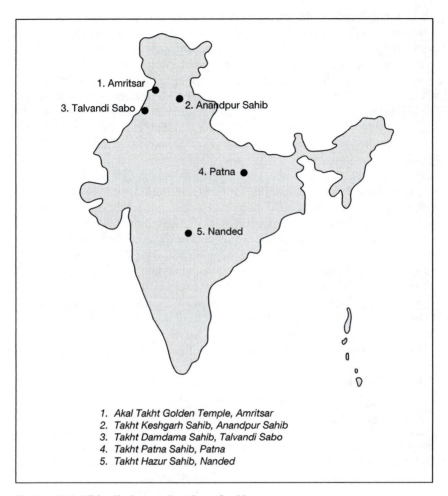

1. Akal Takht Golden Temple, Amritsar
2. Takht Keshgarh Sahib, Anandpur Sahib
3. Takht Damdama Sahib, Talvandi Sabo
4. Takht Patna Sahib, Patna
5. Takht Hazur Sahib, Nanded

Figure 14.1 Sikh pilgrimage: location of *takhts*

In conjunction with a larger study to understand Sikh attitudes, behavior and patterns toward travel and recreation in general (see Jutla 2002 for part of the results), a survey was conducted in 2002 in Toronto and its vicinity where more than 100,000 Sikhs live. Some 120 completed questionnaires were completed and returned to the author. The data from this survey form the empirical basis for the rest of this section.

Sikh participants in the study were asked to identify the activities they undertake on weekends. Seventy percent reported that they regularly attend the *gurdwara* on weekends. Twenty percent attend the *gurdwara* once a month and the remaining 10 percent only visit on special religious occasions such as Guru Nanak's birthday, Guru Teg Bahadur's martyrdom day,

Diwali and Baisakhi. Other occasions include weddings, births and death ceremonies. *Gurdwaras* serve the spiritual need of individuals and also provide an opportunity for social interaction with other members of the Sikh community, to speak their native language of Punjabi and wear their traditional Punjabi clothing. In addition to serving as a place of worship, the *gurdwara* is also a social meeting place for the Sikh community. It provides an important link to Sikh culture and traditions. *Gurdwaras*, especially those outside of India, organize many educational, cultural and charitable events for the local communities they serve. Sikhs raise funds, through the *gurdwara* for hospitals, charities and disaster relief. Punjabi and English classes are also held for adults and children (Minhas 1996: 36).

In response to annual events, 75 percent reported that they attend the annual *Baisakhi* parade in Toronto. All the *gurdwaras* in the vicinity collaborate and organize a colorful parade through downtown Toronto. The parade generally takes place on the second or third weekend of April, and every year thousands of Sikhs participate in it. The *Toronto Star* (2004) reported that about 50,000 Sikhs participated in the 2004 parade. The *Baisakhi* gathering in Toronto has become one of the largest Sikh gatherings in North America. Sikhs from many Canadian provinces and the United States are represented at the event (Gillespie 1999). *Baisakhi* parades are also held in the United States, in New York City, Chicago and Yuba City. Some 20,000 Sikhs participated in the parade of 2004 in New York City.

Many Canadian Sikhs spend their longer vacation times in Canada or travel abroad to the United States, the South Pacific, Asia or Western Europe. Nonetheless, 55 percent of the Sikhs who usually travel abroad visit India. While nearly all Sikh travelers to India disembark in Delhi, the majority (80 percent) head to Punjab after spending a couple of days in Delhi. The remaining 20 percent go to the neighboring states of Haryana, Himachal Pradesh and Uttar Pradesh. Eighty percent of the travelers who visit India make special efforts to visit historic *gurdwaras*, in addition to shopping for Indian products and visiting their ancestral villages.

The results show that some of the most important Sikh pilgrimage places visited by Sikhs who live in India (Jutla 2002) as well as those of the diaspora include *Akal Takht Sahib* at Amritsar (The Golden Temple) and the *Takht Keshgarh Sahib* at Anandpur Sahib, which are mentioned in the *Ardas*. The Golden Temple in Amritsar is the most popular Sikh shrine. It is probably the most well known of the historic buildings associated with Sikh history, culture and tradition (Jutla 2002). This was visited by 80 percent of respondents. The city of Amritsar was founded by Guru Ram Das, the fourth Guru, in 1589. The Golden Temple is the largest and most important Sikh gurdwara, attracting 20,000 visitors daily and up to 200,000 on special occasions associated with Sikh history (*Guinness Book of Records* 2000: 175). It is well known for its display of fireworks on the festival of *Diwali*. This occasion holds a special significance for Sikhs because it marks the return to Amritsar of the sixth Guru, Hargobind, after his release from a Mughal prison. The entire complex is lit with millions of lights that produce a dramatic reflection in the water pool. Pilgrims who visit Amritsar also visit the *gurdwaras* of Taran Taran and Goindwal in the surrounding areas of Amritsar.

Takht Kesgarh Sahib at Anandpur was visited by 70 percent of the study participants. It marks the place where the tenth guru, Gobind Singh, baptized the first Sikhs and created a casteless society. It is known for the celebration of *Baisakhi* and *Hola Mohalla.* April 1999 marked the three hundredth anniversary of the creation of the *Khalsa,* when the Sikh religion was given a formal identity. Millions of Sikh pilgrims not only from various parts of India but from many corners of the world made pilgrimage to Anandpur Sahib that year. On *Hola Mohalla,* Sikhs show their martial traditions by displaying sword fighting, horseback riding, etc.

Takht Damdama Sahib was visited by 20 percent of the respondents. Here, Guru Gobind Singh revised the *Guru Granth Sahib* by adding the verses written by his father, the ninth Guru, Teg Bahadur. *Takht Hazoor Sahib* in Nanded was built by the great Sikh ruler, Maharaja Ranjit Singh, to mark the place where Guru Gobind Singh passed away. It was visited by 10 percent of respondents. *Takht Hazoor Sahib* houses many of the Guru's possessions such as weapons, clothing and jewelry. *Takht Patna Sahib*, which marks the birthplace of Guru Gobind Singh, was also visited by 10 percent of respondents.

In recent years, *Hemkunt Sahib*, the Sikh shrine in the state of Uttarkand, has become a very popular place among Sikh pilgrims, especially in the last 25 years. Five percent of respondents visited this location. *Hemkunt Sahib* is located in a dramatic setting next to a lake of ice and surrounded by seven snow-topped mountain peaks. The importance of this place is marked in the *Dasam Granth*, a compilation of writings by Guru Gobind Singh.

In addition to these historical shrines, respondents who get an opportunity to spend a couple of days in Delhi visit the historic *gurdwaras* of *Sheesh Ganj, Rakab Ganj,* and *Bangla Sahib. Gurdwara Sheesh Ganj,* in the walled city of Delhi, which marks the place where Guru Teg Bahadur was beheaded by the order of the Mughal ruler Aurangzeb; *Gurdwara Rekab Ganj* marks the place where his body was cremated. *Gurdwara Bangla Sahib* is located at Connaught Place, in the heart of New Delhi. Here the eighth Guru, Harikishan, served and cured victims of cholera and smallpox. *Gurdwara Kirtatpur* is located on the Sutlej River, about 10 miles from the town of Anandpur Sahib. Many Sikhs come here to leave the ashes of their deceased loved ones. Although this does not have any religious significance, it is a mark of respect to the deceased person.

There are also several significant Sikh shrines scattered throughout Pakistan. One of the most important is *Nankana Sahib*, the birthplace of Guru Nanak. A large number of Sikh pilgrims from India and other countries cross the Indo-Pakistani border to visit *Nankana Sahib* every year especially on the occasion of his birthday. In the past this has proven to be very difficult owing to the acrimonious relationship between India and Pakistan. Visiting this place for the Sikhs of India requires more planning since passports and visas are needed. A major regional newspaper of Punjab, *The Tribune* (2004), reported that some 3,000 Sikhs left India to participate in the *Baisakhi* festival in Pakistan. There are also hundreds of historic Sikh shrines scattered throughout India and in countries such as Pakistan, Tibet and Bangladesh.

The majority of Sikhs understand that pilgrimage itself does not wash away their sins or purify them. They visit major Sikh shrines during their family vacations or on certain religious holidays that are associated with the birth and martyrdom of the Sikh gurus, as well as events associated with Sikh history. Sikh pilgrims visit major Sikh shrines for a variety of reasons, both spiritual and historical. Many *gurdwaras* were built during the lifespan of the Gurus and some even by the Gurus themselves. These places strongly evoke the past, and pilgrims feel a sense of closeness to the Gurus. Pilgrims also believe that it is important to expose their children to their own cultural tradition and history. This is of particular importance to the pilgrims from not only outside Punjab but also from countries outside India. Sikh pilgrimage centers are crucial to the local economy in terms of transportation, hotels, restaurants and the retail sector. Many of the towns that have developed around these centers thrive on the income generated from pilgrims (Jutla 2002).

Conclusion

Before Sikhism's development, pilgrimage was not a matter of choice but it was one of the mandates of the existing religions of Hinduism and Islam. The motivation for going on pilgrimage was not adventure or vacation, but it was necessary to eradicate sins and achieve salvation (Singh 2004). Sikh scriptures do not endorse the view that pilgrimage is necessary for spiritual enlightenment. The *Guru Granth Sahib* makes many references to these acts and emphasizes that it is fruitless to go on pilgrimage for such purposes. It states that the act of pilgrimage will neither wash away sins nor break the cycle of reincarnation. The holy writ states that a person may make millions of pilgrimages to sacred shrines, but none of these are equal to the worship of God's name (*Guru Granth Sahib*: 973). Sikhism instead stresses the need to immerse oneself in God's name. The true disciple of God should rise at dawn and meditate. This is a symbolic *tirath ishnaan*, "taking a bath in the tank of *Naam* (God's Name). While he repeats and meditates on God's Name he washes away the sins and maladies of his soul" (*Guru Granth Sahib*: 305).

Nowadays Sikh practices have changed. They go on pilgrimage to pray for health, happiness and prosperity, and to thank God for showering them with blessings. Sikhs visit *gurdwaras* associated with their Gurus and their history, such as the Golden Temple at Amritsar and Keshgarh Sahib at Anandpur. *Gurdwaras*, martyrs and heroes in Sikh history are remembered in the *Ardas* at every Sikh gathering, so visiting these places which commemorate the heroism of the past is important to the community. Although pilgrimage is not prescribed by the scriptures, Sikhs are pulled to these places because of the historical significance of events and places that shaped the community. They also go on pilgrimages to connect to their spiritual and cultural traditions. The act of pilgrimage provides spiritual satisfaction and a reaffirmation of their faith. This is particularly true among Sikhs living outside Punjab or abroad. Pilgrimage provides a link to the past and a sense of community among Sikhs from different parts of India and from the diaspora at large. Sikh pilgrimage centers create a sacred geography and play a vital role in creating a sense of community among its members.

References

Barber, R. (1991) *Pilgrimages*, Woodridge: The Boydell Press.

Bharati, A. (1963) "Pilgrimage in Indian tradition," *History of Religions* 3: 135–167.

Bharati, A. (1991) "Grammatical and notational models of Indian pilgrimage," in M. Jha (ed.) *Social Anthropology of Pilgrimage*, New Delhi: Inter-India Publications.

Bhardwaj, S. (1983) *Hindu Places of Pilgrimage in India: A Study in Cultural Geography*, Berkeley: University of California Press.

Bhardwaj, S. (1987) "Hindu pilgrimage," in M. Eliade (ed.) *Encyclopedia of Religions*, New York: Macmillan.

Bhardwaj, S. (1991) "Hindu pilgrimages in America," in M. Jha (ed.) *Social Anthropology of Pilgrimage*, New Delhi: Inter-India Publications.

Bhatnagar, A. (2004) "Devotees crowded out," *The Tribune*, online edition, September 2, 2004, available at www.tribuneindia.com

Bhatti, S. (1986) "An outline of Sikh architecture," in D. Singh (ed.) *The Sikh Art and Architecture*, Chandigarh: Panjab University.

Black, D. (2004) "Sikhs honour their holiest book," *Toronto Star*, online edition, September 27, 2004, available at www.thestar.com

Cohen, E. (1992) "Pilgrim centers," *Annals of Tourism Research* 19: 33–50.

Coleman, S. and Elsner, J. (1995) *Pilgrimage Past and Present: Sacred Travel and Sacred Space in the World Religions*, London: British Museum Press.

Dogra, R. and Mansukhani, G. (1996) *Encyclopedia of Sikh Religion and Culture*, New Delhi: Vikas Publishing House.

Dubey, P. (1987) "Kumbh Mela: origins and historicity of India's great pilgrimage fair," *National Geographic Journal of India* 33: 469–492.

Duggal, K.S. (1989) *Sikh Gurus: Their Lives and Teachings*, Honesdale, PA: The Himalayan International Institute.

Eade, J. (1992) "Pilgrimage and tourism at Lourdes, France," *Annals of Tourism Research* 19: 18–32.

Eliade, M. (1969) *Images and Symbols*, New York: Sheed and Ward.

Farrington, K. (1999) *The History of Religion*, Godalming, Surrey: Quadrillion Publishing.

Fowler, J. (1997) "Hinduism," in J. Fowler, M. Fowler, D. Norcliffe, N. Hill and D. Watkins (eds) *World Religions*, Brighton: Sussex Academic Press.

Gillespie, K. (1999) "Sikhs celebrate historic day," *The Toronto Star*, online edition, April 26, 1999, available at www.thestar.com

Guinness Book of Records (2000) New York: Bantam Books.

Guru Granth Sahib Translated by Singh Sahib Khalsa, Tucson: Hand Made Books, available at www.Sikhnet.com

Hannam, K. (2004) "India and the ambivalences of diaspora tourism," in T. Coles and D.J. Timothy (eds) *Tourism, Diasporas and Space*, London: Routledge.

Hawley, J. and Goswami, S. (1983) *A Play with Krishna: Pilgrimage Drama from Brindavan*, Princeton, NJ: Princeton University Press.

Hudman, L.E. and Jackson, R.H. (1992) "Mormon pilgrimage and tourism," *Annals of Tourism Research* 19: 107–121.

Jackowski, A. and Smith, V. (1992) "Polish pilgrim-tourists," *Annals of Tourism Research* 19: 92–106.

Jutla, R. (2002) "Understanding Sikh pilgrimage," *Tourism Recreation Research* 27: 65–72.

McMullen, C. (1989) *Religious Beliefs and Practices of Sikhs in Rural Punjab*, New Delhi: Manohar Publications.

Michaud, H. (1998) "Walking in the Footsteps of the Guru: Sikhs and Seekers in the Indian Himalayas," unpublished Master's thesis, University of Calgary, Canada.

Minhas, M. (1996) *The Sikh Canadians*, Edmonton: Reidmore Books.

Nolan, M. and Nolan, S. (1992) "Religious sites as tourism attractions in Europe," *Annals of Tourism Research* 19: 68–78.

Rinschede, G. (1992) "Forms of religious tourism," *Annals of Tourism Research* 19: 51–67.

Singh, I. (2004) "Sikh pilgrimage: a study of ambiguity," *Sikh Spectrum* May 16, available at www.sikhspectrum.com (accessed November 1, 2005).

Singh, I.J. (2001) *The Sikh Way, A Pilgrim's Progress*, Guelph, ON: The Centennial Foundation.

Singh, K. and Rai, R. (2001) *The Sikhs*, New Delhi: Roli Books.

Singh, R.P.B. (1997) "Sacred space and pilgrimage in Hindu society: the case of Varanasi," in R.H. Stoddard and A. Morinis (eds) *Sacred Places, Sacred Spaces. The Geography of Pilgrimages*, Baton Rouge, LA: Louisiana State University Press.

Singh, R.P.B. (2005) "Geography of Hindu pilgrimage in India: from trends to prospects," in S. Skiba and B. Domański (eds) *Geography & Sacrum: Festschrift to Antoni Jackowski*, Kraków: Instytut Geografii i Gospodarki Przestrzennej, Uniwersytet Jagielloński.

Smith, V. (1992) "Introduction: the quest in guest," *Annals of Tourism Research* 19: 1–17.

Sopher, E. (1987) "The message of place in Hindu pilgrimage," *National Geographic Journal of India* 33: 353–369.

Sovik, R. (2001) "The Kumbha Mela comes to us," *The Kumb Mela Times* January 24, available at www.kumbhamelatimes.org (accessed November 1, 2005).

The Tribune (2004) Online edition, April 6, available at www.tribuneindia.com

Toronto Star (2004) Online edition, April 26, available at www.thestar.com (accessed June 10, 2004).

Turner, V. (1973) "The center out there: pilgrim's goals," *History of Religion* 12: 191–230.

Turner, V. and Turner, E. (1978) *Image and Pilgrimage in Christian Culture: Anthropological Perspectives*, New York: Columbia University Press.

Vukonić, B. (1992) "Medjugorje's religion and tourism connection," *Annals of Tourism Research* 19: 79–91.

Vukonić, B. (1996) *Tourism and Religion*, Oxford: Pergamon.

15 Pilgrimage in Hinduism
Historical context and modern perspectives

Rana P.B. Singh

Touring is an outer journey in geographical space primarily for the purpose of pleasure seeking or curiosity. Pilgrimage in the traditional sense is an inner journey manifest in exterior space in which the immanent and the transcendent together form a complex spiritual and travel phenomenon. Generally speaking, human beings need both – outward and inward journeys. Hinduism, or more appropriately *Sanatana Dharma* ('the eternal religion'), has a strong and ancient tradition of pilgrimage, known as *Tirtha-yatra* ('tour of the sacred fords'), which formerly connoted pilgrimage involving holy baths in water bodies as a symbolic purification ritual. Faith is central to the desires, vows and acts associated with Hindu pilgrimage, and pilgrimage is a process whereby people attempt to understand the cosmos around them. The number of Hindu sanctuaries in India is so large and the practice of pilgrimage so ubiquitous that the whole of India can be regarded as a vast sacred space organized into a system of pilgrimage centres and their hinterlands (Bhardwaj 1973: 7).

Hinduism, considered by many to be the world's oldest surviving religion, dates back to approximately 5000 BCE. It is the third largest religion in the world after Christianity and Islam, consisting of approximately 13 per cent of the world's population. Hinduism is the majority religion in India, Nepal, Mauritius, and on the island of Bali in Indonesia, as well as a secondary or otherwise major religion in Guyana, Fiji, Suriname, Bhutan, Trinidad and Tobago, Sri Lanka, Bangladesh and Singapore. Many other countries also have large South Asian-based Hindu populations, including Canada, the United States, the United Kingdom, Australia and South Africa.

Hinduism is a polytheistic faith that reveres several gods and goddesses who have control over various elements of creation, life conditions and nature. Hinduism is an inclusive religion in that its adherents accept that all religions are paths to the same goal and are therefore highly tolerant of people of other faiths. Hinduism is unique from other world religions in that there is no messiah, guru or founding prophet. Instead, according to Hindu tradition, the Creator simultaneously formed both the universe and all knowledge about it. Seers, or *Rishi*s, obtained this knowledge directly from the Creator and recorded it in sacred writ known as the *Vedas*, which are comprised of a complex (to the non-Hindu) system of sub-levels of holy writings and epic tales that provide guidelines for achieving harmony in life. Similarly, there is no central religious head-quarters or individual authority to interpret religious canons. Instead, each individual learns what he or she must do to seek his or her own piety and higher level of being.

Travel for pilgrimage purposes is an important part of Hindu doctrine and millions of adherents travel throughout India and from abroad each year to participate in enormous festivals, pilgrimage circuits and ritual cleansings (Kaur 1982, 1985; Singh 2004). Likewise, thousands of people of other religions visit India each year to admire its ancient and beautiful Hindu architecture and important historical sites that are associated with the religion (Gupta 1999; Ichaporia 1983). This chapter focuses specific-ally on Hindu religious tourism in India, with primary stress being placed on forms and functions of sacred sites, pilgrimage events and routes, and places of sacred value.

Hindu pilgrimage: history, place and mythology

According to the *Mahabharata* (13.111.18), a fifth-century BCE epic containing the holy scripture *Bhagavad Gita*, pilgrimage places are auspi-cious for Hindus because of the extraordinary power of their soil, the efficacy of their water, and because they were made holy by visits or proclamations by the sages (holy wise men) (Bhardwaj 1973; Sharpley and Sundaram 2005). By journeying to these powerful places and performing sacred rites, pilgrims obtain what are called 'fruits', or a transformation of themselves or their life situations (Sax 1991: 13). Through the combined processes of sacralization, ritualization and deeper interconnectedness, places become distinct 'scared places' or *sacredscapes*, possessing the characteristic of an eternal bond between the human psyche and the spirit of nature (Singh 1995: 97). For Hindus, pilgrimage (*Tirtha-yatra*) is an act and process of spiritual crossing; to cross the *sacredscape* is to be transformed.

Pilgrimage is a spiritual quest – a guiding force unifying divinity and humanity; it is a search for wholeness. Ultimately the wholeness of land-scape and its sacred and symbolic geography creates a 'faithscape' that encompasses sacred place, sacred time, sacred meanings and sacred rituals, and embodies both symbolic and tangible psyche elements in an attempt to realize humankind's identity in the cosmos. The act of pilgrimage, including the journey, activities and experiences of companionship, is itself a ritual that has transformative value, a reinterpretation of the idea of 'experience'. The Hindu term *Tirtha-yatra* itself denotes this quality. According to Hindu beliefs, by undertaking a pilgrimage, an individual is transformed and begins life anew.

Of all the religious practices associated with Hinduism, pilgrimage is believed to be the most important and meritorious rite of passage. Hindu pilgrimage involves three stages: initiation (from the time one decides to take the journey to the beginning of the journey), liminality (the voyage itself and experiences involved), and re-aggregation (the home-coming). The human quest to find peace and experience sacred space drives faith-building and the desire to travel (Singh 2005). Feelings associated with positive pilgrimage experiences and faith-building causes pilgrims to return to their normal life and share experiences with other members of society. This forms a cyclic frame of travel known as '*pilgrimage mandala*' (Figure 15.1).

An example of Hindu perspectives on pilgrimage (Figure 15.2) would explain this issue. Starting from a believer (person) to the Ultimate (state of reality), at least four layers exist but they are interconnected through

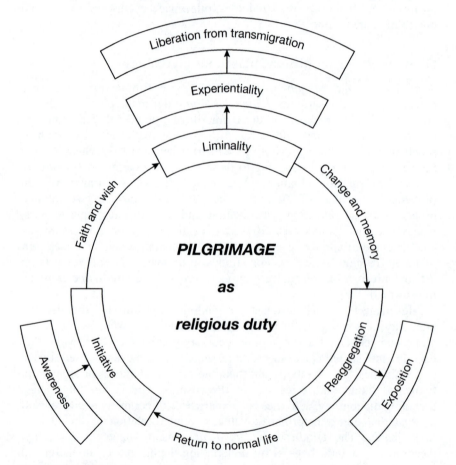

Figure 15.1 Pilgrimage as religious duty

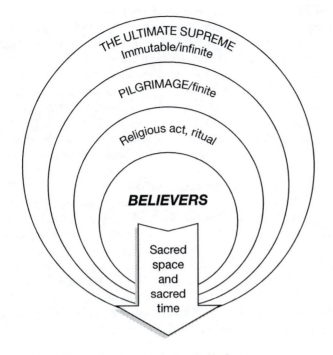

Figure 15.2 The Hindu outlook of pilgrimage

sacred space and sacred time. In theological context this is the eternal will to interconnect a person to the Ultimate while in social context it refers to a march from individual to universal humanity. The act of pilgrimage starts from inner space (*home*) to outer territory, and later in the reverse manner returning to the home. Pilgrimage is a *way* to heal the body and soul by walking and opening the soul to the spirit inherent in Mother Earth.

The notion of *tirtha* symbolizes at least four connotations in ancient Hindu literature: (1) a route to a place where one can receive power (*Rig Veda*, Rg V 1.169.6; 1.173.11); (2) the bank of a river where people can dip in the water as a rite of purification (Rg V 8.47.11; 1.46.8); (3) the sacred site itself which possesses the power of manifestation (Rg V 10.31.3); (4) sacrilized places based upon divine happenings and work of the god(s) that took place there (*Satapatha Brahmana* 18.9). As in many other religions, place and space are an integral part of Hindu pilgrimage.

With the revival of traditional Hinduism during the 1950s, pilgrimages became more popular. Of all domestic travel in India, over one-third is for the purpose of performing pilgrimage (Rana 2003). Some estimates even go so far as to suggest the figure is around 95 per cent (Singh 2001). The growth and importance of pilgrimage tourism may be related to an increased desire among Hindus to assert their identity against an ever more visible Muslim population. Such competition emerged more actively after the destruction of Babri Mosque at Ayodhya on 6 December 1992, by conservative nationalist Hindu groups who wished to build a temple on this sacred site, which is assumed to be the birth place of Lord Rama. This act of aggression resulted in civil disturbances throughout the country. Since then large numbers of Hindus have become more conscious of their Hindu heritage, resulting in increased participation in traditional rituals, celebrations, the construction of temples, and, of course, pilgrimages.

The great epic, *Ramayana* (dated *c*. 1000 BCE), does not directly describe pilgrimages, but it does narrate the routes traversed and the places visited by Lord Rama during his exile. It also draws attention to the natural beauty and inherent powers of important sacred places. These places (e.g. Ayodhya, Prayag, Chitrakut, Panchavati, Nasik, Kishkindha and Rameshvaram) through the course of time developed as important sites of pilgrimage, and many of them are still known throughout all of India as significant locations to visit. The *Mahabharata* epic contains several detailed sections about the 'grand pilgrimages'. The 'Book of the Forest' (III. 82) and the 'Book of the Administration' (XIII. 108) are especially important as they provide descriptions of some 330 places and 12 grand pilgrimage routes covering all corners of India – Kashmir (north), Kamarupa (east), Kanyakumari (south) and Saurashtra (west) (Bhardwaj 1973). According to the *Mahabharata* (XIII.108.16–18):

> Just as certain limbs of the body are purer than others, so are certain places on earth more sacred – some on account of their situation, others because of their sparkling waters, and others because of the association or habitation of saintly people.

These cantos also mention the rules, the ways and the codes of conduct to be followed during pilgrimage, as well as the hierarchy and degree of sanctity of various places.

The *mahatmya* literature (the *puranas*) of the medieval period, dated the eighth to the sixteenth centuries, provides mythological stories as to how, why, to venerate whom, and in what manner pilgrimages should be performed. In total they describe how pilgrim travels symbolize spiritual progress and how pilgrimage is beneficial in being delivered from sins and worldly affairs. These instructions refer to several aspects of spiritual transformation and provide a set of principles associated with pilgrimage:

- Part of religious duty implies being free from other worldly duties.
- One should seek the support of deity to fulfil the journey. By so doing, pilgrims associate more closely with divinities.
- One should seek religious companionship and try to meet other groups of pilgrims while travelling.
- There should be a desire to enhance fellowship in the sect they are associated with.

- Pilgrims should seek to understand the sacred symbols and knowledge of auxiliary shrines and divinities.
- Travellers should try to encounter areas they have not previously visited or known much about.
- Difficult and arduous journeys are a form of penance.
- Pilgrimage is an opportunity to experience earth powers to improve overall well-being, harmony and happiness.

A thorough literary description of mythologies and literature on Hindu pilgrimage can be found in Kane (1974: 11–16).

During the medieval period many digests and treatises were written, all describing the glory of various holy places in different ways and at different magnitudes. These descriptions, in fact, are rearrangements and selections from the *puranas* (18 works of ancient legends) with commentaries, although in many cases some original sources are missing. The earliest among them is Laksmidhara's *Krityakalpataru* (*c.* 1110 CE), where a full canto is devoted to pilgrimages. Other important sources are Vachaspati Mishra's *Tirthachintamani* (*c.* 1460 CE), Narayan Bhatta's *Tristhalisetu* (*c.* 1580 CE), and Mitra Mishra's *Tirthaprakasha* (*c.* 1620 CE). These literatures have yet to be fully interpreted and compared to the field of pilgrimage studies. To this author's surprise, while performing pilgrimage journeys himself he personally found a very close correspondence between the spatial narration given in the *puranic* sources and the present situation in Allahabad, Varanasi, Gaya and Chitrakut.

The first exhaustive and annotated list of about 2,000 Hindu sacred sites, shrines and places was presented by Kane (1974). The other catalogue-type descriptive works on Hindu holy places include Dave's (1957–1961) four-volume work (in English) and Gita Press's (1957) *Kalyana Tirthank*, or short and popular essays on 1820 holy places of India (in Hindi). According to the *Kalyana Tirthank* list, 35 per cent of all sacred places are associated to the god Shiva, followed by Vishnu (16 per cent), and the goddess (12 per cent).

The feminine spirit of nature has received special attention in the books of mythology. There are 51 special sites on the earth that symbolize the dismembered parts of the goddess's body. Every region has its own tradition of varying forms of goddess (Feldhaus 2003). The Tantric tradition symbolized these sites as resting places of pilgrimage by the goddess, resulting in a transformation of energy (Dyczkowski 2004). These 51 goddess-associated sites later increased to 108 (Singh 2000). During the medieval period, all these sites were replicated in Varanasi and are still active points of pilgrimage and other rituals (Singh and Singh 2006).

A taxonomical assessment of Hindu pilgrimage places

Classifying holy places has been an important theme of geographic concern in terms of origin and location, motive, association and manifestation of power. According to the *Brahma Purana* (70.16–19), one of the 18 *purana*s, pilgrimage sites may be classified into four categories: divine sites related to specific deities; demonic sites associated with the mythological demons who performed malevolent works and sacrifices there; sites associated with the lives of important spiritual leaders (sages); and man-perceived sites, which are not believed to be 'chosen' but merely discovered and revered by humans. This taxonomy is not watertight, as some places may overlap categorical lines, being important divine and sage-related sites, for example.

With respect to belief systems and practices as prescribed in the Vedas and as experienced by pilgrims, holy places may be classified into three groups: water-sites, associated primarily with sacred baths on auspicious occasions; shrine sites related to a particular deity and mostly visited by pilgrims who belong to, or are attached to a particular sect; and circuit areas (*Kshetra*), the navigation of which gives special merit based on some system of cosmic *mandala* as in case of Varanasi, Mathura and Ayodhya (Salomon 1979).

Cohn and Marriott (1958) utilized micro- and macro-level acceptance, as well as attractiveness of destination as a way of classifying Hindu sacred places. Bharati (1963, 1970) applied a similar approach. In terms of geographic scale, frequency and routing, Stoddard (1966) proposed a typology of 24 categories. He concluded that factors such as minimal aggregate travel distance, proximity to large urban centres, and social characteristics, such as dominance of a particular cohort of the Hindu population, are not influential upon the distribution of holy places in India. This classification gives less emphasis on the belief systems and phenomenology of religion. In his pioneering study, Sopher (1968) used simple statistical indices to measure pilgrim regions in Gujarat and to classify them. A more detailed and integrated frame of six hierarchical classes of holy places was presented by Preston (1980), a notion that needs more serious attention by geographers in explaining the intricacies of location, institutional foundation, specific characteristics and sacred geography of holy places in India.

Bhardwaj (1973) described and classified Hindu holy places in historical context, albeit without evolution and distribution being properly emphasized together. Since there are several religious traditions and sects within Hinduism, it would be more appropriate to account for the distribution of their sacred places in reference to their development and regional representation, their sacred topography, and perceived and imposed meanings.

In general from the perspective of geographical scale and social coverage, Hindu pilgrimage places may be seen as pan-Indian, those attracting people from all parts of India and glorified in the classical Hindu scriptures; supra-regional, referring to the chief places of the main sects and mostly linked to founders of various shrines (e.g. Pandharpur)

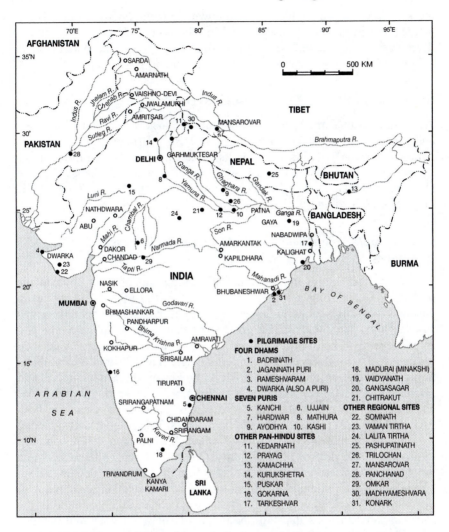

Figure 15.3 Important Hindu places of pilgrimages in India

(Mokashi 1987); regional, connoting the site's dominance in a particular culture or language group and perhaps narrated as representative of pan-Indian places; and local spots associated with ordinary sacred geography, attracting people from nearby villages or towns. Of course there is overlap and transition among these groups, and over time the status of these places may change as well. Moreover, there also exists multi-level places whose identity changes according to sacrality of time and specificity of celebration (Preston 1980). The 7 most sacred cities (*Saptapuris*) are Mathura, Dvarka, Ayodhya, Haridvar, Varanasi, Ujjain and Kanchipuram. Similarly, the 12 most important Shiva (God of Destruction – destruction of evil) abodes are scattered all over India and are known as *Jyotir lingas tirthas*. The 4 abodes of Vishnu (the Preserver God to whom many Hindus pray) in the 4 corners of India comprise another group of popular pilgrimage centers (Figure 15.3) and are representative of pan-Indian pilgrimage places.

Scenarios at 'Three Bridges to Heaven'

According to one of the most authoritative Sanskrit texts on pilgrimage and sacred places, the *Tristhalisetu* (TS – 'Holy Bridge of Three Sacred Cities to Heaven'), dating from the late sixteenth century, the three pillars of the 'bridge to the realm of soul' are Prayaga (Allahabad), Kashi (Varanasi) and Gaya. The first two are located on the Ganga River, while Gaya lies on a tributary of the Ganga.

Kumbha Mela, Allahabad: the world's largest pilgrimage gathering

Sacred site festivals in India (*melas*) are a vital part of Hindu pilgrimage traditions. Celebrating a mythological event in the life of a deity or an auspicious astrological period, *melas* attract enormous numbers of pilgrims from all over the country. The greatest of these, the Kumbha Mela, is a riverside festival held 4 times every 12 years, rotating between Allahabad located at the confluence of the rivers Ganga, Yamuna and the mythical Sarasvati, Nasik on the Godavari River, Ujjain on the Shipra River, and Haridvar on the Ganga (Figure 15.4). Bathing in these rivers during the Kumbha Mela is considered an endeavour of great merit, cleansing both body and spirit. The Allahabad and Haridvar festivals are routinely attended by millions of pilgrims (13 million visited Allahabad in 1977, some 18 million in 1989, and over 28 million in 2001), making the Kumbha Mela the largest religious gathering in the world. It may also be the oldest. There are two traditions that determine the origin/location and timing of the festival. The origins of the location of Kumbha Mela are found in the *Puranas*, ancient texts that tell about a battle between gods and demons wherein four drops of *amrita* (nectar – drink of the gods that gives them immortality) were supposed to have fallen to earth on these *mela* sites (Singh and Rana 2002; Feldhaus 2003). The second tradition establishes the timeframe and is connected to astrological phenomena. The following list demonstrates the astrological periods of the four *melas* and the years of their most recent and near future occurrences:

- Allahabad (Prayaga) – when Jupiter is in Aries or Taurus and the Sun and Moon are in Capricorn during the Hindu month of *Magha* (January–February): 1965, 1977, 1989, 2001, 2012, 2024.
- Haridvar – when Jupiter is in Aquarius and the Sun is in Aries during the Hindu month of *Chaitra* (March–April): 1962, 1974, 1986, 1998, 2010, 2021, 2033.

- Ujjain – when Jupiter is in Leo and the Sun is in Aries, or when Jupiter, the Sun, and the Moon are in Libra during the Hindu month of *Vaishakha* (April–May); 1968, 1980, 1992, 2004, 2016, 2028, 2040.
- Nasik – when Jupiter and the Sun are in Leo in the Hindu month of *Bhadrapada* (August–September): 1956, 1968, 1980, 1992, 2003, 2015.

The antiquity of the Kumbha Mela is shrouded in mystery (Dubey 2001a, 2001b). The Chinese Buddhist pilgrim, Hsuan-tsang, recorded a visit to Allahabad in 643 CE in the company of King Harsavardhana and described a tradition of Magha Mela; however, only around the ninth century did it

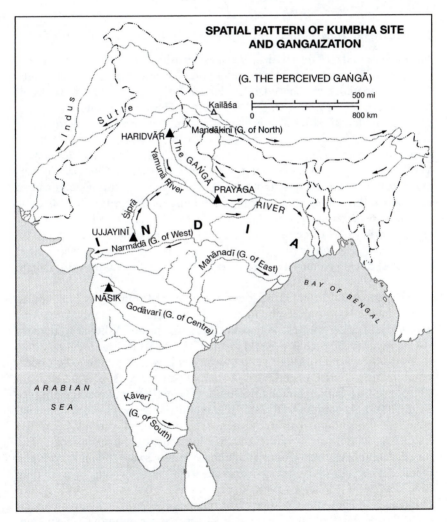

Figure 15.4 Kumbha Mela sites in India

take its present shape under the guidance of the great philosopher Shankaracharya, who had established four monasteries in the north, south, east and west of India, and called upon the Hindu ascetics, monks and sages to meet at these sites for an exchange of philosophical views. Indologists speculate that between the ninth and twelfth centuries other monks and religious reformers perpetuated and reinforced this periodic assemblage of saints and laypeople at sacred places on the banks of the holy rivers to create an environment of mutual understanding among different religious sects. Additionally, the festival gave laypeople the opportunity to derive benefit from their association with the normally reclusive sages and forest yogis. What was originally a regional festival at Prayag thus became the pre-eminent pan-Indian pilgrimage gathering.

Panchakroshi Yatra, Varanasi: experiencing the cosmic circuit

The most sacred city for Hindus, Varanasi (Kashi), has a unique personality possessing all the important pan-India Hindu sacred places in abbreviated form and spatially transposed in its landscape – hence, the city's title of 'cultural capital' of India (Singh 1993, 1997). The sacred territory (*kshetra*) of Kashi is delimited by a pilgrimage circuit, known as Panchakroshi.

In an abbreviated form, the Panchakroshi pilgrimage route of Varanasi symbolizes the cosmic circuit, the centre of which is the temple of Madhyameshvara and radial point at the shrine of Dehli Vinayaka, covering a distance of 88.5km (Figure 15.5). There are 108 shrines and sacred spots along this route, archetypically indicating the integrity of the division of time (e.g. 12 zodiacs) and cardinality of space (9 planets in Hindu mythology, referring to 8 directions and the centre). Among the 108 shrines, 56 are related to Shiva (*linga*). The antiquity of this pilgrimage goes back to the mid-sixteenth century as described in the mythological *puranas* (Singh 2003).

The commonly accepted period for this sacred journey is believed to be the intercalary (thirteenth) month of leap year, commonly known as *malamasa*. During the last Panchakroshi Yatra in the Ashvina Malamasa (18 September–16 October 2001), a total of 52,310 devout local pilgrims and out-of-town pilgrim-tourists performed this sacred journey. To understand pilgrim-tourist experiences better, a survey was conducted with 432 pilgrimage participants by this author during Panchakroshi Yatra (see Singh 2003). According to the study, travel distance, level of faith, mental preparedness, cultural hierarchy, gender context, and various other life conditions, significantly influenced the intensity of the experience. The survey found that small groups (three to six persons) are the most common social setting for performing Panchakroshi Yatra, which is a finding consistent with Sopher's (1968) observations in Gujarat (western India). The data also show the dominance of females (66.2 per cent), which supports the perception that Hindu women are 'more religious' than men.

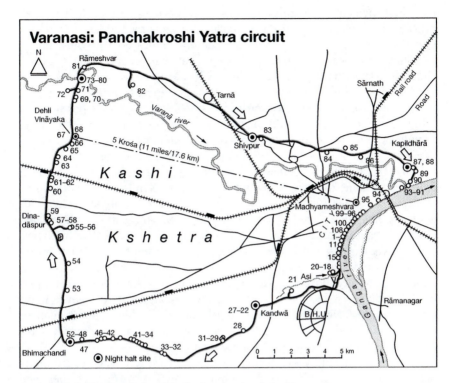

Figure 15.5 The Panchakroshi pilgrimage circuit in Varanasi

This reflects to a large extent, the family-based nature of the pilgrimage experience. The majority of pilgrims were from a proximal area surrounding the city and district of Varanasi. In addition, people from Bengal form a significant cohort owing to the fact that Varanasi has been an important settlement destination for Bengalis since the twelfth century. The religious history of the city and the efficacy of the pilgrimage attract Hindus from all over India, and Nepal. Likewise, in recent years there has been a notable growth in diasporic Hindus from many other countries (e.g. Singapore, United States, Canada, Fiji, South Africa, etc.) travelling to Varanasi to participate in various pilgrimages (Singh 2001).

Well over half of the pilgrim-tourists are older people between the ages of 40 and 60. Adolescent devotees usually accompany their parents and grandparents to support and help them, but they also enjoy the fun of leisure pursuits and sightseeing in addition to the religious rituals of the pilgrimage. Approximately one-fifth of the pilgrims surveyed belong to the lower classes, including peasantry and menial servants. Where education is low and dependency on subsistence farming is high, there is a strong belief in religious and ritualistic activities. Lower educational status is represented by a high percentage of pilgrims and vice versa. More than half (57 per cent) of the foot-pilgrims from the local region claimed to have an education between primary school and graduation (grades 5–10), while among pilgrimage-tourists it is around 70 per cent. The predominance of the Brahmin caste (the highest caste in the Hindu system) is obvious in the observance of Hindu festivals and ritual performances, for by undertaking these rituals, they rejuvenate their professional images, social position and religious status. The hierarchy of higher–lower caste has a positive correspondence with the frequency of devotees. Brahmins and Merchant castes together comprise over half of the pilgrim population (Singh 2003).

Since India's independence in 1947 the upward mobility of the lower caste has become more notable by their adopting symbols and performing religious activities more typically associated with the higher caste. This tendency has encouraged lower caste people to take part in such sacred journeys, as set forth in Sanskrit law books and mythical anthologies. These texts explicitly designate pilgrimage as an appropriate meritorious act for poor people, members of the low caste, and women. However, Hindus of the very lowest caste (e.g. untouchables, such as cobblers, pig-herders, sweepers, basket-makers, and mouse-eaters) almost never make pilgrimages (Morinis 1984: 281). While no noticeable cultural changes have occurred in the Panchakroshi pilgrimage, socio-structural aspects have undergone important changes in the course of time.

Hindu pilgrims enjoy sacred journeys as an earthly adventure from one place to another that entails the combined effects of a spiritual quest and physical hardship – by walking, suffering or avoiding temptation. Believers often speak of the special power of pilgrimage to uplift them (based upon particular qualities of places) and of the compelling effects of various rituals and rites performed by priests at sacred places (Sax 1991).

Gaya: the sacred city of ghostscape

Eulogized as the most sacred place for ancestral rituals, the city of Gaya and its surrounding area claims continuity of tradition at least since the eighth century CE as recounted in the *Vayu Purana*. The ancient writ mentions 324 holy sites related to ancestral rites, of which 84 are presently identifiable and are concentrated in the vicinity of 9 sacred centres (Vidyarthi 1961). At present religious travellers most typically visit only 45 of these sites, although three-quarters of the travellers perform their ancestral rites at only three places: Phalgu River, Vishnupad, and its other associated sacred centres. The cosmogonic hierarchy is marked by three territorial layers: Gaya Mandala, Gaya Kshetra and Gaya Puri, within which there is a complex interweaving of themes of birth, fertility, sun and death (Singh 1999). In the symbolic realm of the cosmic triad, Vishnu's footprints in the Vishnupad temple serve as the *axis mundi*, and the cardinal and solstitial points are marked by hills and other sites of the *mandala* (Figure 15.6).

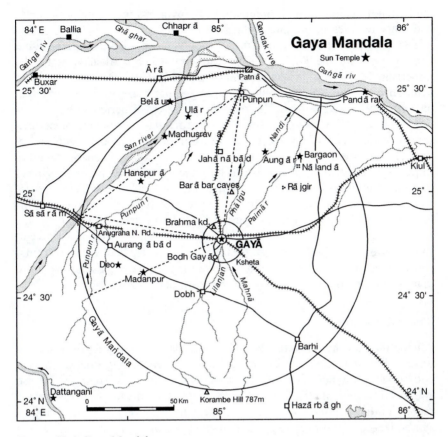

Figure 15.6 Gaya Mandala

The first clear indication of Gaya as a holy place is metaphorically eulogized in the *Rig Veda* (1.22.17). The treatise *Nirukta*, around the eighth century BCE, explains the three most sacred places in Gaya. The glory of Gaya had already been accepted in the period of the *Mahabharata*, especially for ancestral rites. According to inscriptional sources, the antiquity of the site and tradition of ancestral rites in and around the Vishnupad temple goes back to the period of Samudragupta (fifth century CE). The Chinese traveller Hsuan-tsang (seventh century) also mentioned Gaya as a sacred place for bathing, which possesses the power to wash away sins. The name Gaya is derived from a demon-king, Gayasura, who by his arduous austerity, pleased the gods and was blessed that the spirit of all divinities would reside in his body. By the power he gained through deep meditation, the divine spirit met the earth spirit, resulting in the formation of a very powerful matrix. It was this fame that attracted the Buddha to come and perform meditation here. Queen Ahilyabai Holkar of Indore made major sculptural and architectural renovations in Vishnupad temple and other temples in the late eighteenth century.

The three primal objects of nature symbolism described and given ritual connotations are the Phalgu River ('flowing water'), Akshayavata ('the imperishable Banyan') and Pretashila ('the hill of the ghosts'). The river symbolizes fertility by its liquidity ('living water') in which life, strength and eternity are contained. The most common ritual period is the seven-day week (not all weeks have seven days), each day of which is prescribed for particular rituals and ancestral rites, combining sacrality with space, time and function. The texts and traditions of Hinduism persuade devotees to perform ancestral rites at Gaya to help the spirits who, owing to *karma* or an untimely death, have not yet settled down. By doing this, one's forebears can finally achieve a seat in the prescribed abode of manes. This is one of the ideals of Hindus, pursued by the masses, especially in the countryside. As ancestral rites are performed, the spirits of believers' forebears are released from ghost life, which is riddled with suffering, and they are liberated from endless wandering (*moksha*). Each year more than a million Hindus visit Gaya to perform ancestral rites.

The prospects

Among the ancient epics, the *Mahabharata*, dated around the fifth century BCE, is the first source of encouragement for Hindu pilgrimages (*tirtha-yatra*). The mythologies of the medieval period (*puranas*) likewise eulogized sacred places. Many works were written later and encourage sacred journeys as well. According to these holy scriptures, the pilgrimage symbolizes spiritual progress and is encouraged as a way of breaking free of sins and worldly affairs. Pilgrimage travel is prescribed as a duty to earn spiritual advantages and symbolizes different contexts such as routes, riverbanks, shrines and venerated sites associated with wise and respected sages. According to ancient mythology and the Hindu mindset there are many types of hallowed places throughout India, but the most important sacred place is Kashi/Varanasi (Banaras), extolled as one of the three ladders to heaven in company with Allahabad and Gaya.

With the growth of global tourism and a widespread interest in seeing culture in the mirror of history and tradition, religious heritage resource management becomes a critical issue in two primary ways: protection and maintenance of sacred sites and the survival and continuity of pilgrimage ceremonies that preserve centuries-old human interactions with the earth and its mystic powers. Fostering a rediscovery of forgotten or endangered cultural heritage and practices at sacred places that focus on reverence to, and harmony with, the Earth as source and sustainer of life, would be a strong step toward the conservation or preservation of such holy sites. There are many examples of grand Hindu pilgrimages at the regional level, such as Sabarimalai in Kerala (South India), in which even Christians and

Muslims participate (Sekar 1992). Such places are the nexus of cultural integrity. Sopher (1987: 15) has highlighted two contrasting messages in Hindu pilgrimage: a search for roots in places as a basic religious impulse and an ironic form of mental construct of mystical traditions where place has no value. One is free to choose any approach, but for understanding the cultural system in both intrinsic and extrinsic ways, or as an insider or outsider, a human science paradigm would be the best approach of all, as it covers the totality of both, thus attempting to reveal the 'whole' of the culture, human psyche and other functions at play.

References

Bharati, A. (1963) 'Pilgrimage in the Indian tradition', *History of Religions* 3: 135–167.

Bharati, A. (1970) 'Pilgrimage sites and Indian civilization', in J.W. Elder (ed.) *Chapters in Indian Civilization*, Dubuque, IA: Kendall-Hunt.

Bhardwaj, S.M. (1973) *Hindu Places of Pilgrimage in India*, Berkeley: University of California Press.

Cohn, B.S. and Marriott, M. (1958) 'Networks and centres in the integration of Indian civilisation', *Journal of Social Research* (Ranchi) 1(1): 1–4.

Dave, J.H. (1957–1961) *Immortal India* (4 vols), Bombay: Bhartiya Vidya Bhavan.

Dubey, D.P. (2001a) *Kumbha Mela: Pilgrimage to the Greatest Cosmic Fair*, Allahabad: Society of Pilgrimage Studies.

Dubey, D.P. (2001b) *Prayaga: The Site of the Kumbha Mela*, New Delhi: Aryan.

Dyczkowski, M.S.G. (2004) *A Journey in the World of Tantras*, Varanasi: Indica Books.

Feldhaus, A. (2003) *Connected Places: Religion, Pilgrimage, and Geographical Imagination in India*, New York: Palgrave Macmillan.

Gita Press (1957) *Tirthanka Kalyana* [short and popular essays on 1820 holy places of India. In Hindi], Gorakhpur: Gita Press.

Gupta, V. (1999) 'Sustainable tourism: learning from Indian religious traditions', *International Journal of Contemporary Hospitality Management* 11(2/3): 91–95.

Ichaporia, N. (1983) 'Tourism at Khajuraho: an Indian experience', *Annals of Tourism Research* 10(1): 75–92.

Kane, P.V. (1974) *History of Dharmashastra: Ancient and Mediaeval Religions and Civil Law in India*, Pune: Bhandarker Oriental Series.

Kaur, J. (1982) 'Pilgrim's progress to Himalayan shrines: studying the phenomenon of religious tourism', in T.V. Singh, J. Kaur and D.P. Singh (eds) *Studies in Tourism, Wildlife, Parks, Conservation*, New Delhi: Metropolitan.

Kaur, J. (1985) *Himalayan Pilgrimages and the New Tourism*, New Delhi: Himalayan Books.

Mokashi, D.B. (1987) *Palkhi: An Indian Pilgrimage*, Albany: State University of New York Press.

Morinis, E.A. (1984) *Pilgrimage in Hindu Tradition: A Case Study of West Bengal*, Delhi: Oxford University Press.

Preston, J.J. (1980) 'Sacred centres and symbolic networks in South Asia', *Mankind Quarterly* 20(3/4): 259–293.

Rana, P.S. (2003) 'Pilgrimage and ecotourism in Varanasi region: resources, perspectives and prospects', unpublished PhD dissertation, Department of Public Administration & ITS, University of Lucknow.

Salomon, R.G. (1979) '*Tirtha-pratyamnayah*: ranking of Hindu pilgrimage sites in classical Sanskrit texts', *Zeitschrift der Deutschen Morgen Gesellschafft* 129(1): 102–128.

Sax, W.S. (1991) *Mountain Goddesses: Gender and Politics in a Himalayan Pilgrimage*, New York: Oxford University Press.

Sekar, R. (1992) *The Sabarimalai Pilgrimage and Ayyappan Cultus*, New Delhi: Motilal Banarsidas Publishers.

Sharpley, R. and Sundaram, P. (2005) 'Tourism: a sacred journey? The case of Ashram Tourism, India', *International Journal of Tourism Research* 7: 161–171.

Singh, Rana P.B. (ed.) (1993) *Banaras (Varanasi): Cosmic Order, Sacred City, Hindu Traditions*, Varanasi: Tara Book Agency.

Singh, Rana P.B. (1995) 'Towards deeper understanding, sacredscape and faithscape: an exploration in pilgrimage studies', *National Geographical Journal of India* 41(1): 89–111.

Singh, Rana P.B. (1997) 'Sacred space and pilgrimage in Hindu society: the case of Varanasi', in R.H. Stoddard and A. Morinis (eds) *Sacred Places, Sacred Spaces: The Geography of Pilgrimages*, Baton Rouge, LA: Louisiana State University Press.

Singh, Rana P.B. (1999) 'Sacredscape, manescape and cosmogony at Gaya, India: a study in sacred geography', *National Geographical Journal of India* 45(1): 34–63.

Singh, Rana P.B. (2003) *Towards the Pilgrimage Archetype: The Panchakroshi Yatra of Banaras*, Varanasi: Indica Books.

Singh, Rana P.B. (2005) 'Geography of Hindu pilgrimage in India: from trends to prospects', in S. Skiba and B. Domański (eds) *Geography & Sacrum: Festschrift to Antoni Jackowski*, Krakow: Instytut Geografii i Gospodarki Przestrzennej, Uniwersytet Jagielloński.

Singh, Rana P.B. and Rana, P.S. (2002) *Banaras Region: A Spiritual and Cultural Guide*, Varanasi: Indica Books.

Singh, Rana P.B. and Singh, R.S. (eds) (2006) *Sacred Geography of Goddesses in South Asia: Essays in Memory of David Kinsley*, New Delhi: Sundeep.

Singh, R.S. (2000) 'Goddesses in India: a study in the geography of sacred places', unpublished PhD dissertation, Department of Geography, J.P. University.

Singh, Sagar (2004) 'Religion, heritage and travel: case references from the Indian Himalayas', *Current Issues in Tourism* 7(1): 44–65.

Singh, Shalini (2001) 'Indian tourism: policy, performance and pitfalls', in D. Harrison (ed.) *Tourism and the Less Developed World: Issues and Case Studies*, Wallingford: CABI.

Sopher, D.E. (1968) 'Pilgrim circulation in Gujarat', *Geographical Review* 58(3): 392–425.

Sopher, D.E. (1987) 'The message of place in Hindu pilgrimage', *National Geographical Journal of India* 33(4): 353–369.

Stoddard, R.H. (1966) 'Hindu holy sites in India', unpublished PhD dissertation, Department of Geography, University of Iowa.

Vidyarthi, L.P. (1961) *The Sacred Complex in Hindu Gaya*, Bombay: Asia Publication House.

16 Sacred places and tourism in the Roman Catholic tradition

Boris Vukonić

All major religions are composed of stories about gods, beliefs, systematised experiences, rituals, symbols, values, norms, communities, movements, organisations and institutions, and religious leaders or holy people who interpret the meaning of every word and religious act. Christianity is one of the most theologically developed religions and is the largest and most widespread religion in the world, with an estimated two billion adherents, or approximately 32 per cent of the total world population (Rinschede 1999: 45). Of this number, just over one billion people are Roman Catholic – 18 per cent of the total world population or more than half of all Christians.

The Roman Catholic Church has had a profound influence on the course of history in the Western world and the development of Christianity. One of its profound influences has been on the development of pilgrimage sites and travel to them, particularly in Europe. With the blurring of pilgrimage and tourism in recent years and the changes in religious focus brought about by the Vatican in the church to become more engaged in secular matters, leaders and theologians in the Roman Catholic Church have been active in explaining the relationship between tourism and religion, more so than any other religion. Religious tourism in the Catholic tradition most often appears in three forms: pilgrimage with continuous groups and individuals visiting religious shrines; large-scale gatherings on the occasion of significant religious dates and anniversaries; and tours of and visitation to important religious places and buildings within the framework of a tour itinerary.

The purpose of this chapter is to discuss this interface between religion and tourism within Roman Catholic tradition. The chapter begins with an examination of the theological reasoning being the relationship between leaders of the church and tourism. Then, attention is given to the significance of Roman Catholic sacred sites to international tourism, including the types of sites that Catholics hold sacred and often visit. The last section discusses various Catholic sacred sites in relation to tourism.

Roman Catholic understandings of tourism

Christianity is based on the life and teachings of Jesus Christ. Most Christians belong to one of three major sub-sects of Christianity: Roman Catholicism, Protestantism or Eastern Orthodoxy. The differences in the teachings between these groups are relatively few; the differences in their religious rituals are somewhat greater, but they differ most in the organisation of the churches themselves. The head of the Roman Catholic Church is the Pope with a papal seat in Rome, while the head of the Eastern Orthodox Church is the Patriarch, whose seat is in Istanbul. All Christians believe in one god, a god who created the world and continues to care for it. The Son of God was sent to Earth to preach the faith and use his own life as an example for his teachings. For Roman Catholics, Jesus is a divine reincarnation in human form, and is considered to be the Saviour who died to save humankind from sin. In view of this, a key religious teaching is that Christ's resurrection proves that eternal life is possible for those who believe in him. Life on Earth, therefore, is only a journey towards the final goal – eternal life after death. Roman Catholics believe that their status on earth is, theologically speaking, *homo viator*, and the church on Earth, or *eclesia peregrinans*, is actually *in statu viae*. This forms the basis of the theologians' acceptance of tourism's place in the life of modern humankind.

There are several theological tenants that influence the relationship between the Roman Catholic Church and tourism. First, to believe primarily means to be responsible to God. God demonstrated friendship and love to humanity, and in return he expects people to demonstrate responsibility by entirely surrendering themselves to his will. Roman Catholics venerate the places where Christ resided during his time on Earth, but also worship at places associated with the lives of other holy people, such as the apostles and individuals whose life and deeds elevated them to sainthood. As such, Roman Catholics mostly pray to God, but also pray to other holy persons close to him. Special status is therefore given to Jesus' mother Mary, the 'Mother of God'. To Catholics, she is the symbol of Christ's concern for humankind, the symbol of a person and mother, who by definition represents simplicity, the utmost care for the family, the sick and the handicapped, and is believed to bring peace to every home and every region. Apart from the figure of Mary, Catholics worship all the places where she resided, as well as the places where she has later appeared. Mary has a special meaning for Roman Catholics, who have devoted many rituals to her as a way of worshipping everything related to the Mother of God. Pope John Paul II, recently deceased, took as his motto *Totus Tuus* – the Mother of God (partly due to the fact that he had lost his own mother at the age of nine). Such an understanding of the Virgin, evident from the fact that the greatest number of vows are made to her, and that she is seen most often among all heavenly apparitions, led some two centuries ago, when the era of total wars started, to Mary being honoured by Catholics as the Queen of Peace. Owing to this stratified nature of the cult of Mary, it is understandable that shrines dedicated to her are the most numerous among Christian shrines (Vukonić 1996: 124).

Related to this is the fact that Roman Catholics worship relics, as well as works of art (statues, paintings) that depict Christ, Mary, the apostles and other saints or events. These are found on church altars around the world. This is an important feature of the practice of Roman Catholicism, and it has also become an important element in the relationship between tourism and the church; the vast geographical area where Roman Catholic teaching has become established encourages believers to travel, and in this way promotes religious tourism. Theology is clear in its claim that the truth is most moving when it is revealed in a person. This seems to occur most vividly when a tourist experiences God, or Jesus, through people, whether this happens by expressing admiration for the beauty of human creativity, or through an encounter with an actual person. Catholicism holds that when a tourist meets people in the context of a different faith or religion, this too can be a meeting with God, since God did not leave any nation without some knowledge of him. In other words, to meet, become aware of, and experience this holiness outside the place where one grew up and the place one is familiar with is valuable for spiritual growth. In the words of the wise Syrah 'He who has travelled a lot, knows a lot, and who has experienced a lot, speaks wisely' (34: 9). If a tourist is not a Christian and comes among Christians, visits their churches, sees their rituals, gets to know them, is it not likely that he might want to learn the Joyous News of Christ's life? This, theologically speaking, is the basis for justifying the emergence of tourism and its development into a universal social phenomenon.

Second, the fundamental liturgical ritual in Catholicism is the Holy Mass, and next to it, the seven sacraments. These are, in fact, ceremonial markers (signs) of divine activities in a person's life: baptism, confirmation, Eucharist or Mass, confession, holy orders, marriage and the last rites. Many Roman Catholic adherents wish to receive each of the sacraments even when travelling away from home, a desire that church leaders do not ignore, especially in places visited by great numbers of tourists from Catholic countries and in tourism destinations where the local population belong to the church. For these reasons, in almost every country where tourism occurs church leaders encourage touring Roman Catholics to visit congregations to participate in the holy sacraments.

Third, the theological view of tourism is preceded by the placing of humans in a theological context. In the history of humankind there are two acts of creation, two lifetimes and two deaths. The first creation is what God did at the beginning of time. Of note here is the instruction given to the first parents in the Garden of Eden, as written in the Book of Genesis: 'Be fruitful, multiply, fill the earth, and subdue it. Have dominion over the fish of the sea, over the birds of the sky, and over every living thing that moves on the earth.' This commandment remains as a challenge and a goal for the believer. Men and women have taken up the challenge and have tried to achieve this goal. Travelling the world, experiencing it, getting to know it, people perceive themselves as masters of the entire natural world. It is travel that gives them the opportunity to do this.

Fourth, the theological understanding of the relationship between Catholicism and tourism comes from notions of the relationship between God and people. This is the relationship of two persons, one of whom is the Creator, and the other his creation. However, this relationship is not one of a master craftsman and his work, but a relationship of persons, one of whom is perfect, and the other imperfect but created by the will of the Creator and in his own image, and also created in the beginning as a rational and free being with the mission to develop him/herself according to the Creator's plan. People have been given a specific charge by God: to develop their personality to become as much like him as possible. For people to achieve this, God gave them the tools they need. On the physical level this means reason and free will. On the metaphysical level it is the revelations and Christ's redeeming mercy. The Bible reveals to people what they should try to become and how they should behave, in essence shedding light on people's nature and their path through life. Part of this path, which is in a material sense a geographical course, is clearly marked by the path Christ followed during his earthly ministry, which makes theological sense in understanding that taking the same path could and should inspire believers, as should visiting the places that Christ himself visited. In early Christianity these were the first and the most important sacred places, locations of Christian and later Roman Catholic pilgrimage.

All this leads to the theological conclusion that every human activity has found a place for itself, or has a place in the 'book of God's revelation about man'. This attitude also implies the relationship of theology towards tourism. While the Bible makes no mention of tourism per se, Catholic teachings hold that all explanations of phenomena, even modern ones, should be governed by the Bible, and therefore tourism should also be understood in biblical terms when considering the totality of humanity. This vision of totality is the link between theology and tourism. Tourism, therefore, entered the world of theology as one of the components of a unified man, and it is in this context that theology should give its answer. The answer must be consistent with God's will. For this reason, in theological explanations of tourism, the truth is sought in God's texts or thoughts as they form the one basis for correctly evaluating tourism as an individual and social phenomenon. This context enables one to find everything on which the justification of tourism is based and promotes a humane and Christian attitude towards foreigners and tourists.

Fifth, theology also sheds light on tourism from the point of view of labour, or human activity. Perhaps the most famous argument is God's commandment from the Book of Genesis to rest on the seventh day. This order came from God's hand and is permanently inviolable. When the 'One who knew of no rest' entered history, people were given the right and the freedom to rest. 'Foxes have holes and birds of the air have nests; but the Son of man has nowhere to lay his head' (Matthew 8: 20).

Sixth, two other notions are theologically related to tourism, and both of them are found at the very core of the tourism phenomenon. The first notion is that of love. From a theological point of view, this is more than just a pious feeling of affection – it often becomes a sense of duty. The apostle of love, John, said: 'Little children, let us not love in word and speech, but in deed and in truth' (1 John 3: 18). 'Owe no one anything, except to love one another' (Romans 13: 8). Here, then, love is a universal law, just as Paul says: 'Love never ends' (1 Corinthians 13: 8). Love, taken as a 'universal law' is one of the major reasons for Christian pilgrimage and Christian ways of expressing belief. The second notion that is linked to tourism is hospitality. In line with God's word, taking care of guests has always been important in the Roman Catholic Church. For example, Matthew 25: 35 reads: 'I was a stranger/traveller (*ksenos*) and you took me in.' Paul warned the saints in his First Epistle: 'Practise hospitality ungrudgingly to one another' – '*Hospitales invicem sine murmuratione*' (1 Peter 4: 9). Tertullian spoke of '*contesseratio hospitalitatis*' and describes the unique institution '*Tesserae communionis*' of that time, which guaranteed the bearer his participation in the Holy Eucharist and a warm welcome into the Church community. In the Old Testament, God repeatedly warned the Israelites to treat strangers well, and the duty of hospitality is also mentioned several times in the New Testament. 'Practise hospitality' (Romans 12: 13). 'Practise hospitality ungrudgingly to one another' (1 Peter 4: 9). A clear reason comes from the well-known relationship towards fellow human beings and the commandment: 'Love thy neighbour'. The author of the Epistle to the Hebrews wrote: 'Do not neglect to show hospitality to strangers, for thereby some have entertained angels unawares' (13: 2). These are the fundamental messages from the Bible and from sacred Christian texts, which theology today takes as proof of the emergence and duration of tourism in the contemporary conditions of modern civilisation and the life of man, who is seen as a theologically integral being.

In both evangelical and pastoral work, tourism creates new, very complex circumstances that carry great responsibility. An important moral issue is the harmonisation of interpersonal relationships between the guests and the hosts. How can it be ensured that the guests do nothing to hurt the feelings of the natives, and that the natives in turn do not succumb to xenophobia, or, more commonly, the exploitation of their guests? Everyone at some point moves from being the host to being the guest. In different situations the same person can play both of these roles. There exists a dialectical relationship between these roles. Theology reminds us of the

Supper at Emmaus, where the disciples became hosts when they invited Jesus to supper, and at the supper, the traveller became the host.

Entirely separate from the notion of tourism, the Catholic Church has a different attitude towards so-called religious tourism. Simply put, the Roman Catholic Church, though not explicitly in favour of the notion of 'religious tourism', does not negate it altogether either. The reason for not excepting this term lies in the understanding that the religious and the profane cannot both be the content of the same moment of time, because their motivation is completely different. More modern views appear only at the end of the twentieth century, a time when the church probably managed to adapt to contemporary life better and more quickly than some other religions. This is, perhaps, the reason why tourism spread so rapidly and widely in Europe, in the countries dominated by Christianity, including Roman Catholicism, and its teachings. 'So the Catholic Church simply had to react in a certain way to a phenomenon in which an enormous number of believers of this persuasion participated' (Vukonić 1996: 72). Religious tourism, especially as interpreted by the theorists of tourism, appears in three forms:

- as a pilgrimage, a continuous group and individual visit to religious shrines;
- as large scale gatherings on the occasion of significant religious dates and anniversaries;
- as a tour of and visit to important religious places and buildings within the framework of a tourism itinerary and regardless of the time of the tour (Vukonić 1996: 75).

Regardless of the intensity of a person's religious conviction or whether he/she is a religious tourist or a tourist who is religious, the behaviour of both categories is determined to a great extent by the characteristics of the space from which they have arrived and the life they lead there. In general, changes in the lifestyle of tourists, including religious tourists, in the destination are not random, but are always connected with certain general social changes, 'with class competition, prestige hierarchies and the succession of changing lifestyles, as well as to external factors such as the cost and modes of transportation, access to regions and countries, and the state of the economy' (Graburn 1983: 24).

In great world religions such as Christianity, organisation is a very important matter, and believers are born in a world where everything – from how they greet other people, to their burial ceremonies – is set and provided for the believer in some way; it is what precedes his/her actual experience and what will largely be the content of his/her own experience. Christianity, and by extension the Roman Catholic Church, therefore, is not just a collection of teachings, feelings, rituals, a symbolic reality, value and rule, it is also a community, movement and organisation. It is often thought that only ideas bring order to the world, but movement and organisation do that also. Here it is important to understand that a system of ideas, beliefs and values – religion as a code of meaning – is something different from people's specific behaviour – religion as a code of behaviour – which can be clearly seen in the difference between Christian theology and church bureaucracy.

Sacred sites: among the main motives and targets of international tourists

According to theology, sacred places are consecrated or 'illuminated by faith', which because of their religious content, become the inevitable subject of visits. Such places are considered holy, and their holiness is a result of an event that took place there or because they comprise a place of worship for hosting sacred events or for guarding sacred relics. Very rarely is a place held sacred by one religion considered sacred by other religions as well. One such rare place, which accounts for its fame, is Jerusalem, which is sacred to Judaism, Christianity and Islam.

In the Catholic tradition, a distinction exists between pilgrimages and those who participate in them on the one hand, and on the other, 'ordinary' believers who visit churches or places related to religious activities and events for liturgical services, mainly prayer. As a rule, the latter are not considered pilgrims because pilgrimage has the form of an organised group visit to a sacred site. However, the number of individual visits is often higher than the number of believers who travel in groups, especially when it comes to church members who come to sacred sites as tourists.

> Many professional travellers on their various secular journeys have often written in their travel journals about their own religious experiences and their improved understanding of religion, especially after undertaking a casual visit not only to one of the Christian sacred sites, but to the sacred sites of other religions as well.
>
> (Cohen 1992)

Sometimes entire cities are considered to be sacred. The largest sacred site for the Christian, and especially Catholic, world is Rome. This can best be illustrated by visits to Rome for the grand anniversary of Christianity in 2000, when 30 million people visited the city that year. Entire cities from biblical history have become sacred sites, such as Bethlehem, Nazareth and Jerusalem. What is more, the degree of believers' interest in them is higher than for most of the Catholic shrines. Considering the number of church buildings in these towns, the average visit by religious and other tourists is somewhat longer than the one-day or half-day visit which is usual for religious tourism.

Sacred sites are markedly different from secular sites, even though both attract large numbers of visitors. Secular sites are places of special interest to an ethnic, racial or social, but not a religious, group. These places are sacred in a certain way and in a different context, because they too are a source of inspiration and a reason for pilgrimage, and they can be defined as a part of contemporary culture (Shackley 2001). Such places possess specific memorial value, for national history, for instance. They can also be reminders of the creation of a nation state or honour the memory of those who died for national survival. These are very often memorials or everlasting flames, typically honouring an unknown hero or heroes. The Second World War camps of Dachau and Auschwitz-Birkenau have been preserved as reminders of the Holocaust. Here, the religious is strongly linked to the secular, and it is hard to draw a line between the two. In the Catholic tradition it has often been the case that in such historical moments religion and political events were very closely connected, so the pilgrimage to such places honours the religious elements as much as the secular ones. The same is true of the memorial burial sites in Europe and Africa, commemorating soldiers who died in the First and the Second World Wars, because the Catholic tradition treats burial sites as places of immense religious importance because they remind people of humanity's imperfection and life everlasting. In the tourism sense, such places have a powerful attraction and are usually visited on specific dates that evoke the memory of a certain historical event. However, the secular 'sacred' sites cannot be defined as religious sites, because even though there are some spiritual characteristics in common, the religious component of secular place is generally not dominant.

Every sacred site in the Catholic religion is based on some historical event, but often on a myth or a legend also. These are, however, composed of signs, words and symbols. A myth is always more than a system of signs, in the same way that a symphony is more than a simple collection of notes, and a poem more than a collection of words. However, the knowledge of the myth's structure does not allow one to enter the core of religion, because religion is more than just a story. People believe, regardless of their knowledge, which means that when it comes to faith one must distinguish between the myth as a rationalisation of a person's needs, and belief as a personal perception of the myth (myth cannot be reduced to personal faith or belief). A story, a myth or a dogma, as the rational level of faith, is still much more important to the elite, particularly intellectual elite, than to the religious masses. Here, then, lies the difference in approach in determining the significance of a particular sacred site to visitors who are mainly believers, from the significance of sacred sites whose visitors are largely non-believers.

It is especially important to mention the way in which 'the word of God' was spread, and still is being spread, to believers, especially through the medium of speech, where the expressive manner of preaching is almost more important than the content of what is being said. Riessman (1993: 1–3) refers to the ways in which stories create order by translating knowing into telling, drawing on linguistic and cultural resources to persuade the listener/reader that a tale is authentic, and in the process, fashioning a relationship between teller and audience. Private constructions are therefore likely to mesh with a community of life stories. Besides, some believe that, historically speaking, the conceptions of Jerusalem and other holy places as they are recorded by writers and preachers are more a result of the personal experience they took with them to the Holy Land rather than an objective representation of what they saw and experienced in the Holy Land itself. 'Pilgrim writers can be seen as writing less about the Holy City than casting Europe itself in the image of Jerusalem' (Bowman 1992: 155). This is what brings the certainty of our present-day knowledge of pilgrimage into question, particularly of pilgrimage in the Middle Ages, but later as well, and shows that this knowledge is only rarely an accurate representation of what particular holy places were like at a particular time.

For Catholics, a visit to a sacred site is more than just attending a liturgical ceremony. For them it is a personal witnessing of faith, in which the feeling of union with other believers represents an important act, and the repeated shaking of each other's hands is only an outward sign of the wish to preserve and strengthen their faith. These are moments in which the members of the Catholic community find confirmation that all questions of life and death remain open, and that togetherness (usually accompanied with an appropriate ritual) is a way to cope with all difficulties, if not to overcome them. This is what makes attendance of rituals and visits to sacred sites so important to Catholics, as it is the easiest way to bring a large number of believers together and allow them to express their religious devotion. However, for this reason the Catholic Church, for a long time, opposed the notion of religious tourism, and in fact refused any notion that a believer could be anything other than a believer when practicing his/her religious beliefs.

Aside from myth (dogma, doctrine, theory, opinion) as the rational presentation of faith and experience as the personal encounter of faith, ritual is the practical side of faith. Ritual has a meaning only when connected to some form of belief. For this reason, ritual is an important structural part of every sacred site in Catholicism. Without ritual, or during the time that ritual is not performed, the sacred site loses some degree of credibility because the ritual represents faith in action: the turning of faith into visible actions. The ritual is an occasion where speech is performed by the whole body, the whole being, and not just through words. Although this statement cannot always be supported in Roman Catholicism, Catholics nonetheless participate in religious rituals with their entire being. This is especially so with true and deeply devout believers. People in nations that have better systems of education and are more secularised seem to be more eloquent in expressing their religious feelings than people in other nations. This is why the latter have a need to express their faith by performing, dancing and singing. This applies even in cases where the same religion is practised but in areas with different levels of education. The behaviour at liturgical ceremonies of Roman Catholic believers in European countries is different from that of believers in Africa or other areas when they attend the same ceremonies. From a tourism point of view, this gives a certain 'folklore' peculiarity to each of these ceremonies. Roman Catholic ceremonies often take the form of general popular celebrations where even non-Catholic believers, in an ecstatic religious procession for example, experience their own spiritual encounters, stressing that it is the universal religious outlook on life that matters most.

Through service and ritual the strong connection between religion and tourism can be seen. Durkheim's *Elementary Forms of Religious Life* (1912) explains that simple forms of every religion are rituals which are performed to celebrate the society itself, and which increase those societies' social solidarity. From this, MacCannell (1976: 13) concludes that tourism is 'a ritual performed to the differentiations of modern complex society'. Durkheim comments that such rituals, 'generally exhibited effervescence, pleasure, games ... All that creates the spirit that has been fatigued by the too great slavishness of daily work' (Durkheim 1912: 426).

That which society often holds 'sacred', primarily the unquestionable and fundamental structure of beliefs about the world, does not have to be religious, 'but nevertheless may be felt as crucially important and capable of arousing strong emotions' (Graburn 1983: 13). An example is Lenin's grave, now a tourist attraction, which may be considered 'sacred' in its own specific way in the context of atheist Russia. This and other similar examples caused Graburn to derive the phrase 'Tourism: the Sacred Journey' and pose a question for himself and his readers about what can be considered 'sacred' in the modern, non-religious world. The borders of the notion of 'sacred', from the profane and the religious point of view, were studied by Victor Turner, who published the results of his analysis on the ritualistic process in religious and tourist practice in various publications (especially in 1969, 1974) in which he strongly emphasised the structural sociological changes that occur among participants in rituals, regardless of whether the ritual contents are profane or religious (see Turner and Turner 1978). Elaborating on the notion of sacred in the religious sense, Cohen introduced the concept of 'the centre' of every inhabited area, a 'place that is sacred above all' (Cohen 1979: 189–193). Cohen explored the phenomenal modes of tourism according to their relation to the centre of the tourist's cosmos, concluding that recreational, diversional and experimental tourists remain loyal to their centre at home, in their own everyday environment; experimental tourists participate in other people's lives, but never adopt somebody else's centre as their own; only the pilgrim 'is fully committed to an "elective" spiritual centre, i.e., one external to his native society and culture'.

Religious experience and expression generally goes beyond all social and cultural forms by which it is expressed and structured, what religious experiences and forms of expressions are acceptable or considered normal and those which are not. Unacceptable religious experiences or expressions generally fall within the realm of the mystical, prophetic and utopian, and are symbolic of going beyond the religious boundaries of conventional religious experience. People in Eastern European countries where the Roman Catholic Church dominates, were not free to practise their religion, or at least not entirely, during the socialist/communist era. The symbol of freedom, both national and religious, existed in all the religious rituals as a constant invitation to its realisation. It was then that the Roman Catholic Church, despite all the bans, was strongest, as its attitudes also served to defend the general interests of the people. As soon as freedom was obtained in the early 1990s, the religious force weakened considerably. The same is true of shrines. Under communism, a visit to a shrine was a sign of revolt, which was only partly religious, and considerably more national and political. Polish people's visits to Czestochowa, for instance, during the time of communism, were more an expression of defiance and political will than religious need. It was also a symbol of the people's religious and national unity and their desire for a different and better future.

In such a context, certain places from recent history, especially those connected with the inhabitants' suffering, became sacred in a religious sense. The second biggest European shrine, Santiago de Compostela, aside from its great religious significance, also acquired historical significance when the Spanish won independence from the Moors (Graham and Murray 1997: 402). In contemporary European history, the pilgrimage route to Santiago de Compostela came to symbolise the unification of Europe (Slavin 2003; Santos 2002). For the people of Mexico, Guadeloupe represents all that is Mexican, just as the Wailing Wall is symbolic of everything Jewish to Jews. Once a year, Fatima is the meeting place for Portuguese workers who temporarily work abroad. 'All these sacred sites have gradually become national symbols and tourism destinations for visitors of Catholic belief [and] . . . all other faiths' (Vukonić 2002: 65).

Often after a war such places are consecrated through a special ceremony which gives their historical significance an added dimension; their religious meaning becomes partly or entirely equal to that of religious centres and sacred sites. However, this only confirms that the union of religious and secular life in such places is strong. Aside from the traditional sites, new places have been established as sacred sites for the gathering and pilgrimage of Catholic believers. The recent wars in Croatia and Bosnia and Herzegovina, as well as the events that followed, demonstrate this. Many of the church buildings that were destroyed or severely damaged in the war, which had only very limited local significance before the war, have become national religious shrines since their reconstruction. In terms of domestic tourism trends, such buildings or locations have become the destinations of tourism visits, much like traditional religious sacred sites.

Sacred sites as tourist attractions in the Catholic tradition

Long before the emergence of Christianity, other civilisations and religions were familiar with the notion of travel to sacred sites. However, Christianity gave such places special significance in the context of its religious beliefs and teachings, because in the history of Christianity, such places served as a kind of tool for the strengthening and dissemination of Christian beliefs. The meaning of the former pilgrimages was primarily penitence, and the pilgrimage was made in the spirit of the medieval 'penitentials'. Shrines were 'ranked' according to merits that the believers would acquire based on the number of visits. Two visits to St David's in Pembrokeshire 'had the same value' as one visit to Rome, for example (Shackley 2001: 121). But the development of modern methods of transport at the end of the nineteenth century and onwards made travelling more possible, and gave rise to tourism on a larger scale in the modern sense of the word. The church could not ignore this phenomenon, whose significance today is enormous and which ultimately has an impact on the relationship between God and humans. The pastoral mission, according to the teachings of Catholicism, means that the bearers of the church's teachings, the Pope, the council and the bishop, have not only the authority, but also the obligation, to support believers in this area of modern life with guidance and pastoral help.

Therefore, eventually leaders of the Roman Catholic Church recognised pilgrimage as related to tourist visits and tourism contents. In his speech on 1 September 1963, Pope Paul VI referred to pilgrimage as 'a special form of tourism', and then later explained his views: 'The Church cannot ignore such a big and complex phenomenon' ('Osservatore romano' 7April 1964). From the start, pilgrimages have always encouraged the setting up of special institutions and regulations, such as shelters for pilgrims, special monastic orders, and regulations for the preservation of the spiritual significance of this kind of tourism. Similar regulations and guidelines are contained in the decision of the Sacred Core of the Sabor of 1936. The Apostle's Constitution '*Exul familia*' of 1952 made an attempt to solve the problem of providing spiritual guidance to people who experienced the great waves of migration of refugees and other people after the Second World War.

Today, sacred sites derive additional value from the accompanying historical and cultural attractions, which add to the appeal of these places for tourism consumption. Notre Dame Cathedral in Paris has long lost its dominant function as being a place of prayer – the 12 million people who visit each year have turned the remarkable religious edifice into a temple of the new age – a tourist temple. The same can be said of Westminster Abbey in London, Chartres Cathedral in France, Cologne Cathedral in Germany, and in fact, almost all the churches in Rome. Gaudi's Sacrada familia in Barcelona is another, even more extreme example of how sacred space is increasingly becoming tourist (i.e. profane) space.

Undoubtedly, for the majority of tourism theorists dealing with the issues of tourist attractions, sacred religious sites attract classic, traditional believers, as well as other visitors (Inskeep 1991; Middleton 1994). The major changes that occurred in the world tourism market in the 1980s altered conventional conceptions of tourist attractions, and, with the intro-duction of new attractions designed to motivate people to travel, helped form a different understanding of sacred buildings and sacred sites as tourist attractions. It can perhaps be said that the cultural programmes and events organised at sacred sites and in individual sacred buildings have become the sine qua non of tourist visits to such places. This correlation between culture and Roman Catholicism is nothing new or unusual, since this link has existed from the beginnings of Christianity, but the novelty is in the visitors' understanding of the sacred and secular elements of these places. This, of course, does not mean that there are any fewer visitors who come to these places for purely religious reasons or that such people have stopped coming altogether, but it points to a significant increase in secular reasons for the consecration of both sacred Catholic sites and individual sacred buildings.

Sacred sites can be grouped together or divided according to different criteria, but this does not bear much significance on the scientific study of religion or tourism. Therefore, such classifications will not be dealt with here. Instead, attention is devoted to the consequences of visits to sacred sites, in view of their religious, tourism, economic and social significance.

From the perspective of town planning, the religious nucleus was not physically separate from the community itself, nor were any artificial barriers created to make the religious nuclei available solely to believers. Nevertheless, certain codes of behaviour were established to separate such nuclei from settlements. Examples include the regulation of traffic next to a sacred building; the special regulations that govern economic activities performed in the immediate vicinity of a sacred building; and the regula-tions that govern construction near the religious nucleus. This, however, has not prevented the commercialisation of sacred sites and life around them by both commercial and religious entities (Olsen 2003). In special authorised shops in St Peter's Square, Vatican City, papal blessings can be purchased as authentic 'spiritual souvenirs' bearing the Pope's picture. The increasing number of shops and stalls that sell souvenirs of more or less authentic religious content and character, along with a certain amount of kitsch, have forced church authorities in many of the shrines to appeal to city authorities to intervene in order to limit this kind of trade, at least in areas and locations in the vicinity of sacred buildings that are not under the church's jurisdiction. Nowadays, there is not a single Catholic shrine in the world that does not face problems of this kind, including Lourdes, Fatima, Compostela, Medjugorje, Padua (the Basilica of Saint Anthony of Padua), Einsiedeln (Switzerland), Jasna Gora in Czestochowa (Poland), and Loreto (Italy).

Also of interest are the sacred or consecrated 'former religious' places, which are, for particular historical reasons, visited more for secular than religious reasons. Famous examples include the seventeenth- and eighteenth-century Jesuit missions in the Guaranis area on the borders of Brazil, Argentina and Paraguay that were proclaimed World Heritage Sites in the 1980s. Today, these are, in fact, ruins of settlements and sacred buildings: Sao Miguel das Missoes in Brazil, and San Ignacio Mini, Santa Ana, Nuestra Senora de Loreto and Santa Maria la Mayor in Argentina. These places, where the Jesuits established farms with members of the native Guarani tribe, actually had a certain amount of political and religious autonomy. After the establishment of the Spanish-Portuguese peace, the missions were closed down, but they have now been restored as a mix of religious and secular tourist attractions. Add to this the world's largest waterfall, Iguasu, located nearby, and there is a large and interesting tourism complex, which people visit for various reasons. This is becoming a new 'design' or type of tourist destination, formed by combining the different features of tourism resources.

Catholicism has a strong tradition of pilgrimage to the so-called Marian shrines, and their dual significance for both religious and tourist visits requires additional attention. The eyewitness account of young Jacopino from 1484, regarding the apparition of the Virgin Mary, and the consequent miraculous events, marked the beginning of a long period of pilgrimages of Christian believers to places where the Virgin appeared (Turner and Turner 1978: 86). Marian shrines have been established in the places of the Virgin's apparitions, which are often accompanied by miraculous healings and events, as well as messages to the visionaries. This has become a common feature of the Marian shrines. Most of these places have been sites of only one or two apparitions, but at Medjugorje the Virgin has been appearing to the same visionaries for several years at regular intervals. 'The fact that so much tourist accommodation has been built both in Medjugorje and Čitluk, as well as in the nine neighbouring villages and towns, supports the view that Medjugorje has become a well-established tourist destination' (Vukonić 1992: 89). One of the oldest and still most frequently visited Marian shrines in the world, the French town of Lourdes, answers many of the questions usually asked when seeking an explanation for the concept of Marian shrines. Pilgrimages to Lourdes, which began in 1858, when the Virgin appeared to one of the villagers, a girl named Bernadette Soubirous, helped the village develop, in terms of the economy, town planning and population, from a settlement of only 4,000 people to a small town of 18,000 inhabitants (Eade 1992). All this happened in an area where the economic indicators for the neighbouring Pyrenean region were in constant decline. The town grew from the church nucleus through the development of accompanying activities (e.g. shops, hospitals, hotels and catering facilities), and turned into a religious, and regional delopment centre in the Pyrenees (Rinschede 1992; Vukonić 2002). Other typical examples, Loreto and Fatima in Portugal, also grew from village communes into towns thanks to Christian pilgrims. A similar thing happened in Medjugorje, Bosnia and Herzegovina, where the Virgin appeared to a group of young people in 1981. Although Medjugorje is the youngest Marian shrine, it is one of the most visited Catholic pilgrimage centres in the world (Vukonić 1996: 146). Another example is Mont-Saint-Michel, which is packed with shops and visitors, identifying it as both a religious and a tourism centre otherwise.

What are the roots of the cult of Mary and to what can be ascribed the place she holds, reflected in the feast day of the Assumption, which among most Catholics ranks third, after Christmas and Easter, in popular religious observance? The modern history of Christianity and a large part of modern theology, accepts clearly that the present-day honouring of the Virgin has many layers. The various elements of the cult of the Virgin can be distinguished from one another according to historical criteria, by establishing what was introduced into it from the previous cult of virgins or great mothers connected to fertility symbols, which originated from agricultural cults (the dates of Mary's feast days correspond to those of harvest festivals), and what was derived from other beliefs (Vukonić 1996: 124).

The English shrine of Our Lady of Walsingham in Norfolk, where the Virgin Mary appeared in 1061 and ordered the construction of a replica of her house in Nazareth, was forgotten for centuries, but in recent years has been re-established and is now visited by Anglican and Roman Catholic believers alike. In a time of strong secularisation of life, this example demonstrates that, thanks to the political changes and events of the last century, religion in general, and the Catholic religion in Europe in particular, has been re-established, and its significant role in the everyday life of many Europeans has been restored. This is one reason for the increasing number of visitors to European shrines, visitors who are not only tourists and believers from Europe, but members of various religions from around the world.

Conclusion

A general intensification of religious belief seems apparent in many parts of the world, accompanied by a weakening of belief in the established church. So it is with the Catholic religion and the Catholic Church, where to be religious no longer means to follow blindly all that the church says and does. The example of Medjugorje is proof that religion is more than religious dogmas and institutions. The Catholic public has shown a lack of interest in the official view of church authorities who question the authenticity of the visitations of the Virgin Mary in Medjugorje, despite the Vatican's recommendation that they should not, and even in the face of specific prohibition.

Pilgrimages are an essential belief in the Roman Catholic tradition. Because there are no signs that the religious motives for travel among Catholics will weaken, they will remain an important component of tourism wherever pilgrims go. The official attitude of the Catholic Church, as expressed in the 'Declaration on Freedom of Belief', says that humankind has been given reason and free will, so the acceptance of God's law is a question of individual conscience. This view of religious freedom pre-supposes freedom from any religious pressure and supports the view that religious tourism will become increasingly individualised and that visits to traditional sacred sites around the world will develop with the same intensity as they did in the past.

References

Bowman, G. (1992) 'Pilgrim narratives of Jerusalem and the Holy Land: a study in ideological distortion', in A. Morinis (ed.) *Sacred Journeys: The Anthropology of Pilgrimage*, Westport, CT: Greenwood.

Cohen, E. (1979) 'A phenomenology of tourist experiences', *Sociology* 13: 179–201.

Cohen, E. (1992) 'Pilgrimage and tourism: convergence and divergence', in E. Morinis (ed.) *Sacred Journeys: The Anthropology of Pilgrimage*, Westport, CT: Greenwood.

Durkheim, E. (1912) *The Elementary Forms of Religious Life*, New York: G. Allen and Unwin.

Eade, J. (1992) 'Pilgrimage and tourism at Lourdes', *Annals of Tourism Research* 19: 18–32.

Graburn, N.H.H. (1983) 'The anthropology of tourism', *Annals of Tourism Research* 10: 9–33.

Graham, B. and Murray, M. (1997) 'The spiritual and the profane: the pilgrimage to Santiago de Compostela', *Ecumene* 4: 389–409.

Inskeep, E. (1991) *Tourism Planning: An Integrated and Sustainable Development Approach*, New York: Van Nostrand Reinhold.

MacCannell, D. (1976) *The Tourist: A New Theory of the Leisure Class*, New York: Schocken.

Middleton, V. (1994) *Marketing in Travel and Tourism* (2nd edn), Oxford: Butterworth Heinemann.

Olsen, D.H. (2003) 'Heritage, tourism, and the commodification of religion', *Tourism Recreation Research* 28(3): 99–104.

Riessman, C. (1993) *Narrative Analysis*, London: Sage.

Rinschede, G. (1992) 'Forms of religious tourism', *Annals of Tourism Research* 19: 51–67.

Rinschede, G. (1999) *Religionsgeographie*, Braunschweig: Westermann.

Santos, X.M. (2002) 'Pilgrimage and tourism at Santiago de Compostela', *Tourism Recreation Research* 27(2): 41–50.

Shackley, M. (2001) *Managing Sacred Sites*, London: Continuum.

Slavin, S. (2003) 'Walking as spiritual practice: the pilgrimage to Santiago de Compostela', *Body and Society* 9(3): 1–18.

Turner, V. (1969) *The Ritual Process*, Chicago: Aldine.

Turner, V. (1974) *Dramas, Fields and Metaphors*, Ithaca, NY: Cornell University Press.

Turner, V. and Turner, E. (1978) *Image and Pilgrimage in Christian Culture*, New York: Columbia University Press.

Vukonić, B. (1992) 'Medjugorje's religion and tourism connection', *Annals of Tourism Research* 19: 79–91.

Vukonić, B. (1996) *Tourism and Religion*, Oxford: Pergamon.

Vukonić, B. (2002) 'Religion, tourism and economics: a convenient symbiosis', *Tourism Recreation Research* 27: 61–67.

17 Tourism and informal pilgrimage among the Latter-day Saints

Daniel H. Olsen

Most major world religions have some sort of doctrinal basis for and formalized rituals relating to pilgrimage travel. In some cases, pilgrimage is a required element of religious worship, whether it is essential in the quest for a happier afterlife or for initiatory purposes (Morinis 1992). However, not all major faiths embrace the notion of pilgrimage, at least in its traditional sense. The Church of Jesus Christ of Latter-day Saints (informally referred to as the Mormon Church or the LDS Church) does not have formal pilgrimage practices or proscriptions, and yet every year thousands of church members travel to places associated with the birth and growth of their religious faith, their scriptural heritage, and to sacred temples. While much of this travel is a grassroots-inspired movement, church leaders also encourage this informal pilgrimage-like travel to important sites, such as temples, and places related to church history and heritage.

The purpose of this chapter is to describe the relationship between tourism and the LDS Church. In particular, three different aspects of this relationship are discussed. First, the views of the church related to tourism will be addressed. Even though little has been officially stated about tourism by church leadership, it is possible to highlight a number of implicit views by examining the scriptural basis of hospitality and proselytization as they relate to tourism and how LDS Church leaders, historically and contemporarily, interact with tourism at historical and sacred sites. Second, the informal motivations Mormons have for traveling to church history locations and heritage sites associated with holy scriptures are examined. Last, the pilgrimage-like travel patterns of Latter-day Saints are examined in terms of the major destinations they visit when traveling for religious purposes. Before discussing these three aspects, however, a brief introduction to the LDS Church is necessary.

A brief introduction to the LDS Church

The Church of Jesus Christ of Latter-day Saints was founded in 1820 in New York State by Joseph Smith Jr, who claimed to have been told by

God that the existing Christian churches and their associated creeds were not His churches, and that Joseph was to establish a new Christian faith under God's direction (Smith 1976). Converts were attracted to the new religion for many reasons, including the claim to a restoration of divine authority, the bringing forth of *The Book of Mormon*, spiritual manifestations, the primitive simplicity of the doctrines taught by Smith, and the positive influence of missionaries (Grandstaff and Backman 1990). The church grew rapidly, but local antagonism forced Smith to move its headquarters to various successive locations, including Kirtland, Ohio; Independence, Missouri; and Nauvoo, Illinois. As the church continued to grow and its center and membership moved from place to place, the doctrines and religious practices Smith promulgated continued to mature and develop. Some of these practices included the establishment of temples, baptisms on behalf of the deceased, celestial or eternal marriage, the Word of Wisdom (abstinence from tobacco, alcohol and coffee), the identification of Independence, Missouri, as the New Jerusalem, the rejection of the Trinity, and polygamy. In 1884 Joseph Smith was killed by a mob, whose members were not favorable to some of the new doctrines and practices being implemented by Smith. Following his death, several successors vied for control of the church (Quinn 1976; Shields 1987, 1990). His eventual successor, Brigham Young, led most of the Mormons to the valley of the Great Salt Lake where they founded Salt Lake City, the current headquarters of the church (Allen and Leonard 1976; Arrington and Bitton 1979; Bushman 1984; Jackson 2003). Today the church boasts a current membership exceeding twelve million, with the majority residing outside of North America.

Latter-day Saint perspectives on tourism

Unlike many religions, pilgrimage does not hold a place within LDS theology, at least in the sense that leaders of the church have not declared formal doctrines pertaining to any forms of pilgrimage. This is in spite of the fact that the church and its leaders and members recognize the existence of sacred spaces and have long held that certain places are more holy or sacred than others (Jackson and Henrie 1983; Olsen 2002; Bradley 2005). "Neither shrines nor pilgrimages are a part of true worship as practiced by the true saints. . . . [T]here is no thought that some special virtue will attach to worship by performing [pilgrimage to sacred sites]" (McConkie 1966: 574). In other words, travel to sacred sites associated with the forgiveness of sins or miraculous healings does not take place within the travel practices and beliefs of the LDS Church (Hudman and Jackson 1992; Jackson 1995).

Related to the lack of formal doctrine or practice of pilgrimage is the fact that church leaders, while having commented on a number of social and moral issues, have not commented upon the phenomenon of tourism.

One possible reason for the lack of a formal statement is that the church is not in the business of tourism but rather the business of saving of souls (Olsen 20004: 12). While the rationale for how issues to be commented upon are chosen is not clear, it is probably a result of the fact that "the teachings of the Church today have a rather narrow focus, range, and direction" that is focused on "central and saving doctrines" (Millet 2003: 19), including specific moral issues such as gambling, pornography and the disintegration of the family, and that of preaching the gospel to all the world (Matthew 28:19–20 – New Testament). Another reason for that lack of official clarity on tourism may result from the LDS belief that believers are to work some issues out for themselves, rather than depend solely on direction and proclamations by church leadership. In the Doctrine and Covenants 58:26–27 (another book of scripture), the Lord says:

> For behold, it is not meet that I should command in all things; for he that is compelled in all things, the same is a slothful and not a wise servant; wherefore he receiveth no reward.
> Verily I say, men should be anxiously engaged in a good cause, and do many things of their own free will, and bring to pass much righteousness.

According to this perspective, it is unnecessary for the church to provide a position on every aspect of culture or tell adherents how to respond to every social situation. As such, Latter-day Saint theology is primarily experiential rather than systematic or propositional, as most of the church's tenets appear as revealed theology, where the knowledge of truth or reality is based on spiritual experiences and revelations from God rather than on logical deductions and formal reasoned statements of truth (Olsen 2000b). Thus, unlike the Roman Catholic Church, which has made numerous statements specifically on tourism (Vukonić 1996), leaders of the LDS Church have not systematically discussed this issue.

Regardless of the absence of formal statements on tourism and leisure pursuits by church leaders – except for encouraging members to live gospel standards, maintain a strict moral code of behavior at all times and to avoid excessive and unwholesome recreation (McConkie 1966: 622) – the LDS Church is extensively involved in tourism from a couple of perspectives. The first deals with hospitality, which has both a scriptural and otherwise historical basis. While there are several scriptural references to hospitality in the Old and New Testaments (e.g. Romans 12:13; Titus 1:8; 1 Peter 1:9), a couple of key scriptural references give insight to how the church views and interacts with tourism. One of these is in Doctrine and Covenants 124:22–23, wherein a revelation was directed to specific members of the church who were instructed in 1841 to build a boarding house or hotel where visitors to Nauvoo who were interested in learning more about the Mormon movement could rest:

Let my servant[s] . . . build a house unto my name, such a one as my servant Joseph shall show unto them, upon the place which he shall show unto them also.

And it shall be for a house for boarding, a house that strangers may come from afar to lodge therein; therefore let it be a good house, worthy of all acceptation, that the weary traveler may find health and safety while he shall contemplate the word of the Lord; and the corner-stone I have appointed for Zion.

According to Smith and Sjodhal (1978: 772–773), "this [r]evelation proves that the Lord wanted the tourists of the world to visit and become acquainted with the Saints. These were not to be surrounded by a wall of isolation. They had nothing to hide from the world." In accordance with this scripture, a hotel called the Nauvoo House was built at a cost of $10,000, but was left unfinished when the church moved its headquarters to Salt Lake City. However, this same reasoning led the church to build the Hotel Utah in downtown Salt Lake City in the early twentieth century (Arrington and Swinton 1986).

Another scripture is found in Isaiah 2:2–3. The Latter-day Saints viewed their trek west as being analogous to the great Exodus of the Children of Israel from bondage in Egypt. Brigham Young and many of the church leaders saw the settling of the Great Salt Lake Basin as fulfilling Isaiah's prophecy:

And it shall come to pass in the last days, that the mountain of the Lord's house shall be established in the top of the mountains, and shall be exalted above the hills; and all nations shall flow unto it.

And many people shall go and say, Come ye, and let us go up to the mountain of the Lord, to the house of the God of Jacob, and he will teach us of his ways, and we will walk in his paths: for out of Zion shall go forth the law, and the word of the Lord from Jerusalem.

Petersen (1981: 56) suggests that tourists are one of the groups of people from "all nations" who come to Salt Lake City seeking the Lord's house (temple) established "in the tops of the mountains." This implies that the Latter-day Saints have a responsibility to be hospitable and courteous to visitors who, according to this scripture, will actively come to Salt Lake City to see the "house of God," as well as prepare to receive those who seek to learn the "word of the Lord" through interpretation and education.

Historically, leaders and members of the LDS Church have had to act as hospitable hosts to visitors who viewed them as an American curiosity (Jackson 1988) because of the church's unique doctrines and practices. With the discovery of gold in California in 1849, the Mexican War in 1846–1848, the development of stagecoach lines in the 1850s, the Utah

War in 1857–1858, and the construction of the Transcontinental Railway in 1869, thousands of outsiders began to visit Salt Lake City. While tourists did come to Utah to see the natural landscape of the west, including the Great Salt Lake and the Rocky Mountains, the majority were drawn to the city because of their interest in seeing the mysterious and unusual Mormon Church. While the tourism literature of the time promoted Salt Lake City as an "orderly, pleasant, clean, relatively crime-free 'must see' tourist attraction" (Eliason 2001), focusing on the garden-like feel of Salt Lake City with irrigation systems running along its wide streets, the beautiful gardens of Temple Square, and the amazing architectural wonder of the Mormon Tabernacle, the Mormons and in particular the practice of polygamy were the main tourist attraction (Gruen 2002). By the end of the nineteenth century Salt Lake City was inundated with tourists, many of them with strong views and prejudices against the Mormons, thanks in part to many misinformed travel brochures, anti-Mormon books, and negative media and travel accounts (Snow 1991; Bishop 1994; Mitchell 1997; Smith 1998; Morin and Guelke 1998; Eliason 2001).

LDS Church leaders were initially unsure how to handle this influx of non-Mormon tourists on an official level. Some in the church hierarchy were convinced that most visitors to Salt Lake City were there to belittle the Mormon Church and its people, and argued that while they could not stop visitors from coming to Salt Lake City, they could restrict non-Mormons from entering church properties. Many of the lay members of the church also felt that tourists, who, being disappointed at not seeing any public displays of polygamy, became "snoopish" (Limerick 2001) and intruded too far into their private lives in an attempt to gain first-hand contact with polygamist culture (Hafen 1997), creating a growing animosity toward the growing number of tourists visiting Salt Lake City. Others, however, saw tourism as a way of creating a new public image in the face of increasing discrimination and bad press about the theocratic nature of the region and the practice of polygamy. It was thought that by opening church properties to tourists, preconceived negative notions of Mormonism visitors brought with them would be dispelled and they may come to tolerate Mormonism even if they did not believe in its doctrines (Bishop and Holzapfel 1993; Hafen 1997). Church leaders went with the second view, the more hospitable approach, and opened church buildings for public perusal and even met with tourists to explain Mormon beliefs (Hafen 1997). In particular, visitors were guided to Temple Square, the spiritual and administrative center of the LDS Church located in the heart of downtown Salt Lake City, where church officials set up an interpretive booth and assigned the faithful to act as guides for tourists visiting the church's main institutional and religious sites. This was clearly an attempt to promote Mormonism in a more positive light and attempt to dispel the negative myths and ideas eastern tourists had of the church.

Today the LDS faith continues to play the role of host to visitors who come to Salt Lake City by not only providing interpretation for visitors at Temple Square, but also by developing various cultural attractions, including a daily recital on the world-famed Tabernacle Organ and music performances by the Mormon Tabernacle Choir and the Orchestra at Temple Square. In addition, the church has created a number of departments within its bureaucracy that focus specifically on hosting visitors at the Square, including the Visitor Activities Department, a tourism-driven agency whose two main purposes are to build relationships with the local tourism interests as well as the promotion of church events and sites in the downtown area. The Hosting Department is responsible for greeting visitors at Temple Square. The Building Hosting Division oversees welcoming visitors to the various buildings located around Temple Square, and the Temple Square Hospitality Corporation, a church-owned for-profit company is interested in attracting visitors to Salt Lake City and Temple Square in particular. The Hospitality Corporation owns and operates three catering facilities, a hotel and four restaurants around Temple Square, and focuses on four main target markets: wedding receptions, tourism, business functions and families. The church also has a strong interest in seeing the success of tourism in Utah as a whole, as demonstrated by its gifts of money and land to support the 2002 Winter Olympic Games in Salt Lake City (Olsen 2006).

The church also views tourism not just from a hospitality perspective, but also as a medium for spreading the spiritual message of Jesus Christ and Mormonism. Throughout its history the LDS Church has had to weather abuses from various stereotypes and negative publicity because of its unique Christian doctrines and its former practice of polygamy. As alluded to earlier, interpretive centers and a guiding system on Temple Square were developed in part to present a positive image of Mormonism as a way of educating visitors about the basic doctrines, beliefs and practices of the church with the hope that visitors would leave with at least a better, more correct understanding of the tenets of Mormonism. While public relations efforts helped to reshape the positive image of Mormonism over time, the media continues to focus on doctrines that differ from other Christian sects and the long forgone practice of polygamy in articles about the church, depicting Mormons to be admired because of their moral and social convictions but not "truly belong[ing] in mainstream society" (Chen and Yorgenson 1999: 112; Chen 2003). Because of this media fetish, negative stereotypes of the LDS Church continue to exist in the minds of many visitors to Salt Lake City, making Mormonism both a modern-day curiosity and a cultural tourist attraction. In a continuing attempt to build bridges with non-Mormons and enhance the faith's public image, church leaders encourage visits to Temple Square and continually refine their interpretive methods to present the cause in a positive manner to the 4 to 6 million people who visit each year.

Related to the role of tourism to enhance the church's public image is that of using tourism as a means of soft proselytizing. The Mormon Church is well known for its proselytizing efforts. In any given year over 50,000 missionaries actively preach the gospel in almost every corner of the world. The fact that non-Mormons come to Salt Lake City and dozens of other LDS heritage sites and interpretive centers throughout the world (Madsen 2003) provides fertile ground for sharing the basic tenets and history of Mormonism in a way that allows visitors who might be interested in learning more about the church and its teachings to receive more information and potentially convert to the faith later (Olsen 2005).

Tourism also serves a pastoral function. Church history is used to build a cohesive Mormon identity within an increasing geographically and ethnically diverse membership through an increased emphasis on the church's pioneer and spiritual history. Official publications and public discourses emphasize the shared religious roots of church members (Madsen 2003) or a shared "heritage of belief" (Hervieu-Léger 1999: 88–90). While there has been no "official" encouragement to travel to sites related to church history, the church does sponsor special centennial observances of important foundational events, such as Pioneer Day, which celebrates the entrance of Brigham Young and the early pioneers into the Salt Lake Valley, and sponsored events, such as the Hill Cumorah Pageant (New York) held at the site where Joseph Smith was given gold plates from which he translated *The Book of Mormon* (Bitton 1994; Gurgel 1976; Hudman 1994; Olsen 1992). In addition, the Church's main website (www.lds.org) has a special section entitled "Places to Visit," which highlights "a wide variety of Church historic sites, pageants, and visitors' centers." Interactive maps are available that allow users to highlight certain key areas of the United States and specific monuments or sites. Pictures of the sites are shown and a short description of their significance is given. The pageants, events, website and the control and maintenance of church historical sites encourage informal pilgrimage-like travel by believers who wish to participate in these events and to explore the sites and their meanings as they relate to their spiritual roots. This in turn leads to a strengthening of people's religious identity and thereby increases the overall cohesion in the church.

Latter-day Saint travel motivations

While there are implicit reasons why leaders of the LDS Church engage in tourism-related practices and activities, it is also important to examine the travel motivations and patterns of Mormons themselves. Each year thousands of believers visit religious heritage sites associated with the history of the church, as well as sites related to scriptural heritage. As noted earlier, this pilgrimage-like travel occurs despite the lack of formal doctrines or practices related to pilgrimage or the lack of sacred sites associated with forgiveness of sins or miraculous healings (Hudman and Jackson 1992; Jackson 1995). While not formal pilgrimages per se, travel by Mormons tends to fit Hudman and Jackson's (1992: 109) "tourism pilgrimage," which "describe[s] tourism that combines travel for recreation or pleasure with religious beliefs, whether or not church doctrines promote pilgrimage."

While no empirical research has been done on the general socio-economic and motivational variables of this particular market segment, one of the reasons LDS adherents travel to historic sites is, according to Mitchell (2001: 9), because "visiting Mormon historical sites, museums and key buildings is one way in which Mormons are able to participate actively in their theology and cosmology." As a proselytizing faith, the church sends missionaries to preach the gospel of Jesus Christ to those who will listen. They teach interested people the basic beliefs of the church, including faith in God and Jesus Christ, after which they invite their "investigators" to pray and receive their own confirmation from God through prayer (see Moroni 10:3–5 in the *Book of Mormon*). This spiritual confirmation is considered the beginnings of a personal testimony – a gift from God, something that is given after some effort on the part of the individual. As part of a testimony one needs to affirm certain key authentic historic events within the restoration of the Church, such as the divinity of Jesus Christ, God's appearance to Joseph Smith, and that *The Book of Mormon* is in fact the religious history of the people who once inhabited the American continents. Davies (2000: 12–13) observed, "there are many Mormons for whom the primal story of the Restoration does constitute the truth: a basic epistemology that furnishes a template for history and for the stories of family life." Belief, therefore, is not just a rational exercise but also an experiential one, in which belief cannot be separated from experience (Olsen 2000a, 2000b)

In accepting the message of the missionaries and having a confirming experience from God, a convert adopts a new religious identity complete with a religious history that can be geographically and materially located (Hovanessian 1992: 200 cited in Chivallon 2001: 461). The person who receives a testimony must then cultivate or increase that faith and knowledge through study, prayer and service throughout their lives. One way in which this can be accomplished is by visiting religious heritage sites related to the history of the church, where key events of the restoration occurred. Because testimonies come about through intangibly qualitative, spiritual or emotional experiences, many Latter-day Saints desire to visit physically the places where many of the key restoration events took place, to "engage with the material remnants and reminders of the [religious] history through embodied memories of their engagement with the objects, buildings and narratives of their theology" (Mitchell 2001: 9).

Coupled with the emphasis placed on gaining a testimony of the message of Mormonism is the importance placed on the family unit. In particular, because of the church's strong emphasis on families, many Mormons travel with immediate or extended family groups to church heritage sites. As Lee (2001: 231) suggests, ". . . family trips help develop a sense of attachment to a destination and support the notion that childhood travel with family members positively influences an individual's attachment to a destination." In this manner, church members not only gain a better appreciation for their faith and strengthen their testimonies, but also "assure the passage of a given content of beliefs from one generation to another" (Hervieu-Léger 1999: 89–90).

Latter-day Saint travel patterns

Hudman and Jackson (1992) identified four major types of destinations Mormon tourists visit when they undertake pilgrimage-like travel: biblical sites, *Book of Mormon* lands, Church history sites, and temples. These four major destination types are highlighted in this section.

Biblical sites

Like other Christian denominations, Latter-day Saints have an interest in the historical and religious sites related to the birth of Jesus Christ and the rise of Christianity in the Holy Land, particularly in Israel. The importance of the Holy Land to the Mormon Church is deep and can be seen in the establishment of a physical presence in Jerusalem via the church-owned Brigham Young University (BYU) Jerusalem Center for Near Eastern Studies, an educational center used for study abroad programs to the Middle East (Olsen and Guelke 2004). This interest, however, goes beyond just a fascination about the lands where Jesus walked. Latter-day Saint theology teaches that Mormons are "literally adopted into Israel and are thereupon brought into the covenant by virtue of their membership in the tribes of Israel" (Shipps 1985: 75). In addition, the return of the Jews to Israel is one of the signs of the forthcoming millennium (Olsen and Guelke 2004).

The LDS interest in visiting biblical locations has led to the development of a small number of tour companies catering specifically to the Mormon pilgrimage tourism market. Tour companies advertise through a number of commercial media outlets, including popular LDS-oriented magazines and the Internet. These companies focus on the spiritual nature of the trip, advertising spiritual experiences and emphasizing the education and cultural value of the trips, combining biblical sites with other more conventional tourist activities. In addition, BYU (before the political tensions in the Middle East) took students to the Jerusalem Center for educational travel purposes, akin to the Jewish educational experience (Cohen 2003). Both pilgrim-tourists and students visit many of the same attractions as other Christian visitors, such as the Garden of Gethsemane, the Garden Tomb, Nazareth, Bethlehem, and other such sites related to the life of Jesus, but also visit the BYU Jerusalem Center and the Orson Hyde Gardens, built in the 1970s in honour of the first nineteenth-century Mormon missionary to the Holy Land (Olsen 2000; Olsen and Guelke 2004).

Book of Mormon *lands*

One of the things that set Mormonism apart from other Christian sects is its belief in additional scripture. *The Book of Mormon* is a translation of an ancient record Joseph Smith claimed to have received from an angel. This record gives the account of a family who traveled from the Holy Land to the Americas around 600 BCE. Once in the Americas the family divided into two social and political entities, one eventually destroying the other. *The Book of Mormon* is the spiritual and secular history of the group that was destroyed, and, as alluded to in the book's subtitle, "Another Testament of Jesus Christ", is looked upon by Latter-day Saints as a scripture comparable to the Bible. It is believed that the native peoples of the Americas are the descendents of this family.

While the specific lands in which the events depicted in *The Book of Mormon* took place have been a subject of debate by scholars and historians (Sorenson 1992), most LDS scholars believe that the events occurred in and around Central and South America. This locational uncertainty is problematic for tour companies that sell packages to the *Book of Mormon* lands. Therefore, the itineraries are based on approximate locations in Central and South America and best-guesses rather than definitive places. However, Mormons tend not to be so much concerned with the accuracy or exactness of the locations as they are with gaining a greater understanding of the cultural and historical context in which the record was written, particularly since the ancient Maya, Inca and Aztec ruins in Central and South America are believed to be linked directly to the people and events in *The Book of Mormon*.

Church history sites

Most Mormons who travel for religious reasons visit sites related to the founding and growth of the church. This is in part because of the proximity of the historical sites to a large number of church members, making them more accessible by car and bus. Ranging from New York to California, these sites allow for both short and extended vacation trips. Therefore, visits to church history sites are more affordable to LDS travelers than tours to the Holy Land or *Book of Mormon* lands. There are a number of tour companies and operators that cater to this particular niche of the LDS informal pilgrimage market. Some operators develop tours encompassing LDS sites in the Eastern United States, while others develop tours that take travelers from New York to Salt Lake City. Other tour operators merge travel to historic sites with visits to LDS temples or sites related to national identity.

In terms of importance, the Hill Cumorah, where Joseph Smith received *The Book of Mormon* record, and the Sacred Grove, where he received his first revelation, are the most sacred. These two sites, located in Palmyra, New York, are what Jackson and Henrie (1983) characterize as mystico-religious sites, or where a supernatural event occurred. At the Hill Cumorah a pageant is held each summer which dramatizes key aspects of the story-line of *The Book of Mormon*. The Hill Cumorah Pageant is the largest outdoor pageant in the United States, drawing over 100,000 people during its two-week performance (McHale 1985). In addition to the Hill Cumorah and the Sacred Grove, other sites of interest include the Grandin Printing Shop, where *The Book of Mormon* was first published, the Peter Whitmer Log Home, where the church was officially organized in 1830, and the two homes in which Joseph Smith's family lived while in Palmyra.

Nauvoo, Illinois, is seen as the mythical Camelot of Mormonism, where the Church achieved its social and doctrinal maturity under Joseph Smith. The church has undertaken a Williamsburg-like approach to restoring Nauvoo to its original conditions through rebuilding brick and log home replicas to show visitors what life might have been like in the 1840s (Backman 1994). Interpretation methods include first-person interpretative guides and numerous hands-on activities for adults and children (Olsen and Timothy 2002). Nearby Carthage, Illinois, is a popular destination as the place where Joseph Smith was martyred.

Independence, Missouri, is an important stop for Mormon pilgrim-tourists as well, because, as noted earlier, this is where Joseph Smith said the New Jerusalem would be built. While Independence is rich with historic value and future tourism potential, there are few historic sites of interest to LDS tourists. However, the Mormon Missouri Frontier Foundation, a multi-denominational non-profit group, has developed a tour highlighting key sites and events relating to early Mormonism's activities in Independence (Olsen and Timothy 2002). Kirtland, Ohio, is another important destination, as this is where the first Mormon temple was built, and where numerous revelations were received by Joseph Smith (Olsen 2004).

There are a number of additional sites of interest. In Sharon, Vermont, is a 38.5-foot granite monument commemorating the birthplace of Joseph Smith (one foot for every year of Smith's life). Numerous markers line the Mormon Trail, the route Brigham Young took when leading the Saints to Utah. In Salt Lake City, Utah, there are numerous monuments, markers and buildings that direct pilgrim-tourists to sites of importance, including Temple Square, the Church Office Building, the Genealogical Library, the "This is the Place" monument, and various other sites. There are also many lesser-known monuments found in Southern Utah, Mexico, California, Europe and Canada, which represent additional historical events and the colonization phase of early Mormonism in the western mountain region.

Temples

Church members are also encouraged to travel to temples. Temples are deemed the most sacred space in Mormon theology, owing to the belief that the sacred ordinances performed in them are essential for salvation (Talmage 1912; Packer 1980). Temples differ from regular meeting houses in that they are reserved for initiatory-type activities that focus on making sacred covenants with God rather than daily or weekly activities and Sabbath Day worship. In addition, while meeting houses are the most dominant physical symbol of an established Mormon presence in an area (Timothy 1992), the building of a temple changes the status of a city or area in the eyes of LDS members (Hudman and Jackson 1992) and establishes an ideological and physical center of the surrounding Mormon community (Parry 1994). Because temple rites and ceremonies are necessary for exaltation, travel to temples is a semi-obligatory ritual for LDS adherents. The purpose of going to the temple is to perform rites or ordinances that will enhance a person's knowledge of what is necessary to be with God after this life. Eternal marriages are also conducted in temples, whereas only civil marriages can be performed in regular church buildings. Once members perform the ordinances for themselves, they are encouraged to return often to perform the same rites on behalf of their deceased ancestors, who did not have an opportunity during their earthly lives to accept Jesus as their Saviour and perform the temple ordinances. The church's emphasis on eternal families, genealogy and ordinance work in proxy for those who have passed away provides the motivation for many LDS members to make frequent return trips to temples over the course of their lifetimes, particularly if a temple is located close to their place of residence.

Some temple visits, according to Hudman and Jackson (1992: 114), should not be classified as a pilgrimage per se, but rather tourism pilgrimage. Some members of the LDS Church desire to "collect" temples or visit many of the 119 temples currently in operation (www.lds.org), even though rituals do not vary from temple to temple. Tour agencies, especially those based in Utah, Arizona, and other areas of high LDS concentration organize temple tours in conjunction with regular tourist activities. For example, some operators combine visits to temples around Central and South America with visits to *Book of Mormon* sites, or mix European temple visits with cultural events, such as the famous Passion Play in Oberammergau, Germany. Other tour companies provide circuits of various temples in the United States that are in close geographic proximity to each other. It is interesting to note that some of the more recently built temples are located in proximity to major tourist destinations, such as Palmyra, Nauvoo and Winter Quarters, where pioneers spent the winter on their trek to Utah.

Conclusion

Today the Church of Jesus Christ of Latter-day Saints is still a public curiosity. For example, numerous newspaper and television reports were disseminated to a worldwide audience during the Salt Lake City 2002 Winter Olympics (Chen 2003). Being on a global stage, church leaders focused on using the Olympics as a platform for being good hosts and presenting to the world a positive image of the church. Through media curiosity, the church was able to refute old stereotypes, such as polygamy (no longer practiced since 1895), and present a more accurate portrayal of its mission and message to an international audience. While the church did not overtly proselytize to visitors during the Olympics, the event gave the faith international exposure, which indirectly and subtly allowed them to soft-sell their faith through self-promotion and education. This use of tourism (in this case via the Winter Olympics) for proselytizing purposes is consistent with the use of church headquarters and historical sites for both pastoral care and missionary purposes. In so doing, the LDS Church acknowledges the importance of tourism as a social phenomenon as it relates to both publicity for the church and meeting the needs of its members. Therefore, a special emphasis has been placed on the interpretation of Mormon heritage sites with a focus on educating non-LDS visitors about the basic beliefs of Mormonism and creating a positive image of the church in the minds of visitors.

As noted in this chapter, thousands of Mormons travel to many destinations related to their religious and scriptural heritage each year. However, relatively few empirical studies have been conducted on this religious tourism market segment (cf. Jackson and Henrie 1983; Jackson *et al.* 1990; Hudman and Jackson 1992; Jackson 1995; Brayley 1999; Mitchell 2001; Olsen and Timothy 2002). Nonetheless, there is evidence that Mormon tourism will continue to grow in the near future, for although church membership is only some 12.5 million strong, and growing by approximately 300,000 new converts per year in addition to natural growth, Stark (1999) has projected the membership of the LDS Church to reach over 63 million members by 2080, making the Church of Jesus Christ of Latter-day Saints one of the fastest growing religions in the world today. Jackson (1995: 196–197) notes that the publication of guidebooks and other books pertaining to biblical and church heritage sites (e.g. Oscarson and Kimball 1965; Kimball 1988; Allen 1989; Anderson and Anderson 1991; Holzapfel and Cottle 1991a, 1991b; Brown *et al.* 1994; Smith 2003), and the growing number of LDS-oriented tours, suggests that commercial interests have recognized the demand for guidance by Mormon tourists in locating and interpreting sacred and historical sites. A recent development in the LDS pilgrimage-tourism market has been the creation of LDS cruises to various parts of the world that include popular Mormon youth and adult speakers, famous LDS celebrities, and professors from the church-owned BYU. It is certain that many more LDS-themed tourism opportunities will arise in the future, all of which provide yet added evidence of the commodification of religious travel by commercial interests (Olsen 2003).

References

Allen, J.B. and Leonard, G.M. (1976) *The Story of the Latter-day Saints*, Salt Lake City: Deseret Book.

Allen, J.L. (1989) *Exploring the Lands of the Book of Mormon*, Orem, UT: S.A. Publishers.

Anderson, W.C. and Anderson, E. (1991) *Guide to Mormon History Travel*, Provo, UT: Bushman Press.

Arrington, L.J. and Bitton, D. (1979) *The Mormon Experience*, New York: Alfred A. Knopf.

Arrington, L.J. and Swinton, H.S. (1986) *The Hotel: Salt Lake's Classy Lady: The Hotel Utah 1911–1986*, Salt Lake City: Publisher's Press.

Backman, M.V. (1994) "Nauvoo tourist sites," in S.K. Brown, D.Q. Cannon and R.H. Jackson (eds) *Historical Atlas of Mormonism*, New York: Simon and Schuster.

Bishop, M.G. (1994) "The Saints and the captain: the Mormons meet Richard F. Burton," *Journal of the West* 33(4): 28–35.

Bishop, M.G. and Holzapfel, R.N. (1993) "The 'St. Peter's of the New World': the Salt Lake Temple, tourism, and a new image for Utah," *Utah Historical Quarterly* 61(2): 136–149.

Bitton, D. (1994) "The ritualization of Mormon history," *Utah Historical Quarterly* 43(1): 67–85.

Bradley, M.S. (2005) "Creating the sacred space of Zion," *Journal of Mormon History* 31(1): 1–30.

Brayley, R.E. (1999) "Finding sacredness in the profane – the Niagara Falls experience," *Visions in Leisure and Business* 18(2): 4–8.

Brown, S.K., Cannon, D.Q. and Jackson, R.H. (eds) (1994) *Historical Atlas of Mormonism*, New York: Simon & Schuster.

Bushman, R.L. (1984) *Joseph Smith and the Beginnings of Mormonism*, Chicago: University of Chicago Press.

Chen, C.H. (2003) "'Molympics?' Journalistic discourse of Mormons in relation to the 2002 Winter Olympic Games," *Journal of Media and Religion* 2(1): 29–47.

Chen, C.H. and Yorgason, E. (1999) "'Those amazing Mormons': the media's construction of Latter-day Saints as a model minority," *Dialogue: A Journal of Mormon Thought* 32(2): 107–128.

Chivallon, C. (2001) "Religion as space for the expression of Caribbean identity in the United Kingdom," *Environment and Planning D: Society and Space* 19: 461–483.

Cohen, E.H. (2003) "Tourism and religion, a case study: visiting students in Israel," *Journal of Travel Research* 42: 36–47.

Davies, D.J. (2000) *The Mormon Culture of Salvation*, Aldershot: Ashgate.

Eliason, E.A. (2001) "Curious Gentiles and representational authority in the city of the Saints," *Religion and American Culture: A Journal of Interpretation* 11(2): 155–190.

Grandstaff, M.R. and Backman, M.V. (1990) "The social origins of the Kirtland Mormons," *Brigham Young University Studies* 30(2): 47–66.

Gruen, J.P. (2002) "The urban wonders: city tourism in the late-19th-century American west," *Journal of the West* 41(2): 10–19.

Gurgel, K.D. (1976) "Travel patterns of Canadian visitors to the Mormon culture hearth," *Canadian Geographer* 20(4): 405–418.

Hafen, T.K. (1997) "City of Saints, city of sinners: the development of Salt Lake City as a tourist attraction 1869–1900," *Western Historical Quarterly* 38: 343–377.

Hervieu-Léger, D. (1999) "Religion as memory," in J.G. Platvoet and A.L. Molendijk (eds) *The Pragmatics of Defining Religion: Contexts, Concepts and Contests*, Leiden, Netherlands: Koninklijke Brill.

Holzapfel, R.N. and Cottle, T.J. (1991a) *Old Mormon Kirtland and Missouri: Historic Photographs and Guide*, Santa Ana, CA: Fieldbrook Productions.

Holzapfel, R.N. and Cottle, T.J. (1991b) *Old Mormon Nauvoo and Southeastern Iowa: Historic Photographs and Guide*, Santa Ana, CA: Fieldbrook Productions.

Hovanessian, M. (1992) *Le Lien Communautaire: Trois Générations d'Arméniens*, Paris: Armand Collins.

Hudman, L.E. (1994) "Historic sites and tourism," in S.K. Brown, D.Q. Cannon and R.H. Jackson (eds) *Historical Atlas of Mormonism*, New York: Simon & Schuster.

Hudman, L.E. and Jackson, R.H. (1992) "Mormon pilgrimage and tourism," *Annals of Tourism Research* 19: 107–121.

Jackson, R.H. (1988) "Great Salt Lake and Great Salt Lake City: American curiosities," *Utah Historical Quarterly* 56(2): 128–147.

Jackson, R.H. (1995) "Pilgrimage in American religion: Mormons and secular pilgrimage," in D.P. Dubey (eds) *Pilgrimage Studies: Sacred Places, Sacred Traditions*, Allahbad, India: Society of Pilgrimage Studies.

Jackson, R.H. (2003) "Mormon Wests: the creation and evolution of an American region," in G.J. Hausladen (ed.) *Western Places, American Myths: How We Think About the West*, Reno: University of Nevada Press.

Jackson, R.H. and Henrie, R. (1983) "Perception of sacred space," *Journal of Cultural Geography* 3(2): 94–107.

Jackson, R.H., Rinschede, G. and Knapp, J. (1990) "Pilgrimage in the Mormon church," in G. Rinschede and S.M. Bhardwaj (eds) *Pilgrimage in the United States*, Berlin: Dietrich Reimer Verlag.

Kimball, S.B. (1988) *Historic Sites and Markers along the Mormon and Other Great Western Trails*, Urbana: University of Illinois Press.

Lee, C.C. (2001) "Predicting tourist attachment to destinations," *Annals of Tourism Research* 28: 229–232.

Limerick, P.N. (2001) "Seeing and being seen: tourism in the American west," in D.M. Wrobel and P.T. Long (eds) *Seeing & Being Seen: Tourism in the American West*, Lawrence: University Press of Kansas.

McConkie, B.R. (1966) *Mormon Doctrine*, Salt Lake City: Bookcraft.

McHale, E.E. (1985) "'Witnessing for Christ': the Hill Cumorah Pageant of Palmyra, New York," *Western Folklore* 44(1): 34–40.

Madsen, M.H. (2003) "Mormon Meccas: the spiritual transformation of Mormon historical sites from points of interest to sacred space," PhD dissertation, Syracuse University.

Millet, R.L. (2003) "What is our doctrine?," *The Religious Educator* 4(3): 15–33.

Mitchell, H. (2001) "'Being There': British Mormons and the history trail," *Anthropology Today* 17(2): 9–14.

Mitchell, M. (1997) "Gentile impressions of Salt Lake City, Utah, 1849–1870," *The Geographical Review* 87(3): 334–352.

Morin, K.M. and Guelke, J.K. (1998) "Strategies of representation, relationship, and resistance: British women travelers and Mormon plural wives, ca. 1870–1890," *Annals of the Association of American Geographers* 88(3): 436–462.

Morinis, E.A. (1992) "Introduction: the territory of the anthropology of pilgrimage," in A. Morinis (ed.) *Sacred Journeys: The Anthropology of Pilgrimage*, Westport, CT: Greenwood Press.

Olsen, D.H. (2000) "Contested heritage, religion, and tourism," unpublished Master's thesis, Bowling Green State University, Bowling Green, Ohio.

Olsen, D.H. (2003) "Heritage, tourism, and the commodification of religion," *Tourism Recreation Research* 28(3): 99–104.

Olsen, D.H. (2006) "Religion, tourism and the management of the sacred," unpublished PhD dissertation, University of Waterloo, Waterloo, Ontario.

Olsen, D.H. and Guelke, J.K. (2004) "Spatial transgression and the BYU Jerusalem Center controversy," *Professional Geographer* 56(4): 503–515.

Olsen, D.H. and Timothy, D.J. (2002) "Contested religious heritage: differing views of Mormon heritage," *Tourism Recreation Research* 27(2): 7–15.

Olsen, S.L. (1992) "Centennial observations," in D.H. Ludlow (ed.) *Encyclopedia of Mormonism*, New York: Macmillian.

Olsen, S.L. (2000a) "Remembering and witnessing at church historic sites," paper presented at the *Symposium on "Remembering,"* Brigham Young University, October 6.

Olsen, S.L. (2000b) *Theology as a Cultural System*, available at www.aliveonline.com/ldspapers/StevesTalk.html

Olsen, S.L. (2002) *The Mormon Ideology of Place: Cosmic Symbolism of the City of Zion, 1830–1846*, Provo, UT: Joseph Fielding Smith Institute for LDS History.

Olsen, S.L. (2004) "A history of restoring historic Kirtland," *Journal of Mormon Studies* 30(1): 120–128.

Oscarson, R.D. and Kimball, S.B. (1965) *The Travelers' Guide to Historic Mormon America*, Salt Lake City: Bookcraft.

Packer, B.K. (1980) *The Holy Temple*, Salt Lake City: Bookcraft.

Parry, D.W. (1994) "Introduction," in D.W. Parry (ed.) *Temples of the Ancient World*, Salt Lake City: Deseret Book.

Petersen, M.E. (1981) *Isaiah for Today*, Salt Lake City: Deseret Book.

Quinn, M.D. (1976) "The Mormon succession crisis of 1844," *Brigham Young University Studies* 16(2): 187–233.

Shields, S.L. (1987) *The Latter Day Saint Churches: An Annotated Bibliography*, New York: Garland Publishing.

Shields, S.L. (1990) *Divergent Paths of the Restoration: A History of the Latter-day Saint Movement*, Bountiful, UT: Restoration Research.

Shipps, J. (1985) *Mormonism: The Story of a New Religious Tradition*, Chicago: University of Illinois Press.

Smith, B.C. (2003) *The LDS Family Travel Guide: Independence to Nauvoo*, Orem, UT: LDS Family Travels.

Smith, C.S. (1998) "The curious meet the Mormons: images from travel narratives, 1850s and 1860s," *Journal of Mormon History* 24(2): 155–181.

Smith, H.M. and Sjodahl, J.M. (1978) *Doctrine and Covenants Commentary*, Salt Lake City: Deseret Book.

Smith, J.S. (1976) *History of the Church of Jesus Christ of Latter-day Saints, Period I: History of Joseph Smith the Prophet, by Himself: Volume IV*, Salt Lake City: Deseret Book.

Snow, E.J. (1991) "British travelers view the Saints," *Brigham Young University Studies* 31: 63–81.

Sorenson, J.L. (1992) *The Geography of Book of Mormon Events: A Source Book*, Provo, UT: Foundation for Ancient Research and Mormon Studies.

Stark, R. (1999) "Extracting social scientific models from Mormon history," *Journal of Mormon History* 25(1): 174–194.

Talmage, J.A. (1912) *The House of the Lord*, Salt Lake City: Deseret Book.

Timothy, D.J. (1992) "Mormons in Ontario: early history, growth and landscape," *Ontario Geography* 38: 20–31.

Vukonić, B. (1996) *Tourism and Religion*, Oxford: Elsevier.

18 Conclusion

Whither religious tourism?

Dallen J. Timothy and Daniel H. Olsen

Billions of people throughout the world associate with various socio-structural systems that venerate higher sources of power and authority, including a Creator, God, Goddess, mythical deity, Earth Mother, the Sun, the Moon, a Savior, a prophet, or an infinite and intangible power source. For most believers, their terrestrial existence or mortal phase of life is by its very nature a personal quest for understanding about, and the reasons for, their being. Many people read books and holy scripts, exercise, eat certain foods, pray or chant, and physically travel to designated destinations as a way of materially augmenting their spiritual journeys. For millions of faithful, these elements are part of larger belief systems that provide the principles by which they live their lives and become an integral part of their individual or collective social identity.

In purely quantitative terms, the majority of people who undertake travel as a way of connecting with someone or something greater than themselves at locations understood to be sacred, holy, or otherwise reverenced, are devout adherents to some formal belief system or network of believers. However, even non-believers, or atheists, can experience something outside themselves that whispers acknowledgment of something beyond. Thus, as this book has attempted to demonstrate, there can be a significant difference between being religious and being spiritual. For instance, someone who does not necessarily subscribe to the tenets of the faith that owns a site being visited, or even someone who denies the existence of God, can have a spiritual experience there, because spiritual refers more to "a personal belief in, or a search for a reason for one's existence; a greater or ultimate reality, or a sense of connection with God, nature, or other living beings" (Ibrahim and Cordes 2002: 18). Thus, even pleasure-seeking, hedonic tourists who, amid their sightseeing schedules stop in to holy places may feel a heightened inner awareness or connection to something beyond themselves.

Despite the pervasiveness of religious tourism and spiritual connections to place, relatively few scholars have explored the multitudinous and multifarious relationships between religion, spirituality, and tourism. This book has aimed to advance this nascent discussion in the field of tourism studies by examining the conceptual implications of religious tourism and how these play out experientially among various world belief systems. As the contributors to this volume have made abundantly clear, the convergence of religion, spirituality and tourism manifests in countless perspectives on personal and social identity, subjective and objective reality, dissonance, inclusion, exclusion, harmony, discord, compulsion, commercialization/commodification, persuasion, and spiritual connection to place. Although the book focuses on many dynamics of religious and spiritual mobility, several other ideas may be drawn from its contents.

The tourist-pilgrim debate noted in the introductory chapter has yet to be solved, but perhaps a solution is not as important as understanding the roots of contention. Scholars are at odds over how to define this growing phenomenon of religious tourism, assigning labels such as pilgrimage tourism, religious tourism, spiritual tourism, and tourism pilgrimage (Jackowski 1987; Hudman and Jackson 1992; Tyrakowski 1994; Vukonić 1996; Santos 2003). While religious tourism seems to be the most common choice, there has been no consensus in defining it. This is in part because the term is generally used by tourism promoters and researchers to describe the phenomenon in two different ways: those whose impetus to travel combines both religious (dominant) and secular (secondary) motives and people who visit sacred sites during their journeys to other attractions and destinations (Smith 1992; Bywater 1994; Russell 1999). On the one hand, from a religious perspective, religious tourism is separate from other forms of tourism because it is characterized by its aims, motivations, and destinations (Liszewski 2000), as equating pilgrimage with tourism or even religious tourism would make pilgrimage co-equal with more hedonic types of undertakings, such as wine tourism or sex tourism (Ostrowski 2000: 57). From an industry or academic perspective, definitions of tourist types are typically not based on motivations but rather on the activities tourists engage in while traveling. For example, a person traveling to participate in a cultural festival would be considered a cultural tourist, while a person wanting to experience the thrill of skydiving may be considered an adventure tourist. From this classificatory perspective, pilgrimage can be seen as a form of tourism categorized by pilgrim activities, and by virtue of this, a pilgrim may be seen as a type of tourist who travels for religious or spiritual reasons (Timothy 2002; Timothy and Boyd 2003). From the religious organizational perspective, however, this designation would be inadequate, for tourists are unscrupulous adventurers with low moral standards. Thus, at its very core, the roots of debate are all about perspective and interpretation. From the perspective of tourism planners, promoters, and scholars, pilgrims are simply tourists, for tourists are not defined by motives, rather by the simple fact that they travel away from home.

Because of this confusion the term religious tourism is generally used in an undifferentiated or uncritical way (Santos 2003). For example, Aziz (2001: 155), in discussing tourism and travel in an Arab/Islamic context, writes that "it is a moot point as to whether or not the *Hajj* is a form of tourism, and debate on this topic is ongoing." However, Russell (1999: 40) has given a starting point in defining religious tourists as those "who set out to visit a destination of religious significance for a specifically religious purpose." The religious tourist, according to Russell, is different from the religious heritage tourist, who visits the same sites but for cultural and historical interests rather than a search for spiritual meaning. However, with the modern-day separation of spirituality from religion, and with the multiple perspectives from which religious tourism can be considered, it is likely that defining religious tourism and who is a religious tourist will continue to stimulate much debate in the future.

A subject that has received only scant attention in the literature, primarily by critics of religious (mass) tourism, is the negative social and ecological impacts of this type of travel (Sizer 1999). While Gupta (1999) suggests that religious tourism is more innocuous than other types of tourism, the evidence noted in this book shows otherwise, and in many cases pilgrim-tourists may be even more injurious that others. Likewise, some authors have noted the negative impacts of religious tourism on the commercialization of places and artifacts that were once held as sacred locations and icons imbued with deep devotional value. The commodification of religious symbols and the economic implications of selling them should become of increased interest to researchers as the world becomes more consumption-oriented.

There are several heritage tourism issues that can also be gleaned from the previous 17 chapters. Clearly, pilgrimage places are heritage sites, even if their meanings differ from person to person. As such, it would be important to examine matters related to power and empowerment through heritage in the context of religious travel. This has yet to surface in the literature. Likewise, problems associated with interpretation, conservation, funding, authenticity of place and experience, planning, and heritage contestation should factor strongly in future research endeavors (Digance 2003; Rana and Singh 2000; Singh and Rana 2001; Timothy and Boyd 2003) and would reveal a great deal about the junction between religion and tourism and the role of heritage meanings.

One of the hottest issues today in commentary on heritage is the overlapping claims to sacred space by dissimilar religious groups and their adherents. Jerusalem is perhaps the very best example of this, although there are many more in Islam, Christianity, and other spiritual traditions. In recent years considerable bloodshed has come from disputes over sacred space, and many other less-violent challenges have emanated from two or more religious institutions having to share a common place-bound heritage. Olsen and Timothy (2002) identified three types of contested heritage in the realm of religion: multiple groups sharing the same heritage places; factions within a group interpreting a common heritage differently; and parallel heritages, where the history of two groups unfolded at the same place and time. In all of these cases, the situation leads to questions about whose heritage is more important and which should be preserved or interpreted. Social amnesia about unpleasant events that might have occurred, and divergent versions of the past – all with religious and tourism implications, have important conceptual bearings for understanding religious heritage in conflict (Timothy and Boyd 2003).

Likewise, the up-and-coming focus of heritage tourism specialists on "personal heritage" travel sheds some light on things of a spiritual nature. As noted previously in this volume, some heritage themes and patterns that might at first appear to be secular are now being seen as spiritual, such as visiting national monuments erected to fallen war heroes, cemeteries, and traveling to places with which an individual has a personal heritage connection. For Americans, hearing the national anthem being played at the Tomb of the Unknown Soldier in Virginia can be a very spiritual experience, whether or not one has any official religious affiliation. Likewise, discovering one's roots in the lands of one's ancestors can be a thrilling spiritual experience that makes people feel close to their forebears (Meethan 2003; Timothy 1997). This personal attachment to place then has strong spiritual connotations that negotiate an individual's identity as a hyphenated person between a spiritual motherland and a secular homeland (Mazumbar and Mazumbar 2004).

Future research might also be brought to bear on how religious groups venerate places of spiritual importance. Among most indigenous peoples, including Native Americans, the entire earth is sacred, and if some places are deemed more holy than others, it is because some kind of mythological event was said to have occurred there (Dobbs 1997; Ball 2000). Similarly, how religions borrow elements of traditional faith systems or how various sects borrow from each other, given the obvious barriers between religions in recognizing that others might possess some elements of truth, heretofore has not been a major area of study. This is unfortunate, as there is undoubtedly a great deal to learn from religious diffusion and how this might play out in pilgrimage terms.

Many world religions were not systematically highlighted in this volume. This is not an acknowledgment of lack of importance, rather a function of space and time constraints, and a dearth of a critical mass of published knowledge about various religions' approaches to tourism. While it was not a focus of a particular chapter, a great deal has been written about indigenous groups and tourism, including various implications for native spiritual traditions (e.g. Barnett 1997; Butler and Hinch 1996; Hall and McArthur 1993; Johnston 2003). Native Americans are a common focus of commentators on indigenous spirituality and tourism. American Indian traditions are animist in nature, wherein they see the earth and indeed the entire universe as being alive and human beings as an integral part of nature, not its master. In the Native American worldview, "all life forms and natural elements are connected, interrelate, and interplay; no part of nature is considered more important than another" (Ibrahim and Cordes 2002: 19). Indigenous people were mystified by the European tendency to see the earth as savage, untamed, wild, and primitive. From their perspective, the world is a "fantastic and beautiful creation engendering extremely powerful feelings of gratitude and indebtedness, obliging us to behave as if we are related to one another" (Forbes 2001: 284). American Indian, and other indigenous peoples, traditionally celebrated nature through primitive forms travel. For instance, young males were required to leave the clan to find themselves and participate in initiation ceremonies. Dances and other celebrations were carried out to worship the earth and their role in nature (Ibrahim and Cordes 2002). New Age travel borrows a great deal from indigenous religion, but little is known about commonalities between animist beliefs and other spiritual worldviews that may have major bearings on travel behavior.

The study of various Protestant groups may also shed light on new perspectives on religious tourism. Within most Protestant sects, there is no theology regarding sacred places and pilgrimage because of the belief that God can be found within individuals and

> in the midst of the praises of his people. Linking God to a place is too constraining and routinized, and represents the very opposite of a "movement". God is a mind-heart-body experience in most strands of Protestantism; little value is put on buildings, aesthetics or "shrines". The church is people, not places.
>
> (Percy 1998: 285)

This view results in many evangelicals seeing pilgrimage as a debased superstition. The whole idea of "sacred ground" is abhorrent to them, and many regard statue worship (i.e. the Virgin Mary) as a form of idolatry (Tyler 1990: 16). While members of various Protestant sects do travel to places of historical significance, especially in the Holy Land (Collins-Kreiner and Kliot 2000), they do not consider this a form of pilgrimage (Tweed 2000). With the exception of this shallow glance at Protestantism and Jutla's chapter on Sikh tourism, this volume has focused overwhelmingly on religious bodies that encourage travel and pilgrimage. Thus, we still know very little about faiths that discourage travel for spiritual purposes.

Many other religions have practices and doctrines related to pilgrimage and other forms of travel, yet little is known about these situations. Examples include Jainism (Banarjee 1995; Chapple 2001; Mukherji 1989; Salgia 2002), Baha'i (McGlinn 1999), and Eastern and Greek Orthodox (Jotischky 2000; Raivo 2002). There is also the matter of religion as a tourist attraction. Comparative investigations to explore religion and tourism in a wider range of religious traditions would make a valuable and long-overdue contribution to the study of religious tourism.

Finally, the issue of change deserves mention in this context, particularly change in religion and change in religious forms of travel in recent years. As people move toward finding deeper spiritual meaning than what traditional religious bodies could offer in a fast-paced, post-modern world, new forms of spirituality have begun to supersede many traditional religious organizations and practices. These transformations are manifest in people beginning to realize that each individual is imbued with some degree of divine power, and that the ability to exercise that power rests within each person (Cooper 1999). This growing trend is manifest in tourism terms as people travel to places around the world that have previously been marked as sacred, but they also have begun to create their own sacred space where they can find solace and a personal connection to God or other higher existence (Tresidder 1999).

Changes in human mobility and modernization have also significantly affected traditional pilgrimage. Efficient modes of transportation, such as flights and air-conditioned motorcoaches, have replaced traditional means of arriving at sacred destinations, thereby changing the very nature of pilgrimage, which in most religious traditions required some degree of penitent travail in arriving at their destinations (Jackson and Davis 2000). There is also the matter of virtual pilgrimages, in which anyone can now participate, thereby allowing access to celebrations and rituals that heretofore have been off-limits to non-believers or too far away to visit. Despite these modern influences, some observers believe pilgrimages still have the ability to function as a rite of passage, to change the status of pilgrims from simple adherents to strong believers (Utterback 2000).

It is clear from the contributions in this volume that the relationships between tourism and religion are multifaceted and worthy of current and future investigation. Religious or spiritually motivated journeys were among the oldest forms of travel. In contemporary terms, religious tourism, wherein people are motivated to visit hallowed places by desires for deep soul-searching, intimacy with deity, forgiveness of sins, educational experiences, patriotic stirrings, a chance to admire magnificent architectural or natural wonders, or simply out of curiosity about different cultures and religions, is one of the most significant types of tourism in the world today by volume and prevalence. With the passing of Pope John Paul II in the spring of 2005, over 4 million pilgrim tourists during a two- or three-day period were said to have traveled to the Vatican to pay their last respects. Regardless of the motives for experiencing sacred places and events, there are clear indications that this form of travel will continue to grow far into the future. It may undergo significant changes as the world becomes more mechanized, modernized, and liberalized, but it will assuredly continue to expand.

References

Aziz, H. (2001) "The journey: an overview of tourism and travel in the Arab/Islamic context," in D. Harrison (ed.) *Tourism and the Less Developed World: Issues and Case Studies*, Wallingford: CABI.

Ball, M. (2000) "Sacred mountains, religious paradigms, and identity among the Mescalero Apache," *Worldviews* 4(3): 264–282.

Banarjee, S.R. (1995) "Jainism through the ages," *Jain Journal* 29(4): 129–166.

Barnett, S. (1997) "Maori tourism," *Tourism Management* 18: 471–473.

Butler, R.W. and Hinch, T. (eds) (1996) *Tourism and Indigenous Peoples*, London: International Thomson Business Press.

Bywater, M. (1994) "Religious travel in Europe," *Travel & Tourism Analyst* 2: 39–52.

Chapple, C.K. (2001) "The living cosmos of Jainism: a traditional science grounded in environmental ethics," *Daedalus* 130(4): 207–224.

Collins-Kreiner, N. and Kliot, N. (2000) "Pilgrimage tourism in the Holy Land: the behavioural characteristics of Christian pilgrims," *GeoJournal* 50: 55–67.

Cooper, J. (1999) "Comprehending the circle: Wicca as a contemporary religion," *The New Art Examiner* 26(6): 28–33.

Digance, J. (2003) "Pilgrimage at contested sites," *Annals of Tourism Research* 30(1): 143–159.

Dobbs, G.R. (1997) "Interpreting the Navajo sacred geography as a landscape of healing," *Pennsylvania Geographer* 35(2): 136–150.

Forbes, J.D. (2001) "Indigenous Americans: spirituality and ecos," *Daedalus* 130(4): 283–300.

Gupta, V. (1999) "Sustainable tourism: learning from Indian religious traditions," *International Journal of Contemporary Hospitality Management* 11(2/3): 91–95.

Hall, C.M. and McArthur, S. (eds) (1993) *Heritage Management in New Zealand and Australia: Visitor Management, Interpretation and Marketing*, Auckland: Oxford University Press.

Hudman, L.E. and Jackson, R.H. (1992) "Mormon pilgrimage and tourism," *Annals of Tourism Research* 19: 107–121.

Ibrahim, H. and Cordes, K.A. (2002) *Outdoor Recreation: Enrichment for Lifetime*, Champaigne, IL: Sagamore.

Jackowski, A. (1987) "Geography of pilgrimage in Poland," *The National Geographical Journal of India* 33(4): 422–429.

Jackson, R.H. and Davis, J. (2000) "Pilgrimage in western arid China," *Pilgrimage Studies* 8: 155–182.

Johnston, A.M. (2003) "Self-determination: exercising indigenous rights in tourism," in S. Singh, D.J. Timothy, and R.K. Dowling (eds) *Tourism in Destination Communities*, Wallingford: CABI.

Jotischky, A. (2000) "History and memory as factors in Greek Orthodox pilgrimage to the Holy Land under crusader rule," *Studies in Church History* 36: 110–122.

Liszewski, A. (2000) "Pilgrimages or religious tourism?," in A. Jackowoski (ed.) *Peregrinus Cracoviensis*, Krakow: Institute of Geography, Jagiellonian University.

McGlinn, S. (1999) "A theology of the state from the Bahā'ī teachings," *Journal of Church and State* 41(4): 697–724.

Mazumbar, S. and Mazumbar, S. (2004) "Religion and place attachment: a study of sacred places," *Journal of Environmental Psychology* 24: 385–397.

Meethan, K. (2003) "'To stand in the shoes of my ancestors': tourism and genealogy," in T. Coles and D.J. Timothy (eds) *Tourism, Diasporas and Space*, London: Routledge.

Mukherji, S.C. (1989) "Cultural heritage of Bengal in relation to Jainism," *Jain Journal* 24(1): 25–34.

Olsen, D.H. and Timothy, D.J. (2002) "Contested religious heritage: differing views of Mormon heritage," *Tourism Recreation Research* 27(2): 7–15.

Ostrowski, M. (2000) "Pilgrimages or religious tourism," in A. Jackowski (ed.) *Peregrinus Cracoviensis*, Krakow: Institute of Geography, Jagiellonian University.

Percy, M. (1998) "The morphology of pilgrimage in the 'Toronto Blessing'," *Religion* 28: 281–288.

Raivo, P. (2002) "The peculiar touch of the East: reading the post-war landscapes of the Finnish Orthodox Church," *Social and Cultural Geography* 3(1): 11–24.

Rana, P.S. and Singh, Rana P.B. (2000) "Sustainable heritage tourism: framework, perspective and prospect," *National Geographical Journal of India* 46: 141–158.

Russell, P. (1999) "Religious travel in the new millennium," *Travel & Tourism Analyst* 5: 39–68.

Salgia, T. (2002) "Jainism: a solution for world peace," *Dialogue and Alliance* 16(1): 46–51.

Santos, M.G.M.P. (2003) "Religious tourism: contributions towards a clarification of concepts," in C. Fernandes, F. McGettigan, and J. Edwards (eds) *Religious Tourism and Pilgrimage*, Fatima, Portugal: Tourism Board of Leiria.

Singh, Rana P.B. and Rana, P.S. (2001) "The future of heritage tourism in Varanasi: scenario, prospects and perspectives," *National Geographical Journal of India* 47: 201–218.

Sizer, S.R. (1999) "The ethical challenges of managing pilgrimages to the Holy Land," *International Journal of Contemporary Hospitality Management* 11(2/3): 85–90.

Smith, V.L. (1992) "Introduction: the quest in guest," *Annals of Tourism Research* 19: 1–7.

Timothy, D.J. (1997) "Tourism and the personal heritage experience," *Annals of Tourism Research* 24: 751–754.

Timothy, D.J. (2002) "Sacred journeys: religious heritage and tourism," *Tourism Recreation Research* 27(2): 3–6.

Timothy, D.J. and Boyd, S.W. (2003) *Heritage Tourism*, Harlow: Prentice Hall.

Tresidder, R. (1999) "Tourism and sacred landscapes," in D. Crouch (ed.) *Leisure/Tourism Geographies: Practices and Geographical Knowledge*, London: Routledge.

Tweed, T.A. (2000) "John Wesley slept here: American shrines and American Methodists," *Numen* 47(1): 41–68.

Tyler, C. (1990) "Spreading the word," *Geographical Magazine* 62(7): 12–19.

Tyrakowski, K. (1994) "Pilgrims to the Mexican highlands," *Geographica Religionum* 8: 193–246.

Utterback, K.T. (2000) "Saints and sinners on the same journey: pilgrimage as ritual process," *Medieval Perspectives* 15: 120–129.

Vukonić, B. (1996) *Tourism and Religion*, Oxford: Elsevier.

Index